Inferential Models

Reasoning with Uncertainty

MONOGRAPHS ON STATISTICS AND APPLIED PROBABILITY

General Editors

F. Bunea, V. Isham, N. Keiding, T. Louis, R. L. Smith, and H. Tong

Monographs on Statistics and Applied Probability 147

Inferential Models
Reasoning with Uncertainty

Ryan Martin
University of Illinois at Chicago, USA

Chuanhai Liu
Purdue University, West Lafayette, Indiana, USA

CRC Press
Taylor & Francis Group
Boca Raton London New York

CRC Press is an imprint of the
Taylor & Francis Group, an **informa** business

A CHAPMAN & HALL BOOK

CRC Press
Taylor & Francis Group
6000 Broken Sound Parkway NW, Suite 300
Boca Raton, FL 33487-2742

First issued in paperback 2020

© 2016 by Taylor & Francis Group, LLC
CRC Press is an imprint of Taylor & Francis Group, an Informa business

No claim to original U.S. Government works

ISBN-13: 978-1-4398-8648-9 (hbk)
ISBN-13: 978-0-367-73780-1 (pbk)

Visit the Taylor & Francis Web site at
http://www.taylorandfrancis.com

and the CRC Press Web site at
http://www.crcpress.com

Dedication

While Chuanhai and I were finalizing this manuscript for publication, my grandfather, Jack Martin, passed away. In his more than 90 years, he faced many challenges, including U. S. Army combat service during World War II, dangerous and physically demanding steel mill labor to support his family, and caring for his beloved wife, Frances Martin, during her battle with cancer. These experiences made him tough, but the Jack I knew was joyful, loving, and inquisitive. In particular, though he had no formal training in statistics, he was always genuinely interested to know what kind of work I was doing. I would tell him that I was working on something new and that I wasn't sure if it would ever amount to anything. Fortunately, before his passing, he had a chance to see a draft of this book and know that the work did eventually amount to something. Grandpa was loved by many and, for the positive influence he had on my life, I dedicate this book to him.

Ryan Martin

Contents

Preface

The conversion of experience to knowledge is the cornerstone of the scientific method and, arguably, the most valuable of mankind's intellectual activities. Despite the importance of this activity, the field of statistics, which is concerned with how experience-to-knowledge conversions should be made, is still surprisingly underdeveloped. To be clear, there has been substantial development in statistical methodology, driven in part by rapid scientific and technological advances, but the subject lacks a solid theory for reasoning with uncertainty. This book grew out of the authors' recent attempts to understand the basic foundations of statistical inference, with an emphasis on developing sound methods for inference in modern and challenging high-dimensional problems. No doubt, developing a new and satisfactory framework for statistical inference is an ambitious goal. Our intention in writing this book is to share our experience in developing some deep intuition about statistical inference. Our hope is that this book is just the beginning, rather than the end, of a long and interesting story about the development of tools for sound scientific reasoning.

The authors recognize the difficulties the reader will find in studying this book. Unlike many other advanced texts in statistics, the reader's main obstacle to overcome is not the mathematics. Instead, the reader is asked to adjust to a new way of thinking compared to that of existing schools of thought on statistical inference. It is fair if readers from the Neyman–Pearson school of thought view our contribution as a framework for constructing confidence intervals and testing procedures with *exact* error rate control. Beyond these basic sampling distribution properties, we hope the reader will also recognize that our approach provides alternative meaningful probabilistic interpretations of some common inferential summaries, such as *p*-values, which have been criticized by many. Our new perspective also provides insights on the interesting and challenging problems of conditional and marginal inference. Readers having a slant toward the Bayesian school of thought will also find some obstacles in this book, in particular, our construction of posterior probabilistic inferential summaries without a prior and Bayes formula. We explain in the book what we believe to be the key shortcomings with the existing schools of thought but the point is not to be critical; rather, we hope that by identifying some difficulties with standard thinking, the reader will be motivated to think carefully about the correct approach for scientific inference, even if he/she ultimately does not align with our proposed approach.

The main topic of the book is our recently developed *inferential model* (IM) framework. At a high level, an IM produces meaningful prior-free probabilistic inference; by "meaningful" we mean that the inferential output is properly calibrated so that the numerical values can be interpreted both within and across experiments/experimenters, and not just in the limit as sample size increases. Our claim to achieve exact prior-free probabilistic inference is a bold one, but we believe that the basic idea of IMs is quite simple. Like the fiducial and the Dempster–Shafer theory of belief functions, we introduce an unobservable auxiliary variable to characterize uncertainty. IMs diverge from these existing methods by seeking to *predict* the particular unobserved value of this auxiliary variable, rather than forcing a particular sort of posterior distribution—via a "continue to regard" operation—on the auxiliary variable itself. The prediction is carried out using a suitably chosen random set, which yields a meaningful measure of uncertainty, in particular, a belief function for the unknown parameters given the observed data.

This book describes, in detail, the motivation for and the general construction of an IM. The specific content comes mostly from various papers we have published recently on the subject, but we have put forth a serious effort to make clear both (i) the motivation for the new approach, and (ii) how the new approach actually differs from existing methods, such as fiducial. Chapters 1–3 present the background and motivation for IMs, and Chapter 4 lays out the basic IM framework. Random sets play an important role in this new framework, and Chapter 5 provides some background on these not-so-familiar objects. Chapters 6 and 7 consider two important dimension-reduction strategies, namely, conditioning and marginalization, which are closely connected to the efficiency of the corresponding IM. The conditioning operation concerns the combination of information coming from different sources, and our results cast some new light on the familiar combination rules based on sufficient statistics and Bayes formula. Our proposed marginalization strategy provides a precise description of when exact marginal inference is possible, providing some new insights on several challenging marginalization problems, including the Behrens–Fisher problem. Applications of the IM machinery are considered in Chapters 8 (linear models) and 9 (prediction). Chapter 10 is a first attempt at constructing an IM for challenging high-dimensional problems. There, we construct a valid IM for a collection of assertions of interest, simultaneously, and in addition to expected connections to various multiplicity adjustments in the literature, we also find an unexpected insight concerning post-selection inference. Chapter 11 presents an extreme generalization of the IM framework, one that does not require a complete description of the connection between data, parameters, and auxiliary variables but still provides exactly valid prior-free probabilistic inference. Finally, in Chapter 12 we give some concluding remarks and some ideas for new theoretical, computational, and methodological problems to be considered. To help researchers who find IMs interesting and want to make their own contributions, Chapter 12 also provides a list of ten open problems that we think are particularly important and challenging. We also have a companion website[1] for the book with links to relevant papers and collaborators' websites, as well as R codes for some of the numerical examples.

It was our goal to keep the mathematical level of the book relatively low. For the most part, students with a solid background in calculus and calculus-based prob-

---

[1]Go to either of the authors' professional websites for the link.

ability, along with basic knowledge about set theory, linear algebra, and real analysis, will be able to follow our calculations. Occasionally some more sophisticated mathematical arguments, based on, say, measure theory, are needed but these are not essential to understanding the main ideas. To grasp the motivation behind the proposed IM approach, it is ideal that the reader be familiar with the key concepts and arguments of the frequentist and Bayesian schools of thought. A brief review of frequentist and Bayesian inference, along with references, is presented in Chapters 1 and 2. There we also provide a summary of the main ideas behind the fiducial and Dempster–Shafer approaches, with references. The later chapters include some applications of the IM machinery to some specific problems but, in these cases, all the models are relatively simple. A reader familiar with generalized linear models will have sufficient background to follow all the examples presented in the book.

A project of this magnitude requires help from many different people and organizations, and here we take this opportunity to acknowledge a few of them. We would especially like to thank Professor Arthur P. Dempster for sharing his broad knowledge on the historical development of statistical inference, along with his deep thoughts on statistics in general and the Dempster–Shafer theory of belief functions in particular. Thanks also go out to Professor Donald Rubin for his constant encouragement, saying to us on many occasions: "I think you're onto something...". David Grubbs, Laurie Oknowsky, Karen Simon, and Katy Smith at Taylor & Francis have been of invaluable help in putting the final version of this book together, and we very much appreciate their efforts. During the time when this book was written, we were supported in part by the U. S. National Science Foundation, grants DMS–1007678, DMS–1208833, and DMS–1208841, and we are grateful for the Foundation's investment in outside-the-box thinking about statistical inference. We should also acknowledge the many grant proposal and journal paper reviewers whose critical comments forced us to think more carefully about the fundamental ideas. Finally, we would like to thank our families—Ryan's wife Cheryl and Chuanhai's wife Shu and daughters Jane and Melinda—for their support and encouragement, without which this book would not have been possible.

Ryan Martin
Chicago, Illinois

Chuanhai Liu
West Lafayette, Indiana

Chapter 1

Preliminaries

1.1 Introduction

Researchers would agree that statistical methods are of fundamental importance to their scientific investigations, which explains the substantial developments in statistical methodology and computing in recent years. On the other hand, there has been very little progress on the foundations of statistical inference. These foundational considerations are of the utmost importance, as they distinguish Statistics as a branch of Science and not just a collection of methods for scientists to use. Moreover, there are now alternatives—machine learning, data science, etc—poised to challenge Statistics, so setting a solid theoretical foundation is an urgent matter. This book attempts to lay out a new set of foundations that satisfy the goals of statistical inference, as we understand them, along with an approach, called *inferential models* (IMs), that fits within this new framework. Before we can get into these new details, we must first set some preliminary notation and concepts, which is the goal of this chapter. Section 1.2 fixes some basic ideas from probability theory and statistics that the reader is assumed to be familiar with. Section 1.3 attempts to lay out what, in our opinion, is the fundamental goal of scientific inference, and Section 1.4 gives a preview of one of the main ideas of the book, namely, that prediction plays a fundamental role in probabilistic inference. It is taken as the basic technique for propagating uncertainties specified in the sampling model to the space of quantities of interest, summarizing the knowledge gained about these quantities through observations. Finally, Section 1.5 provides an outline of the rest of this book.

1.2 Assumed background

1.2.1 Probability theory

A standard prerequisite for any text in statistical inference is a solid background in probability theory. For the most part, calculus-based probability theory will be sufficient for our purposes, though we will occasionally bring up some more technical matters concerning measurability, etc, where some background in measure-theoretic probability theory would be needed for complete understanding. To fill in this background, we refer the reader to some standard textbooks on probability theory, for example, [24, 86, 210]. For the most part, we will assume the reader is familiar with the basic concepts in these texts but, to fix some notation, terminology, and concepts, we will give a brief review of probability theory next.

The mathematical starting point for a probability problem is a specification of a triple, $(\Omega, \mathscr{A}, \mathsf{P})$, where Ω is a set, called the sample space, $\mathscr{A}$ is a collection of events, or subsets of Ω, of interest, and P is a set-function called a probability measure, that maps sets in $\mathscr{A}$ to values in $[0,1]$. The idea is that the number $\mathsf{P}(A)$ for an event $A \in \mathscr{A}$ represents the probability that this event will occur when the underlying experiment is performed. The simplest case is where the sample space Ω consists of finitely many outcomes, so that $\mathscr{A}$ can be taken as the collection of all possible subsets of Ω, called the power set, and denoted by 2^{Ω}. For infinite sample spaces, it is generally not possible to define P for all possible subsets, so the notion of a σ-algebra of subsets is required; see Exercise 1.1 for a definition of σ-algebra. We will assume throughout the book that the reader is familiar with the case of infinite (countable or uncountable) sample spaces, as presented in [24].

The simplest example of a probability space is that of a Bernoulli trial, where there are only two possible outcomes. For a coin-tossing example, let $\Omega = \{H, T\}$. Then $\mathscr{A}$ consists of $\varnothing$, Ω, $\{H\}$, and $\{T\}$. Of course, there is nothing special about a sample space with only two possible outcomes, and it is easy to extend these ideas to the case of a sample space $\Omega = \{1, \ldots, K\}$ with K possible outcomes.

The probability measure P cannot be defined arbitrarily—there are some rules or axioms that it is assumed to satisfy. The three axioms are as follows:

1. (*Non-negativity*) $\mathsf{P}(E) \geq 0$ for all $E \subseteq \Omega$,

2. (*Additivity*) for any collection of mutually exclusive events, $E_1, \ldots, E_n$, i.e., for each $j \neq i$, $E_i \cap E_j = \varnothing$, P satisfies $\mathsf{P}(\bigcup_{i=1}^{n} E_i) = \sum_{i=1}^{n} \mathsf{P}(E_i)$, and

3. (*Normalization*) $\mathsf{P}(\Omega) = 1$.

Some further properties of P, which are consequences of these three axioms and are relevant to our discussions in later chapters, are left as exercises in Section 1.6.

The definition of conditional probability of A given B, i.e., the updated probability of event A when event B is known to have occurred, follows intuitively:

$$\mathsf{P}(A \mid B) = \frac{\mathsf{P}(A \cap B)}{\mathsf{P}(B)}, \quad \text{provided that } \mathsf{P}(B) > 0, \tag{1.1}$$

where $A \cap B$ is the intersection of A and B, which is the event that both A and B occur. The resulting identity

$$\mathsf{P}(A \cap B) = \mathsf{P}(B)\mathsf{P}(A \mid B) \tag{1.2}$$

is called the *multiplicative property* of probability. This is useful to define formally the concept of independence, that is, events A and B are said to be independent if

$$\mathsf{P}(A \cap B) = \mathsf{P}(A)\mathsf{P}(B). \tag{1.3}$$

Since $A \cap B = B \cap A$, it follows from (1.2) that

$$\mathsf{P}(A \cap B) = \mathsf{P}(B)\mathsf{P}(A \mid B) = \mathsf{P}(A)\mathsf{P}(B \mid A).$$

The last two symmetric factorizations of the probability $P(A \cap B)$ gives immediately the well-known *Bayes formula*, which states that

$$P(A \mid B) = \frac{P(A)P(B \mid A)}{P(B)}, \quad \text{provided that } P(B) > 0. \tag{1.4}$$

Conditional probability plays an important role in statistical inference. If the statistical model has a joint probability distribution specified for all unknown quantities of interest, as in the Bayesian setting (Section 1.2.3), then inference reduces to conditional probability calculations. While this is operationally simple, there are challenges in specifying a complete joint distribution in real-world problems.

A useful way to specify a probability measure P is via a random variable representation. Loosely speaking, functions defined on sample space Ω and taking values in a Euclidian space $\mathbb{R}^d$ are called s. Using random variables to represent well-specified probabilities on quantities on sample spaces is useful in application of probability theory as they stand for meaningful quantities. For example, the outcome of a Bernoulli trial can be represented by the random variable X, which is commonly defined with $X = 1$ for "success" and $X = 0$ "failure." In this case, the probability space is represented by the sample space $\Omega = \{0, 1\}$ and the probability mass function on 2^Ω that can be simply defined by specifying the probability of "success"

$$P(X = 1) = \theta, \quad \theta \in [0, 1],$$

because the probability of any other event is uniquely determined according to the three axioms. For example, the probability of "failure" is $P(X = 0) = 1 - \theta$. For a random variable that takes only finitely or countably infinitely many values, the function $x \mapsto P(X = x)$ is the *probability mass function*, or pmf. The pmf uniquely determines the probability measure P. Quantities having continuous values can be represented by continuous random variables. Continuous random variables are often characterized by a *probability density function*, or pdf. The expectation and variance operators, written as E and V, are assumed familiar and will be used throughout. We review some of the concepts and terminologies with a few examples.

Example 1.1. The standard uniform random variable U takes values in the interval $(0, 1)$, with the key feature that the probability that U falls into a sub-interval is the length of the sub-interval. In particular,

$$P(U \leq u) = u, \quad u \in (0, 1).$$

This probability model is uniquely determined by what is called the cumulative distribution function (cdf) on the real line $\mathbb{R} = (-\infty, \infty)$:

$$F(u) = \begin{cases} 0 & u \leq 0, \\ u & 0 < u < 1, \\ 1 & u \geq 1. \end{cases} \tag{1.5}$$

The derivative of $F(u)$ with respect to u,

$$f(u) = \begin{cases} 0 & u \leq 0, \\ 1 & 0 < u < 1, \\ 0 & u \geq 1, \end{cases} \tag{1.6}$$

is the pdf. Note that although F is not differentiable at the two end points of $[0, 1]$, its continuity means that $P(U = 0) = P(U = 1) = 0$ and thus can be ignored. Alternatively, one can first specify the pdf (1.6) and then derive the cdf as

$$F(u) = \int_{-\infty}^{u} f(t)dt,$$

which is the same as that in (1.5). In what follows, we write $U \sim \text{Unif}(0, 1)$ to mean that the random variable has the *uniform distribution* on the interval $(0, 1)$. Easy computations reveal that $E(U) = 1/2$ and $V(X) = 1/12$.

Example 1.2. Suppose that X is a measurement of a physical quantity θ, subject to a random error. Assume that the random error follows a Gaussian or normal distribution with standard deviation parameter $\sigma > 0$. That is, $E(X) = \theta$ and $V(X) = \sigma^2$. Then the probability model for X is fully described by the pdf

$$f(x) = \frac{1}{\sqrt{2\pi}\sigma} e^{-\frac{(x-\theta)^2}{2\sigma^2}}, \quad -\infty < x < \infty, \tag{1.7}$$

with the real line $\mathbb{R}$ as its sample space. The corresponding cdf is

$$F(x) \equiv P(X \leq x) = \int_{-\infty}^{x} f(t)dt, \quad -\infty < x < \infty. \tag{1.8}$$

This *normal distribution* model is denoted by $N(\theta, \sigma^2)$, and we write $X \sim N(\mu, \sigma^2)$. This probability calculation above can be reduced to those involving only the standard normal distribution, $N(0, 1)$ using a change-of-variable. Let

$$Z = \frac{X - \theta}{\sigma}$$

be a new random variable, a function of the random variable X. Denote by $F_Z(z)$ the cdf of Z. Then

$$F_Z(z) = P(Z \leq z) = P(X \leq \theta + \sigma z) = \int_{-\infty}^{\theta + \sigma z} \frac{1}{\sqrt{2\pi}\sigma} e^{-\frac{(x-\theta)^2}{2\sigma^2}} dx.$$

With this change-of-variable method, the cdf can be written as

$$F_Z(z) = \int_{-\infty}^{z} \frac{1}{\sqrt{2\pi}} e^{-\frac{x^2}{2}} dx = \int_{-\infty}^{z} \phi(x)dx,$$

where

$$\phi(x) = \frac{1}{\sqrt{2\pi}} e^{-\frac{x^2}{2}}, \quad -\infty < x < \infty$$

denotes the standard normal pdf. The standard normal cdf is given by

$$\Phi(z) = \int_{-\infty}^{z} \phi(x)dx, \quad -\infty < z < \infty.$$

Scientists are familiar with the specification of probability in terms of random variables because observed data are treated as realizations of random variables. It may be less familiar to some that specified probability can be also represented by random variables that may be purely imaginary and have no physical meaning. In other words, random variables can be introduced for simplicity of mathematical operation and as a tool for reasoning with uncertainty already described by probability models. This is illustrated by the Bernoulli model.

Example 1.3. Consider the binary random variable X from the coin-flipping experiment. If θ is the probability of "success," then we can write $X \sim \text{Ber}(\theta)$ for the *Bernoulli distribution*, where the probability mass function of X is given by

$$P(X = x) = \begin{cases} \theta & \text{if } x = 1, \\ 1 - \theta & \text{if } x = 0. \end{cases}$$

This Bernoulli model can be represented equivalently (Exercise 1.6) by

$$X = I_{[0,\theta]}(U) \quad \text{where} \quad U \sim \text{Unif}(0,1),$$

and $I_{[a,b]}(x)$ is the indicator function for the interval $[a,b]$, i.e., it takes value 1 if $x \in [a,b]$ and value 0 otherwise. In this example, U is not a quantity that lives in the "real world" as it is not a part of the actual experiment that generates the observable X. However, such purely imaginary quantities can be introduced for the purpose of reasoning with probability models.

It should be noted that the imaginary variable U in the previous example is of a different nature compared to the missing data or latent variables introduced in popular computational methods such as the Expectation–Maximization (EM) and Gibbs sampler algorithms, which are designed to simplify certain computations; for detailed discussion of these methods, see [160] and references therein. Missing data introduced in such problems do not change uncertainty specified in the inferential problems. The representation of uncertainty using imaginary quantities, which we call *auxiliary variables*, shall be used to help reason with uncertainty. For example, the auxiliary variable Z in the unit normal problem $X = \theta + Z$ with unknown mean θ is not treated as a missing value in Bayesian and frequentist approaches, but rather as a device for specifying the sampling model or likelihood. Auxiliary variables play a fundamental role in the approach to inference taken in this book.

As a last point in this section about probability, we should mention that probabilities can have different interpretations. One is a frequency interpretation, which is the most common, and maybe the only interpretation presented in standard textbooks. In this context, the *frequency probability* of an event E is defined as the limiting relative frequency, or proportion of instances, of E in independent but otherwise identical experiments, such as tossing a coin. That is, under the frequency setup, when we say that the probability a coin lands on Heads is θ, then the understanding is that, in an infinite sequence of coin tosses, the limiting proportion of Heads observed equals θ. Mathematically, this can be written as

$$\theta = \lim_{n \to \infty} \frac{|\{i = 1, \ldots, n : X_i = 1\}|}{n},$$

where $X_1, \ldots, X_n$ are independent binary random variables, where $X_i = 1$ corresponds to the coin landing on Heads, and $|B|$ denotes the cardinality of a finite set B. It is straightforward to extend this idea to any other setting, not just Bernoulli trials. The other interpretation is a subjective one, and we call this *belief probability*. In this case, the belief probability assigned to an event E need not correspond to a limiting proportion in a sequence of experiments. The common setting in which this is introduced is in a betting context; a nice presentation is in Kadane [142]. Fortunately, the probability calculus is the same regardless of the interpretation, but we claim that the interpretation is important. In particular, as we argue in this chapter and the next, the conversion of frequency probabilities, which come from the posited model for the observable data, to belief probabilities about the parameter of interest is the essential but often overlooked step in statistical inference.

1.2.2   Classical statistics

The statistics problem is almost a reversal of a probability problem. In the latter, everything about the experiment is known and the goal is to make some statement about realizations of that experiment. In the former, only the realizations are known, and the goal is to make statements about the experimental settings that produced them. There are very good textbooks on classical statistics, including [40] and [22] and, at a bit higher level, [158] and [159]. We assume that readers are familiar with the standard material in these texts. Our goal in this section is just to introduce the notation to be used throughout and briefly review some basic but relevant concepts.

Let X represent the observable data, a random variable, taken values in a sample space $\mathbb{X}$, and write the distribution of X as $\mathsf{P}_{X|\theta}$. Here X can be a scalar or a vector. The idea behind this notation is that there is actually a collection of possible distributions for the observable data, and we call this the *sampling model*. That is, the postulated sampling model is that one of the distributions in the class $\{\mathsf{P}_{X|\theta} : \theta \in \Theta\}$ is responsible for producing X, and it is the goal of the statistician to identify which one. We refer to θ as the parameter of interest, and Θ the parameter space.

In classical statistics, there are essentially three different problems that one can encounter, and we discuss these three briefly next.

- *Point estimation.* Here the goal is to identify some function $\hat{\theta} : \mathbb{X} \to \Theta$ such that the *estimator* $\hat{\theta}(X)$ has desirable properties, such as small mean square error, consistency, etc. This is a standard approach in the frequentist framework, but point estimators derived from Bayesian posterior distributions (discussed in Section 1.2.3) are also frequently encountered. We do not consider point estimation to be genuine "statistical inference" as we discuss in Chapter 2, so we will not have much to say about it in the book. In fact, point estimation can be treated as a special case of the two problems to be discussed next; see Exercise 1.12.

- *Hypothesis testing*. This problem begins with a pair of disjoint subsets Θ_0 and Θ_1 of Θ and the goal is to decide which of the two subsets is more likely to contain the true θ. Formally, the two hypotheses are written as $H_0 : \theta \in \Theta_0$, the null hypothesis, versus $H_1 : \theta \in \Theta_1$, the alternative hypothesis. Often, Θ_1 is the complement of Θ_0, i.e., $\Theta_1 = \Theta_0^c$, and in such cases we can write $H_1 : \theta \notin \Theta_0$. The classical goal is to define a function $\delta : \mathbb{X} \rightarrow \{0,1\}$, called a testing rule, such that $\delta(X)$ has desirable properties in terms of Type I and Type II error rates; see Exercise 1.8. In our view, hypothesis testing is the fundamental problem in statistical inference. However, as will be made clear in the book, our objectives are different than those in the classical testing context.

- *Set estimation*. This problem is based on the idea that point estimation is not really "inference," so the goal is to identify a function $C : \mathbb{X} \rightarrow 2^\Theta$ such that $C(X)$ is a set of reasonable values of θ based on the observed data, where "reasonable" is measured in terms of sampling properties, i.e., coverage probability; see Exercise 1.11. This problem is a dual to the hypothesis testing problem since, given a test δ_{θ_0} for testing $H_0 : \theta = \theta_0$, a set $C(X)$ can be defined as $\{\theta_0 : \delta_{\theta_0}(X) = 0\}$. Then the set $C(X)$ inherits its properties from the test. In this book, we can similarly derive set estimates from our basic inferential output but, as will be made clear, our sets have a simpler interpretation.

Two important and related concepts that arise in classical statistics are information and likelihood. Loosely speaking, the concept of information is important because those data summaries discussed above should not throw away any relevant features in the data. A way to make this precise is to identify those functions of the data X that retain all the relevant features. A statistic $T = T(X)$ is a *sufficient statistic* if the conditional distribution of X, under the sampling model $\mathsf{P}_{X|\theta}$, given T, does not depend on θ. In other words, knowing the value of T is sufficient to generate a new set X' of probabilistically similar data.

Sufficiency is a desirable property, but the definition is not helpful for identifying a sufficient statistic, nor does it really explain what is the "relevant information." For this, we need to introduce the *likelihood function*. Suppose that the distribution $\mathsf{P}_{X|\theta}$ admits a pmf or pdf, which we write as $p_\theta(x)$. Then the likelihood function is $L_X(\theta) = p_\theta(X)$, the pmf or pdf at the observed X but treated as a function of θ. The intuition behind the choice of name is that, given X, a θ for which $L_X(\theta)$ is large is "more likely" to be true value compared to a θ' for which $L_X(\theta')$ is small. The name "likelihood" was coined by Fisher [96]:

> What has now appeared is that the mathematical concept of probability is ... inadequate to express our mental confidence or indifference in making ... inferences, and that the mathematical quantity which usually appears to be appropriate for measuring our order of preference among different possible populations does not in fact obey the laws of probability. To distinguish it from probability, I have used the term "likelihood" to

designate this quantity; since both words "likelihood" and "probability" are loosely used in common speech to cover both kinds of relationship.

Fisher's point is that $L_X(\theta)$ is a measure of how *plausible* θ is, but that this measure of plausibility is different from our usual understanding of probability. We understand p_θ, for fixed θ, as a pre-experimental summary of our uncertainty about where X will fall, but the likelihood $L_X(\theta)$ gives a post-experimental summary of how likely it is that model $P_{X|\theta}$ produced the observed X. In other words, the likelihood function provides a ranking of the possible parameter values.

The likelihood helps us to deal with the issue of sufficient statistics and a measure of information. The first is that the celebrated Neyman–Fisher factorization theorem which says roughly, that the shape of the likelihood function depends on data X through a sufficient statistic only, which provides a tool for identifying a sufficient statistic in a given problem. The second is the notion of *Fisher information*, which, for scalar θ, assuming certain regularity conditions, is defined as

$$I(\theta) = \mathsf{E}_{X|\theta}\left\{\left(\frac{\partial}{\partial \theta}\log p_\theta(X)\right)^2\right\} = -\mathsf{E}_{X|\theta}\left(\frac{\partial^2}{\partial \theta^2}\log p_\theta(X)\right).$$

The definition can be extended in a straightforward way for vector θ, at the expense of more complicated notation. The Fisher information provides a measure of information in data and/or in a function of data. Let $T = T(X)$ be a statistic, and write $I_T(\theta)$ for the Fisher information in an observation of T; similarly, write $I(\theta) = I_X(\theta)$ to emphasize that it is based on an observation of X. Then

$$I_T(\theta) \leq I_X(\theta),$$

with equality if and only if T is a sufficient statistic; see Schervish [215], Theorem 2.86, for a proof of this claim. This provides a clear explanation of in what sense a sufficient statistic $T = T(X)$ retains all the information in the full data X. We will see another application of the Fisher information in Chapter 2.

Finally, we briefly address the question of how to construct point estimators, tests, etc. For this, the likelihood function is again a useful tool. Following Fisher's intuition, a natural choice for an estimator $\hat{\theta}(X)$ is the maximizer of the likelihood function, called the *maximum likelihood estimator*, or MLE, which can be interpreted as the most plausible value of θ for the given X. The exact distribution of $\hat{\theta}(X)$ generally is not known, however, in many problems, a reasonable approximation is

$$\hat{\theta}(X) \sim \mathsf{N}(\theta, \{nI(\theta)\}^{-1}),$$

where $X = (X_1, \ldots, X_n)$ is a n-vector of independent and identically distributed (iid) observations and n is large. From this approximate distribution, the construction of tests and set/interval estimates is straightforward. The statistical software widely used by practitioners usually produces output based on this approximation.

1.2.3   Bayesian statistics

The Bayesian approach starts with the basic premise that probability is the appropriate tool for describing uncertainty. So, since the parameter θ is unknown, our description of the sense in which it is unknown should be described via probability. So, the Bayesian approach begins with a *prior distribution* Π for θ that encodes one's beliefs about the whereabouts of θ before data is observed. Suppose that Π has a pdf $\pi(\cdot)$ on Θ. The basic assumption is that the prior distribution is meaningful, so that $L_X(\theta)\pi(\theta)$ is a justifiable full probability model for (X, θ). In this case, statistical inference is simply a probability exercise, as discussed above. In particular, the Bayes theorem provides a tool for producing a *posterior distribution* Π_X, which encodes one's updated beliefs about θ based on the data X. In particular, the posterior distribution Π_X has a pdf, denoted by π_X and is given by

$$\pi_X(\theta) = \frac{L_X(\theta)\pi(\theta)}{\int_\Theta L_X(\theta)\pi(\theta)\,d\theta}, \quad \theta \in \Theta. \tag{1.9}$$

The result in (1.9) looks similar to that simple Bayes formula given in Section 1.2.1 but the proof is a bit more involved when working with pdfs; see, for example, [215], Theorem 1.31 and its proof, for the technical details. The use of the posterior distribution in (1.9) for statistical inference dates back to Bayes and Laplace; see [231]. We should also mention that the above calculations are valid even if the prior π is *improper*, i.e., $\int \pi(\theta)\,d\theta = \infty$. In that case, the assumption behind the formula (1.9) is that the posterior π_X is *proper*, i.e., $\theta \mapsto L_X(\theta)\pi(\theta)$ is integrable for the given X. The use of improper priors is common, particularly in cases where there is limited prior information available; see Chapter 2.3.1.

We will assume that the reader is familiar with the use of the posterior distribution for constructing point estimators, hypothesis tests, and credible regions as discussed in the early chapters of [110] and [111]; see Exercises 1.15–1.17 for some examples.

It is clear that the prior distribution has an impact on the posterior distribution in a number of ways. One is in terms of the interpretation of posterior probabilities. A question is if the posterior distribution should be viewed as a frequency or belief probability. The answer depends on the interpretation of the prior. That is, if Π has a frequency interpretation, then the posterior does too, since the joint distribution (X, θ) naturally inherits the frequency interpretation. However, if Π has a belief interpretation, then the Bayes formula above is understood as an updating of beliefs. Examples of priors with a belief interpretation are the subjective approaches of deFinetti [58, 59] and Savage [212]. More details on the interpretation of priors can be found in [91] and the references therein.

Given the importance of the prior in both the calculation and interpretation of the posterior probabilities, the specification of a good prior is essential, and also challenging. The main focus of this book is the case where limited or no prior information is available, and we will defer our discussion of modern "objective Bayes" using non-informative priors to Chapter 2. Here we discuss some classical issues, motivated by a sort of "keep it simple" strategy. In early years, due to limitations in computing power, analytic simplicity played an important role in the choice of prior [29]. An important class of prior distributions for which the posterior computations are simple are called *conjugate priors*. A prior π is conjugate (relative to the sampling model) if the posterior π_X has the same functional form as π.

Example 1.4. Suppose X has a binomial distribution, i.e., $X \sim \text{Bin}(n, \theta)$, where n is a known integer and $\theta \in [0, 1]$ is unknown; see Exercise 1.4. The pmf for X is

$$p_\theta(x) = \binom{n}{x} \theta^x (1 - \theta)^{n-x}, \quad x = 0, 1, \ldots, n.$$

Consider $\text{Beta}(a, b)$ prior distribution for θ, with pdf

$$\pi(\theta) \propto \theta^{a-1} (1 - \theta)^{b-1}, \quad \theta \in [0, 1];$$

the quantities $a > 0$ and $b > 0$ are specified numbers, called hyper-parameters. It is easy to see (Exericse 1.14) that the posterior distribution has pdf

$$\pi_X(\theta) \propto \theta^{a+X-1} (1 - \theta)^{b+n-X-1}, \quad \theta \in [0, 1],$$

so that the $\text{Beta}(a, b)$ prior distribution is conjugate for the binomial model.

Conjugate priors can be used to construct other priors which are simple, in the sense that certain posterior computations can be done analytically. For example, mixtures of conjugate priors are simple, as are certain hierarchical models with conjugate priors at the top layers of the hierarchy. In particular, in nonparametric problems, Dirichlet process priors [88, 190] are used to model probability distributions, and employing a conjugacy structure at the higher levels of a hierarchical model will greatly simplify computations [83, 171, 170].

What justification is there for choosing a prior simply because it provides relatively simple posterior computations? An answer to this question comes from asymptotic theory, in particular, a result that is often referred to as a Bernstein–von Mises. An easily accessible discussion can be found in [110] and a precise statement and proof is given in [111]. Roughly, the Bernstein–von Mises theorem states that, under some mild regularity conditions, when the sample size n is large, the posterior distribution for θ closely resembles a normal distribution, $N(\hat{\theta}_n, \{nI(\hat{\theta}_n)\}^{-1})$, where $\hat{\theta}_n = \hat{\theta}_n(X)$ is the MLE of θ based on data X of size n, and $I(\theta)$ is the Fisher information. This closely agrees with the standard first-order asymptotic theory for the MLE presented above. There are two interpretations of this result. First, since the posterior approximation does not depend on the choice of prior, it means that the prior does not have much of an impact asymptotically, which provides some justification of a prior motivated by simplicity. Second, the connection with the limiting distribution of the maximum likelihood estimator suggests that the posterior probability of certain events can be assigned a frequency interpretation, at least asymptotically.

An important point, in the context of interpretation of probability, is that a prior that has a belief probability interpretation is updated, via Bayes formula, to a posterior distribution that also has a belief probability interpretation. Thus, the frequency probability coming from the sampling model has been successfully converted into a belief probability for inference. However, how can we interpret posterior probabilities if the prior has no belief (or frequency) interpretation? This is a relevant question in the context of this book, where no prior information is available. In such cases, the modern Bayesian approach is to introduce a sort of "default" prior, often improper, that has no probability interpretation. Then, except for the limited large-sample cases mentioned in the previous paragraph, there is no clear interpretation of the posterior probabilities, which is a problem. See Chapter 2.3.1.

1.3 Scientific inference: An overview

Science can be defined as the systematic search for knowledge about the physical world based on observation and experimentation. Statistical inference plays a fundamental role in the scientific method, as it provides a framework by which experience, in the form of observed data, can be converted into knowledge. We define statistical inference as follows:

> statistical inference *provides meaningful probabilistic summaries of evidence available in the observed data concerning the truthfulness and falsity of any assertion or hypothesis about the unknown quantities of interest.*

The remainder of this section is devoted to clarifying what we mean here, within the context of the simple location estimation problem.

Suppose that the available data consists of a collection of observations, $X_1, \ldots, X_n$, where each X_i represents a measurement of a common physical quantity, θ, but subject to an unobserved random error, which we write as $Z_i = X_i - \theta$, $i = 1, \ldots, n$. Suppose further that the observations are independent. We shall assume here that an appropriate model for these data is

$$X_1, \ldots, X_n \overset{iid}{\sim} N(\theta, \sigma^2), \tag{1.10}$$

where θ is the mean and σ^2 is the variance. We have a bit more to say about the modeling step below.

The goal is to make inference on the mean θ, the quantity of interest. That is, we want to assign belief probabilities to any assertion A of interest about θ. Examples of such assertions include $A = [\theta_0, \theta_1]$ for some fixed $\theta_0 \le \theta_1$. Since the sampling model (1.10) implies some uncertainty about θ given the observed $X = (X_1, \ldots, X_n)$, at least for the case $\sigma > 0$, our belief probabilities are generally strictly between 0 and 1. Moreover, since the uncertainty is controlled by the sampling model, our belief probabilities should have some connection to the sampling model as well. We will write $\mathsf{bel}_X(A)$ for the *belief* in the truthfulness of the assertion A given data X. As will be demonstrated below, the belief is specified by working with auxiliary variables whose probability distribution is associated with (1.10). Therefore, although our beliefs will have different mathematical properties than an ordinary probability, see below, we will use the two terms—belief and probability—interchangeably.

Besides the belief assigned to A, we will also assign a belief to A^c, which corresponds to evidence supporting the falsity of the assertion A. So, for a given assertion A, we would report the pair $(\mathsf{bel}_X(A), \mathsf{bel}_X(A^c))$ or, equivalently, the pair $(\mathsf{bel}_X(A), \mathsf{pl}_X(A))$, where

$$\mathsf{pl}_X(A) = 1 - \mathsf{bel}_X(A^c)$$

denotes the *plausibility* of the assertion A, a measure of the evidence in data X that does not support the falsity of A. Here is the point where belief diverges from the usual probability. Evidence that does not support the falsity of A need not support the truthfulness of A. As a simple example, consider a criminal trial in court. For the assertion $A = \{$defendant is guilty$\}$, witness testimony may conflict with the defendant's claimed alibi, which is evidence that does not support the falsity of A. However, this testimony does not provide direct evidence for the truthfulness of A. Mathematically, this property can be represented as

$$\mathsf{bel}_X(A) \leq \mathsf{pl}_X(A), \quad \text{for all assertions } A. \tag{1.11}$$

From the definition of plausibility, we can now see that bel_X is not a probability measure in the usual sense, for it does not satisfy the familiar complementation rule, $P(A^c) = 1 - P(A)$, which is a simple consequence of the additivity axiom of probability (Exercise 1.3). We believe that a break from the usual theory of probability is required for objective scientific inference; see Section 2.4. In some contexts, the belief and plausibility are referred to as lower and upper probabilities, respectively. The difference $\mathsf{pl}_X(A) - \mathsf{bel}_X(A)$ which, in some sense, describes how different bel is from probability, is called the "don't know" probability [67].

This book is focused on formalizing the above procedure, which can be described algorithmically as follows:

$$(\text{model, data, assertion}) \mapsto (\mathsf{bel}(\text{assertion}), \mathsf{pl}(\text{assertion})). \tag{1.12}$$

That is, in our view, statistical inference boils down to processing the model, which describes the uncertainty, and the observed data to produce, for any assertion of interest, an associated pair of numbers that represent the belief and plausibility of that assertion. Then the relative magnitudes of the belief and plausibility can be used for decision-making, etc. Each of the items appearing in (1.12) will be discussed further below. Although the ideas presented throughout this book are new and perhaps unfamiliar, we want to emphasize that our goal is to keep the formulation of the scientific inference problem, e.g., (1.12), and its solution, as simple as possible. That is, throughout the book we try to stick to the following principle:

everything should be made as simple as possible, but no simpler,

a quote attributed to Albert Einstein.

Data. The data component in (1.12) stands for the observable quantities; in the normal mean illustration above, $X = (X_1, \ldots, X_n)$ is the data, the observable and eventually observed output of the experiment or observational study.

Model. The model component in (1.12) stands for the posited rule that describes how the data component was generated which, in turn, describes how the observable data and the unknown quantities of interest are related. In the above illustration, we assumed a normal model, i.e., $X_1, \ldots, X_n \overset{\text{iid}}{\sim} N(\theta, \sigma^2)$. More generally, the model for the observable data, written as $P_{X|\theta}$, is a family of probability distributions indexed by a parameter θ, taking values in a space Θ, called the parameter space, and usually a subset of some Euclidean space. The quality of the inferential output is closely tied to the adequacy of the model, so the choice of model is of considerable importance, particularly when the sample size is small. In this book, however, the focus is on the inference step, so, for the most part, we will take the model to be given. Although later chapters will say a bit about the model-building step in some applications, we refer the reader to [46, 238] for general details on model-building.

Assertion. Now that we have observed/observable data, X, and a model, $P_{X|\theta}$, that links the data to the unknown parameter θ of interest, we can say that the assertion component in (1.12) stands for a subset A of the parameter space Θ. In the normal mean illustration above, the parameter space for the mean θ is $\Theta = \mathbb{R}$, so, for example, any interval would be an assertion. Usually, the assertion (or collection of assertions) of interest will be determined by the context, as in the more familiar case of hypothesis testing. Moreover, as our goal is to provide meaningful belief and plausibility measures for *assertions of interest*, in some cases our description of the "$\mapsto$" step, which is the main emphasis in this book, will depend on those assertions. It should be pointed out that the assertion could be *any* subset of Θ, that is, as we will see later, A being non-measurable or having measure zero will not be of any concern.

Inferential output. The inferential output component in (1.12) is the pair of mappings, bel and pl, which map 2^{Θ} to $[0, 1]$. As discussed above, the subadditivity (1.11) means that our "degrees of belief" in A, given data, is represented by an interval. A main premise of this book is that, in general, this interval cannot be made simpler without sacrificing validity. By the nature of the scientific process, it is essential that conclusions made are verifiable and reproducible, and this demands that statistical degrees of belief can be interpreted on a meaningful and scientifically agreed upon scale. Our *validity* criterion guarantees that the belief and plausibility functions are calibrated to that scale, and can therefore admit a common interpretation across researchers and experiments. As we will demonstrate, Dempster's "don't know" probability, namely the non-zero width of our degrees of belief, is the driving force behind our validity results.

To end this section, we will make two comments on the interpretation—in fact, dual interpretation—of the inferential output in (1.12), which is based primarily on the use of "auxiliary variables" in the construction; see Section 1.4.2.

- The inferential output should be interpreted as a belief probability, or a range of belief probabilities. That is, there is no notion of frequencies in its interpretation, it is simply a measure of the evidence in data supporting the truthfulness and falsity of the assertion in question. However, as will be made clear in what follows, these belief probabilities are obtained using the usual (measure-theoretic) probability, albeit on a different auxiliary space to be defined later. The point is that we can carry out well-defined probability calculations on one space and then interpret the results of those calculations as belief probabilities on a different space.

- In order for the belief probabilities to have any meaning in the statistical problem at hand, they must be connected in some way to both the observed data and the postulated sampling model. This, again, is carried out through the use of auxiliary variables. The key point to be made here is that these auxiliary variables have a distribution that can be interpreted on a frequency scale, and the use of these auxiliary variables in constructing our inferential output makes it possible to calibrate our belief probabilities for a common interpretation by intelligent minds. As discussed above, this calibration, or validity, property is essential to the scientific process. Moreover, this use of auxiliary variables with associated frequency probability to construct inferential output with a belief probability interpretation, makes our earlier point of converting frequency probability to belief probability for the purpose of scientific inference more concrete.

1.4 Prediction and inference

1.4.1 *Predicting realizations of random variables*

We will see that the prediction of realizations of a random variable is fundamental to the goal of inference. By "prediction" here we mean simply reinterpreting the frequency probability of an event such as $\{X \in B\}$ as a belief probability. The numerical values of these two probabilities are equivalent, but their interpretation is different. The latter assumes that the experiment that generates X is yet to be performed, and the probability is understood as a measure of the chance that the to-be-observed X will take value in B. The former, on the other hand, can cover the case that the experiment has been performed—hence X is not really "random"—but the realization has not been observed, and the probability is understood as a measure of our belief that X has taken value in B. Clearly, the distinction rests on whether the experiment that generates X has taken place or not, but the main point is that there is no practical difference between the two cases when the probability has a belief interpretation. The statistical inference problem, as we describe below, can be viewed as one where the experiment has taken place so the goal is prediction of an unobserved auxiliary variable that is required. Here we discuss just two simple examples of this transition of frequency probabilities to belief probabilities in the context of predicting the realization of a random variable.

Example 1.5. Suppose that a Bernoulli trial is performed and results in the realization of a Bernoulli random variable $X \in \{0,1\}$, where the success probability θ sits in the interval $\Theta = [0,1]$. The frequency interpretation of probabilities associated with this model has been discussed before. In this case, suppose the trial has been performed but the resulting X is unobserved. Then the probabilities associated with the events $\{0\}$ and $\{1\}$ cannot be interpreted as a frequency, but as a measure of our belief in the respective events. There is no trouble in assigning our belief probabilities to be the same as the frequency probabilities, i.e., $\mathrm{bel}(\{1\}) = \theta$ and $\mathrm{bel}(\{1\}) = 1 - \theta$.

Example 1.6. Suppose that one is interested in predicting a realization of the standard normal random variable $X \sim \mathrm{N}(0,1)$. An assertion on X is a subset A of $\mathbb{R}$. If A is measurable, i.e., is contained in the Borel σ-algebra $\mathscr{B} = \mathscr{B}(\mathbb{R})$, of subsets of

$\mathbb{R}$, then the belief probability for the event $\{X \in A\}$ can be set equal to the frequency probabilities: $\mathrm{bel}(A) = \mathrm{P}(X \in A) = \int_A \phi(x)\,dx$. However, an interesting point is that A need not be measurable to assign a belief probability. For example, define an "inner approximation" of A by measurable sets

$$A_\uparrow = \bigcup_{B \in \mathscr{B}, B \subseteq A} B.$$

Then the belief probability $\mathrm{bel}(A)$ can be assigned the value $\mathrm{P}(X \in A_\uparrow)$. Similarly, the belief probability $\mathrm{bel}(A^c)$ can be defined as $\mathrm{P}(X \in A_\uparrow^c)$, where $A_\uparrow^c$ is the inner approximation of A^c.

1.4.2 Statistical inference via prediction

It was seen that, for the prediction of realizations of a random variable, it is straight-forward to convert the available frequency probabilities to belief probabilities. A simple and important consequence of this conversion is that the belief probabilities inherit the calibration properties of their frequency counterparts, which is relevant for the validit of the resulting inference. The more interesting case is where there is not a full probability model for all unknowns. In particular, we consider the case where the sampling model for the observable data depends on an unknown parameter, and the goal is to make inference on that parameter. As we will see, there is an associ-ated auxiliary random variable, and its prediction, in the sense of Section 1.4.1, is equivalent, in a certain sense, to the inference problem.

For simplicity, we will stick with the normal mean illustration. That is, X is a measure of an unknown physical quantity θ, subject to a random error Z, i.e., $X = \theta + Z$, and Z is assumed to satisfies $Z \sim \mathsf{N}(0,1)$. The main point, which is the driving force behind all of the developments in this book, is the following obvious but important observation:

> if Z were observed, then we could solve for $\theta = X - Z$ exactly, which corresponds to the best possible inference.

The problem, of course, is that the auxiliary variable Z cannot be observed, so this "best possible inference" is unattainable. The next best thing is to accurately predict the unobserved value of Z, which is related to what was presented in Section 1.4.1. The remainder of the book is devoted to describing how a framework for statistical inference is built around the idea of accurate and efficient prediction of unobservable auxiliary variables. Here, as a sort of preview of the upcoming chapters, we elaborate on some of the key steps.

A first point is that, among X, θ, and Z, only Z is (marginally) predictable. By "predictable" we mean that there is a fully specified probability distribution whose frequency probabilities can be converted, as in Section 1.4.1, to belief probabilities for prediction and ultimately inference. On the other hand, X is not predictable in this sense because its probability distribution depends on θ, which is unknown. An important observation is that there are two ways that we can attempt to predict Z, depending on what frequency probability distribution we assign to it. One idea is to use the "conditional distribution" of Z, given $X = x$, for the prediction problem. However, given $X = x$, the distribution of Z is degenerate, i.e., $Z = x - \theta$ with probability 1. Since θ is unknown, this distribution is unknown, so Z is *not* conditionally predictable. See the relevant section on fiducial inference in Chapter 2 for more discussion of this point. The remaining option is to use the "marginal distribution" of Z for the prediction problem. That is, perform the prediction step based on the $N(0, 1)$ model, ignoring X entirely. This is the approach we advocate throughout the book.

Toward inference on θ, consider an interval assertion of the form

$$A = A_{\theta_0, \delta} = [\theta_0 - \delta, \theta_0 + \delta],$$

where $\theta_0 \in \mathbb{R}$ and $\delta \geq 0$ are specified values. From the connection among (X, θ, Z), there is a corresponding interval on the Z-space, namely,

$$[X - \theta_0 - \delta, X - \theta_0 + \delta], \tag{1.13}$$

so we could assign our belief probability for A to be the frequency probability, with respect to the distribution $N(0, 1)$ for Z, of the interval (1.13). The problem with this setup is that the event above depends on X and, hence, indirectly on Z, the quantity to be predicted; see [81] for some relevant discussion. This inconsistency is a result of the link between Z and X that depends on an unknown θ. The idea is to break this link, so that meaningful prediction of Z can be carried out independent of X.

Here we describe one simple idea to break the link between X and Z, which appeared early in the development of the inferential models; see [186, 269]. The idea begins with a set-valued mapping $z \mapsto \mathcal{S}_z$, a map from the Z-space $\mathbb{Z} = \mathbb{R}$ to the power set $2^{\mathbb{Z}}$, with the constraint that $\mathcal{S}_z$ contains z. If we consider the random variable Z, then $\mathcal{S}_Z$ is a *random set*, with distribution determined by that of Z and the structure of the mapping $z \mapsto \mathcal{S}_z$, such that $\mathcal{S}_Z$ contains Z with probability 1. We will have a lot more to say about random sets in this book; see Chapters 4–5.

To facilitate the discussion, write Z' for a random variable with the same distribution as Z, i.e., $Z' \sim N(0, 1)$. The idea is to use $\mathcal{S}_{Z'}$, for $Z' \sim N(0, 1)$, to predict Z independent of X, that is, we interpret a realization of $\mathcal{S}_{Z'}$ as containing Z, and we refer to this as a *predictive random set*. A consequence of the belief that $Z \in \mathcal{S}_{Z'}$ is that $\theta \in X - \mathcal{S}_{Z'}$, where

$$X - \mathcal{S}_{Z'} = \{\theta' : X = \theta' + z' \text{ for some } z' \in \mathcal{S}_{Z'}\}.$$

Based on this representation, we can write the belief probability for $A = A_{\theta_0, \delta}$ as

$$\text{bel}_X(A) = P\{X - \mathcal{S}_{Z'} \subseteq A\} \tag{1.14}$$

and, similarly,

$$\text{bel}_X(A^c) = P\{X - \mathcal{S}_{Z'} \subseteq A^c\}. \tag{1.15}$$

These two are belief probabilities defined to represent knowledge on the status of A and A^c, so (1.14) and (1.15) are subject to evaluation so that they are meaningful. From the viewpoint of scientific reproducibility, suppose we have n scientists who

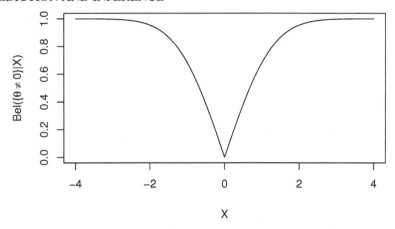

Figure 1.1 *The belief of $A = \{\theta \neq 0\}$ given X.*

have performed the same experiment independently to gain knowledge about the same physical quantity under the same postulated sampling model $N(\theta, 1)$. Scientist i reports a belief probability $\mathrm{bel}_{X_i}(A)$ based on his/her observation X_i, $i = 1, \ldots, n$. These belief probabilities should be consistent with one another in some sense, relative to the choice of predictive random set. Since the sampling model for $X_1, \ldots, X_n$ is iid, a sensible evaluation criterion is to consider the distribution of the belief probabilities as functions of the observables, $X_1, \ldots, X_n$. Ideally, $\mathrm{bel}_X(A)$ should be large and $\mathrm{bel}_X(A^c)$ should be small when A is true; likewise, $\mathrm{pl}_X(A)$ should be small and $\mathrm{bel}_X(A^c)$ should be large when A is false. However, the predictive random set has a fixed distribution, so only one of these cases can be controlled, and we opt to force $\mathrm{bel}_X(A)$ to be stochastically small, as a function of X, when A is actually false and, similarly, $\mathrm{bel}_X(A^c)$ is stochastically small when A is actually true. See Exercise 1.18 for details on stochastic ordering. More specifically, we require that

$$\sup_{\theta \notin A} P_{X|\theta}\{\mathrm{bel}_X(A) \geq 1 - \alpha\} \leq \alpha, \quad \forall\, \alpha \in (0,1), \qquad (1.16)$$

and

$$\sup_{\theta \in A} P_{X|\theta}\{\mathrm{bel}_X(A^c) \geq 1 - \alpha\} \leq \alpha, \quad \forall\, \alpha \in (0,1). \qquad (1.17)$$

The criteria (1.16) and (1.17) are called the *validity criteria* and their frequency interpretation is further explained in Chapter 4. We say that inferential methods that do not satisfy (1.16) and (1.17) are "approximate inferential methods." Accordingly, valid inference is also said to be *exact*.

It is shown in Chapter 4 that it is easy to construct predictive random sets such that the criteria (1.16) and (1.17) hold. For example, the use of the centered predictive random set

$$S_{Z'} = \{z' : |z'| \leq |Z'|\}, \quad Z' \sim N(0,1),$$

leads to valid belief probabilities. For an illustrative example, consider the assertion

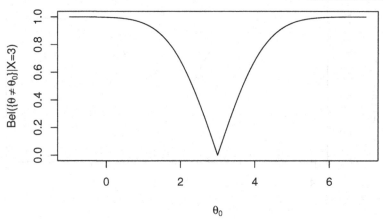

Figure 1.2 *The belief of $A = \{\theta \neq \theta_0\}$ given $X = 3$.*

$A = A_{\theta_0,\delta}$ with $\theta_0 = 0$ and $\delta = 0$, i.e., the point assertion that $\theta = 0$. The belief probability $\text{bel}_X(\{0\})$ equals zero for all X. The belief probability $\text{bel}_X(\{0\}^c)$ for the negation of $\{0\}$ is

$$\text{bel}_X(\{0\}) = \text{P}\{|Z| < |X|\} = 1 - 2\Phi(-|X|).$$

As a function of X, the belief probability $\text{bel}_X(\{0\}^c)$ is shown in Figure 1.1.

Similarly, for a sequence of point assertions of the form $A = A_{\theta_0,0}$, as θ_0 varies, the belief probability for each assertion based on the observed value $X = 3$ is depicted in Figure 1.2. These functions, in particular, the belief function for fixed X and varying θ_0 will be important tools for our inferential framework to be described in more detail in the upcoming chapters.

1.5   Outline of the book

This book focuses on reasoning towards exact probabilistic inference in the cases where a probability model is not specified for all unknowns in the model. Chapter 2 provides a brief review of existing schools of thought, including Bayes, frequentist, fiducial, Dempster–Shafer theory, and other approaches for "distributional inference." Our main thesis is that producing exact prior-free or prior-less probabilistic inference is difficult using these existing frameworks, thus motivating our proposed approach. Our discussion there is critical of these existing methods, but we do believe that these methods are useful and important in the sense that they provide simple alternatives or approximations to the best possible inference. Bayes is perhaps the simplest, provided that a prior (or collection of priors) is available for the unknown parameters that produces sensible inferential results. In addition to its conceptual simplicity, Bayesian inference is also computationally doable thanks to recent computational developments, in particular, Markov chain Monte Carlo [160].

Chapter 3 discusses two simple principles, namely, the validity principle and the efficiency principle, which lay the foundation for the IM framework. Details of the IM framework are presented in Chapter 4, and the construction of convenient predictive random sets is discussed in Chapter 5. Chapters 6–7 describe two advanced IM tools, namely, conditional and marginal IMs, which are based on auxiliary variable dimension-reduction techniques, designed for the purpose of efficiency. Chapters 8–9 describe important application of these two dimension-reduction techniques, respectively, inference in linear models and prediction of future observables. The case of designing optimal predictive random sets for multiple simultaneous assertions is discussed in Chapter 10, with applications to variable selection in regression. Those two dimension-reduction methods are attractive in the sense that they lead to exact and efficient inferential methods, but it appears that they may be difficult to apply in general settings, in particular, settings without a special structure like that in the normal mean example above. To overcome such difficulties, Chapter 11 introduces a simpler approach, called *generalized IMs*. A proposed computational algorithm discussed there fills in the hole that was previously thought difficult, i.e., to compute (exact) confidence regions via Monte Carlo methods. As we explain there, this proposal goes deeper than the well-known bootstrap methods [54, 74]. Finally, in Chapter 12, we make some concluding remarks and suggest some further research topics to be considered.

1.6   Exercises

Exercise 1.1. For a sample space Ω, a σ-algebra $\mathscr{A}$ of subsets of Ω satisfies:

- $\Omega \in \mathscr{A}$;
- $A \in A$ implies $A^c \in \mathscr{A}$; and
- $A_1, A_2, \ldots \in \mathscr{A}$ implies $\bigcup_{i=1}^{\infty} A_i \in \mathscr{A}$.

Use DeMorgan's law to show that if $A_1, A_2, \ldots \in \mathscr{A}$, then $\bigcap_{i=1}^{\infty} A_i \in \mathscr{A}$.

Exercise 1.2. Let $\mathscr{C}$ be a generic collection of subsets of a space Ω. Let $\{\mathscr{A}_\beta : \beta \in B\}$ be a collection of σ-algebras that contain $\mathscr{C}$. Show that

(a) the collection $\{\mathscr{A}_\beta : \beta \in B\}$ is not empty;

(b) $\bigcap_{\beta \in B} \mathscr{A}_\beta$ is a σ-algebra.

[The σ-algebra in part (b), denoted by $\sigma(\mathscr{C})$, is the smallest σ-algebra containing $\mathscr{C}$, called the σ-algebra generated by $\mathscr{C}$. An important example is the *Borel* σ-algebra of the real numbers, denoted by $\mathscr{B} = \mathscr{B}(\mathbb{R})$, which is the σ-algebra generated by the collection of open subsets of $\mathbb{R}$.]

Exercise 1.3. Use the basic axioms of probability to prove the complementation rule, i.e., $\mathsf{P}(A^c) = 1 - \mathsf{P}(A)$ for any measurable A.

Exercise 1.4. The *binomial distribution*, written as $\mathrm{Bin}(n, \theta)$, is the distribution of the number of successes in n independent Bernoulli trials with success probability $\theta \in [0, 1]$. The pmf for a random variable $X \sim \mathrm{Bin}(n, \theta)$ is

$$p_\theta(x) = \binom{n}{x} \theta^x (1 - \theta)^{n-x}, \quad x = 0, 1, \ldots, n.$$

(a) Derive this pmf formula from the definition.

(b) Show that $X \sim \text{Bin}(n, \theta)$ has the same distribution as $Z_1 + \cdots + Z_n$, where $Z_1, \ldots, Z_n \overset{iid}{\sim} \text{Ber}(\theta)$. [*Hint:* Use moment generating functions.]

Exercise 1.5. An important model for discrete data is the *Poisson distribution*, written as $\text{Pois}(\theta)$, where $\theta > 0$ is the mean parameter. The pmf for $X \sim \text{Pois}(\theta)$ is

$$p_\theta(x) = e^{-\theta}\theta^x/x!, \quad x = 0, 1, 2, \ldots.$$

(a) Show that $\text{Pois}(\theta)$ can be understood as a limit of $\text{Bin}(n, \theta_n)$, where $n \to \infty$ and $\theta_n \to 0$ in such a way that $n\theta_n \to \theta \in (0, \infty)$. [*Hint:* Show that the Poisson pmf is obtained as the limit of the binomial pmf under the stated conditions.]

(b) Show that if $X \sim \text{Pois}(\theta)$, then $E(X) = V(X) = \theta$.

(c) Show that if $X_i \sim \text{Pois}(\theta_i)$, independent, $i = 1, \ldots, n$, then $\sum_{i=1}^n X_i$ is distributed as $\text{Pois}(\sum_{i=1}^n \theta_i)$. [*Hint:* Use moment generating functions.]

Exercise 1.6. Show that $X = I_{[0,\theta]}(U)$, for $U \sim \text{Unif}(0, 1)$, has a Bernoulli distribution with parameter θ.

Exercise 1.7. Let F be a continuous and strictly increasing cdf.

(a) Show that X has cdf F if and only if $F(X) \sim \text{Unif}(0, 1)$.

(b) Explain the implications of the result in (a) for simulating random variables.

Exercise 1.8. Let $X_1, \ldots, X_n \overset{iid}{\sim} N(\theta, 1)$, and suppose the goal is to test $H_0 : \theta = \theta_0$ versus $H_1 : \theta > \theta_0$, where θ_0 is some fixed number. Consider a test that rejects H_0 if and only if $\bar{X} > k$ for some suitably chosen k. That is, the test function δ is given by

$$\delta(X_1, \ldots, X_n) = \begin{cases} 1 & \text{if } \bar{X} > k \\ 0 & \text{if } \bar{X} \leq k. \end{cases}$$

The Type I error of the test δ is the probability of rejecting H_0 when it is true, i.e., $P_{\theta_0}(\delta = 1)$. Similarly, the Type II error of the test δ is the probability of not rejecting H_0, i.e., $P_\theta(\delta = 0)$. Note that the latter error probability is actually a function of θ.

(a) For given $\alpha \in (0, 1)$, find $k = k(\theta_0, \alpha)$ such that the above test has Type I error probability equal to α.

(b) Find the Type II error probability function.

(c) Suppose the goal is to have Type I error probability at most α and, at a given θ_1, the Type II error probability at most β. How large does n have to be?

Exercise 1.9. The error probabilities defined in Exercise 1.8 can be computed before data is observed, so they cannot be understood as measures of the evidence in the observed data concerning the truthfulness of H_0 or H_1. A *p-value* does, however, provide some evidence; see [181]. For the test described in the previous problem, the *p*-value is defined as

$$\text{pval}_{\theta_0}(\bar{x}) = P_{\theta_0}(\bar{X} \geq \bar{x}),$$

where $\bar{x}$ is the observed value of $\bar{X}$ in the actual sample.

(a) Derive the *p*-value for this test.

(b) Show that, if H_0 is true, then $\text{pval}_{\theta_0}(\bar{X})$, as a function of $\bar{X}$, is $\mathsf{Unif}(0,1)$.

(c) For given $\alpha \in (0,1)$, show that the test δ', with

$$\delta'(X_1,\ldots,X_n) = \begin{cases} 1 & \text{if } \text{pval}_{\theta_0}(\bar{X}) \le \alpha \\ 0 & \text{if } \text{pval}_{\theta_0}(\bar{X}) > \alpha, \end{cases}$$

is equivalent to the test δ derived in Exercise 1.8.

Exercise 1.10. A general method for constructing tests (and also confidence regions, see Exercise 1.11) is based on *likelihood ratios*. Consider testing $H_0 : \theta = \theta_0$ versus $H_1 : \theta \ne \theta_0$, where θ_0 is some fixed value. Let $L(\theta)$ be the likelihood function, depending on the observed data X, and let $\hat{\theta}$ be the maximum likelihood estimator. Then the likelihood ratio statistic is

$$T = L(\theta_0)/L(\hat{\theta}).$$

Then the likelihood ratio test rejects H_0 iff T is less than a specified cutoff k.

(a) If the sampling model is $X \sim \mathsf{N}(\theta,1)$, write down the likelihood ratio statistic.

(b) Fix $\alpha \in (0,1)$ and find the cutoff $k = k(\theta_0,\alpha)$ such that the likelihood ratio test has Type I error probability α. How does the resulting test compare to that in Exercise 1.8?

(c) Find the p-value of the likelihood ratio test.

Exercise 1.11. Given data/model $X \sim \mathsf{P}_\theta$, consider a X-dependent interval $C(X)$ and define its coverage probability

$$\mathsf{P}_\theta\{C(X) \ni \theta\}. \tag{1.18}$$

We say that $C(X)$ is a $100(1-\alpha)\%$ confidence interval for θ if its coverage probability (1.18) equals $1-\alpha$ for all θ.

(a) Let X be an exponential random variable with mean θ. Derive the likelihood ratio test for $H_0 : \theta = \theta_0$ versus $H_1 : \theta \ne \theta_0$ that has Type I error probability α.

(b) Invert the likelihood ratio test above to construct a set of θ_0 values such that the above likelihood ratio test *does not* reject. Show that this set is a $100(1-\alpha)\%$ confidence interval for θ

Exercise 1.12. Use the experience from the previous exercise to justify the claim that point estimation is a special case of the testing or set estimation problem. For example, consider taking intersections of $100(1-\alpha)\%$ confidence intervals over α, and/or maximizing the p-value function over θ_0.

Exercise 1.13. Calculate the Fisher information for

(a) $\mathsf{N}(\theta,1)$;

(b) $\mathsf{Bin}(n,\theta)$;

(c) $\mathsf{Pois}(\theta)$;

(d) $\mathsf{Exp}(\theta)$, with pdf $(1/\theta)e^{-x/\theta}$ for $x > 0$ and $\theta > 0$;

(e) $\mathsf{N}(\mu,\sigma^2)$;

(f) Gamma(α, β), with pdf $\{\Gamma(\alpha)\beta^{\alpha}\}^{-1}x^{\alpha-1}e^{-x/\beta}$, for $x > 0$ and $\alpha > 0$, $\beta > 0$.

(g) Does Unif$(0, \theta)$ admit a Fisher information?

Exercise 1.14. Show that

(a) a beta prior is conjugate to a Bin(n, θ) sampling model (Example 1.4);

(b) a gamma prior is conjugate for a Pois(θ) sampling model;

(c) a normal prior is conjugate for a N$(\theta, 1)$ sampling model;

(d) and a normal–inverse gamma prior, with hyper-parameters $(\eta, \lambda, \alpha, \beta)$ and pdf

$$\pi(\mu, \sigma^2) \propto (\sigma^2)^{-(\alpha+3/2)} \exp\left\{-\frac{2\beta + \lambda(\mu - \eta)^2}{2\sigma^2}\right\}$$

is conjugate for a N(μ, σ^2) prior.

Exercise 1.15. Let $X = (X_1, \ldots, X_n)$ be an iid sample from N$(\theta, 1)$, and consider testing $H_0 : \theta = \theta_0$ versus $H_1 : \theta \neq \theta_0$. Let the prior be $\theta \sim N(\theta_0, 1)$, i.e., the prior is centered at the null value and scaled the same as the sampling model, which we take to be 1 without loss of generality. The z-statistic is $Z_n = n^{1/2}(\bar{X} - \theta_0)$, where $\bar{X} = n^{-1}\sum_{i=1}^{n} X_i$ is the sample mean.

(a) Argue that the posterior probability for H_0 is useless for testing.

(b) As an alternative, consider the *Bayes factor*, or ratio of marginal likelihoods. For a general point-null, the Bayes factor for H_0 versus H_1 is defined as

$$B_{01}(X) = \frac{L_n(\theta_0)}{\int L_n(\theta)\pi(\theta)\,d\theta},$$

where π is the prior density and $L_n(\theta)$ is the likelihood function. Show that, for this normal mean problem, the Bayes factor simplifies to

$$B_{01}(X) = (1+n)^{1/2}\exp\left\{-\frac{nZ_n^2}{2(1+n)}\right\}.$$

(c) The interpretation of $B_{01}(X)$ is that large values provide support for H_0 and small values provide support for H_1. Show that, in this case, on a sequence of data such that Z_n remains constant as $n \to \infty$, then $B_{01}(X) \to 0$, no matter the value of Z_n, and explain why this is counter-intuitive. (This is called the *Jeffreys–Lindley paradox* [163, 207].)

Exercise 1.16. One way to construct a Bayesian $100(1 - \alpha)\%$ interval estimate— called a *credible interval*—is by returning the middle $100(1 - \alpha)$ posterior probability interval. That is, define the interval $[\ell(X), u(X)]$ such that

$$\int_{\ell(X)}^{u(X)} \pi_X(\theta)\,d\theta = 1 - \alpha.$$

In other words, $\ell(X)$ and $u(X)$ are the lower and upper $\alpha/2$ quantiles of the posterior distribution.

(a) Consider the $\text{Bin}(n, \theta)$ model in Example 1.4. Describe computation of the 95% credible interval.

(b) In general, a 95% credible interval is *not* a 95% confidence interval. Simulate data for different values of n and plot the empirical coverage probability of the 95% credible interval as a function of θ. Does the choice of the beta prior hyper-parameters (a, b) have an effect?

Exercise 1.17. The credible interval construction in Exercise 1.16, based on middle posterior probability, only works for one-dimensional parameters. A more general strategy is to use a highest posterior density region, which is obtained by thresholding the posterior density function. Consider a slice

$$H_X(c) = \{\theta : \pi_X(\theta) \geq c\},$$

and, for a given $\alpha \in (0, 1)$, the $100(1 - \alpha)\%$ highest posterior density credible region is obtained by choosing $c = c_\alpha$ such that $\int_{H_X(c)} \pi_X(\theta) d\theta = 1 - \alpha$.

(a) Describe computation of the highest posterior density region.

(b) Explain how the highest posterior density region might not be connected or, in the one-dimensional case, might not be an interval.

(c) For a one-dimensional θ, argue that the highest posterior density is the same as the central credible interval described in Exercise 1.16 if and only if the posterior is symmetric.

(d) Show that the highest posterior density region is not invariant to transformation. That is, show that the highest posterior density for $g(\theta)$ is not necessarily the image of the highest posterior density region for θ under the mapping g.

Exercise 1.18. A random variable X is said to be stochastically no less than a random variable Y, written $X \leq_{st} Y$, if

$$P(X > x) \geq P(Y > x), \quad \forall x \in \mathbb{R}.$$

(a) Express the stochastic ordering in terms of the cdfs of X and Y.

(b) Express the stochastic ordering in terms of the quantiles of X and Y.

(c) Show that $\text{Bin}(n, \theta_1) \leq_{st} \text{Bin}(n, \theta_2)$, with $\theta_1 \leq \theta_2$.

(d) Show that $N(\mu_1, \sigma^2) \leq_{st} N(\mu_2, \sigma^2)$, with $\mu_1 \leq \mu_2$.

(e) Are $N(0, \sigma_1^2)$ and $N(0, \sigma_2^2)$, for $\sigma_1 \leq \sigma_2$, stochastically ordered?

Exercise 1.19. Let $U \sim \text{Unif}(0, 1)$. For each of the random sets S below, defined via set-valued functions of U, derive the function $\gamma_S(u) = P_S(S \ni u)$, which represented the probability that S contains the fixed point u.

(a) $S = [U, 1]$;

(b) $S = \{u' \in (0, 1) : |u' - 0.5| \leq |U - 0.5|\}$.

Exercise 1.20. As a follow-up to the previous exercise, show that, for both of the random sets defined there, $\gamma_S(U) \sim \text{Unif}(0, 1)$ as a function of U.

Chapter 2

Prior-Free Probabilistic Inference

2.1 Introduction

It is difficult to discuss various attempts to solve the problem of statistical inference, in part because they all have slightly different goals and there is no agreed-upon criteria that can allow one to conclude that one approach is "correct" or better than all the others. The main goal of this chapter is to introduce, in Section 2.2, what, in our opinion, are the essentials for prior-free probabilistic inference. A first conclusion is that the classical frequentist methods, which are focused on the design of procedures with good sampling properties, are not "inference" in our sense. Then, in Section 2.3, we delve into a critical survey of the existing methods which are both prior-free, in the sense that no real prior information is needed to carry out the necessary calculations, and probabilistic, in the sense that the inferential output is a sort of probability distribution. This includes Bayesian methods based on default non-informative priors, Fisher's fiducial inference, the Dempster–Shafer theory of belief functions, and a number of other approaches. Having identified the shortcomings of these existing approaches, we can then address two important questions. First, in Section 2.4, we discuss the role of probability in statistical inference. Our conclusion is that the familiar probability measures are not the appropriate tool for summarizing evidence in data for or against an assertion of interest, and we propose the use of the distribution of a random set, i.e., a *belief function*, as a more appropriate tool. Second, in Section 2.5, we give a high-level explanation of how our new way of thinking can lead to improvements to the existing approaches.

2.2 Probabilistic inference

2.2.1 Definition

Let us begin by recalling the problem setup. We have a sampling model $P_{X|\theta}$, $\theta \in \Theta$, and observable data X. The goal is to learn what we can about θ from the given data, relative to the posited sampling model. In this book we advocate for inference which is "prior-free" and "probabilistic." The first adjective is easy to understand, and is consistent with the stated context in Chapter 1, that is, we assume there is no reliable prior information available about θ. So, our inferential output should not require such prior input. The second adjective is more difficult to understand, and we state our explanation formally in the following definition.

Definition 2.1. A method produces *probabilistic inference* if, given data X, its output is a function $b_X : 2^\Theta \to [0, 1]$ such that the number $b_X(A)$ is a measure of the evidence in data X supporting the truthfulness of the claim "$\theta \in A$" for any $A \subseteq \Theta$

The output b_X in Definition 2.1 is a set-function similar to a probability measure, but we do not assume b_X satisfies the same properties as a probability; see Section 2.4. This definition makes clear what we mentioned briefly in Chapter 1 concerning the important step where frequency probabilities, coming from the sampling model for X given θ, are converted into belief probabilities for inference about θ given X; in fact, we use the notation "b" to emphasize this belief interpretation. However, the definition alone does not imply that the conversion has been carried out such that b_X is meaningful in the sense that it accurately reflects the uncertainty about θ based on a sample X from the posited sampling model. That is, some conditions on the mapping $X \mapsto b_X$ are required to insure that the inferential output is meaningful in a certain sense. We defer discussion of this point, namely, calibration of the inferential output, to Section 2.5 and Chapter 3.

2.2.2   Implications about classical frequentist statistics

By introducing a notion of prior-free probabilistic inference, we are implicitly assuming that any framework that does not satisfy the conditions of Definition 2.1 is not appropriate. Of course, one can certainly question the weight we assign to those conditions, in which case, the consequences we reach are equally questionable. We will defer our explanation of why we require our inferential output to be probabilistic to the end of this chapter and into the next. Since some of the conclusions we reach based on this requirement are provocative, we hope that the reader will delay his/her judgement till after we have made our complete case.

A first conclusion we can reach is that the classical frequentist methods are not "probabilistic inference" in the sense of Definition 2.1. That is, a frequentist method is, essentially, just a rule for decision-making, designed to have certain properties across applications in repeated samples. Take a hypothesis test for example. Given the null hypothesis H_0, a test is a function $\delta(X)$, taking values in the set $\{0, 1\}$, with value 0 meaning "do not reject H_0" and value 1 meaning "reject H_0." It is standard practice to fix the Type I error probability at some level α, a small number in $(0, 1)$, and then select δ among those tests of size α, with low Type II error probability or, equivalently, high power. Although the size and power are probabilities, they can be computed without seeing X, so they do not depend on data. Therefore, they cannot be interpreted as a measure of evidence in data supporting the truthfulness or falsity of H_0. The classical confidence intervals can be handled similarly, as the confidence level is a computation that can be carried out without data. So, the conclusion is that frequentist methods do not produce probabilistic inference in the sense of Definition 2.1; see Section 2.3.4 below for some relevant discussion of confidence distributions. Therefore, we will not consider frequentist methods any further in our survey of existing inferential methods.

Before moving on, a few comments are in order. First, there are certain tools, which fall under the umbrella of classical or frequentist methods, which can be given a probabilistic inference interpretation. One example is the p-value, which is colloquially understood as a measure of evidence in data supporting the null hypothesis and, if the p-value is small, then, naturally, the suggested decision is to reject the null hypothesis. Martin and Liu [181] demonstrated that the p-value can be interpreted as an output resulting from a suitable inferential model (see Chapter 4), therefore, corresponding to probabilistic inference. The authors expect that other frequentist output, such as confidence intervals, can similarly be given a probabilistic inference interpretation, but further work is needed. Second, the repeated sampling properties of frequentist methods are important for scientific inference, in particular, through the notions of reproducibility, etc. Therefore, we do not completely abandon this kind of frequentist calibration; see Chapter 3 and the various validity results that appear throughout the book. In summary, our opinion is that procedures with good properties should be a consequence of quality inferential output. So, we choose to carry out our work at the higher level of providing meaningful and efficient probabilistic inference, and then easily-derived procedures will automatically inherit the desirable validity and efficiency properties. Examples in later chapters will make this clear, and our comparisons of IM-based procedures with existing methods based on frequentist criteria will demonstrate that there is no loss of efficiency in taking this position, in fact, in many cases we will see an efficiency gain due, at least in part, to our careful handling of the underlying uncertainty.

2.3   Existing approaches

2.3.1   Objective Bayes

2.3.1.1   Motivation, Jeffreys prior, and properties

Bayesian methods were reviewed in Chapter 1.2.3. Here we want to focus primarily on the modern situation where a Bayesian analysis is to be carried out in the absence of any meaningful prior information. In other words, we are "ignorant" about the whereabouts of the true parameter θ. However, when a prior is introduced, it inserts a certain amount of information into the problem which, if off the mark, will have a harmful effect on the analysis. Even though that effect will vanish for large samples, there is a desire to work with a prior distribution that, in some sense, has a little effect on the analysis as possible. It turns out that there is no probability distribution that can describe our position of "ignorance" (see Section 2.4) so other ideas are needed. A notion of "prior indifference" was used by Bayes and Laplace to motivate the use of a $\mathsf{Unif}(0,1)$ prior for θ in a $\mathsf{Bin}(n,\theta)$ model. The modern theory of *non-informative priors* tries to generalize this notion of uniformity in a certain way. We refer the reader to [13] and [111] for a discussion of the different viewpoints on non-informativeness and the definition and construction of non-informative priors. Here we will focus mainly on the standard non-informative prior, namely, *Jeffreys prior*. This kind of analysis is often called *objective Bayes*, though we believe that a Bayesian analysis with a default prior is a more appropriate name.

Definition 2.2. Jeffreys rule says to take the prior pdf π for θ to satisfy

$$\pi(\theta) \propto |\det I(\theta)|^{1/2},$$

where $I(\theta)$ is the Fisher information matrix and "det" is the matrix determinant. This prior is called the Jeffreys prior and may be written as π^J, depending on the context.

The Jeffreys rule has the appealing property that it is invariant under transformations of the parameter; see Exercise 2 in Chapter 5 of [111]. However, in most cases, the Jeffreys prior will be improper, i.e., $\int \pi^J(\theta) d\theta = \infty$, but the posterior will generally be proper with a large enough sample. The following example demonstrates the calculation of the Jeffreys prior and reveals that it is a sensible choice, at least for one-dimensional parameters.

Example 2.1. Suppose $X \sim N(\theta, 1)$ and inference on θ is required. It can easily be shown (see Exercise 1.13(a)) that $I(\theta) = 1$, so $\pi^J(\theta) \propto 1$, which is improper. Then, by the symmetry of the $N(\theta, 1)$ density function, it is clear that the posterior distribution of θ, given X, is $N(X, 1)$. Recall the Bayesian credible intervals discussed in Exercise 1.16. In the present case, an upper 95% credible interval for θ is $(-\infty, X + 1.645]$, which is identical to the classical 95% confidence interval and, therefore, has coverage probability exactly equal to 0.95.

That the Jeffreys prior led to a posterior with properly calibrated credible intervals is not a coincidence. In fact, perhaps the strongest theoretical justification for the use of the Jeffreys prior is its "probability matching" property. As first shown in [258], the Jeffreys prior for a scalar parameter is such that one-sided $100(1 - \alpha)\%$ credible intervals $C_\alpha(X) = (-\infty, \theta_\alpha(X)]$, constructed to satisfy

$$\Pi_X^J\{\theta \in C_\alpha(X)\} = 1 - \alpha,$$

also satisfy

$$P_{X|\theta}\{C_\alpha(X) \ni \theta\} = 1 - \alpha + O(n^{-3/2}), \quad n \to \infty,$$

where $X = (X_1, \ldots, X_n)$ is an iid sample of size n, and $O(a_n)$ represents a sequence z_n such that $a_n^{-1} z_n$ is bounded as $n \to \infty$. For some recent surveys of probability matching properties, we refer the reader to [111, 113]. Calibrated Bayes [55, 164, 211] has similar motivations.

Despite the success of the Jeffreys prior in scalar parameter problems, care must be taken in vector parameter problems. Consider the normal model $N(\mu, \sigma^2)$ where $\theta = (\mu, \sigma^2)$ is the unknown parameter of interest. It can be shown (see Section 2.3.1.2) that the Jeffreys prior is

$$\pi^J(\mu, \sigma^2) \propto (\sigma^2)^{-3/2}, \quad \mu \in \mathbb{R}, \ \sigma^2 > 0.$$

However, the Jeffreys prior is not probability matching in this case, and it is more common to use the prior

$$\pi(\mu, \sigma^2) \propto (\sigma^2)^{-1}, \quad \mu \in \mathbb{R}, \ \sigma^2 > 0, \tag{2.1}$$

i.e., independent flat priors for both μ and $\log \sigma^2$, because it has better large-sample frequentist properties. Incidentally, this prior in (2.1) is the "right Haar prior" while the Jeffreys prior is the "left Haar prior," where the names come from the mathematical theory of invariant measures, summarized in [72]. In general, the right Haar priors have better frequency properties, e.g., [111, 145].

2.3.1.2   Reference priors and related ideas

An alternative to specifying a default or non-informative prior on a vector parameter directly is to first do a sort of dimension reduction step, making it a problem of specifying non-informative priors for several one-dimensional quantities.

A now-standard approach is the specification of *reference priors*. In this case, one assumes there is an ordering of the individual parameters in the vector $\theta = (\theta_1, \theta_2)$; we consider only the two-component case, but the idea generalizes to dimensions greater than two. Reconsider the $N(\mu, \sigma^2)$ example above, where $\theta = (\mu, \sigma^2)$ is the parameter of interest. The Fisher information matrix in this case is

$$I(\theta) = \begin{pmatrix} I_{11}(\theta) & I_{12}(\theta) \\ I_{21}(\theta) & I_{22}(\theta) \end{pmatrix} = \begin{pmatrix} (\sigma^2)^{-1} & 0 \\ 0 & 2(\sigma^2)^{-2} \end{pmatrix}.$$

Then the reference prior takes $\pi(\sigma^2 \mid \mu) \propto I_{22}(\theta)^{1/2} \propto (\sigma^2)^{-1}$. Let

$$I_{11.2}(\theta) = I_{11}(\theta) - I_{12}(\theta)I_{22}(\theta)^{-1}I_{21}(\theta)$$

which, in our case, simplifies to $I_{11.2}(\theta) = (\sigma^2)^{-1}$. Since neither $I_{22}(\theta)$ nor $I_{11.2}(\theta)$ depend on μ, a formal calculation, integrating $\log(\sigma^2)^{-1/2}$ with respect to $(\sigma^2)^{-1}$, gives a constant prior for μ. Therefore, the reference prior for $\theta = (\mu, \sigma^2)$ is the same as the left Haar prior in (2.1), which leads to a posterior with desirable frequency properties; see Exercise 2.1. Details about the above calculations, as well as generalizations, are provided in [14, 19, 44, 45], and in the recent review [113]. A substantial generalization of the definition of reference priors was given recently in [15] and work is still ongoing, e.g., [168]. Questions about how to construct a good prior when the parameter of interest is some scalar function of a vector parameter θ have been considered; see, for example, [53, 235].

An alternative approach which, to our knowledge, has not been explored is to assign reasonable non-informative priors for each full conditional prior distribution, and then combine to a full joint prior. Consider again that normal example, where $\theta = (\mu, \sigma^2)$ is the parameter of interest. The proposed idea is to take the conditional priors based on the Jeffreys rule, i.e.,

$$\pi(\sigma^2 \mid \mu) \propto I_{22}(\theta)^{1/2} \propto (\sigma^2)^{-1} \quad \text{and} \quad \pi(\mu \mid \sigma^2) \propto I_{11}(\theta)^{1/2} \propto 1;$$

the latter prior is flat because the upper-left entry in the Fisher information matrix is constant in μ. Then the combined prior matches exactly the reference prior (2.1). This approach can be extended to allow for other specifications of the full conditional priors. A challenge of course is that not all sets of full conditional distributions would correspond to a genuine joint distribution. In such a case, where the full conditionals could be called "incompatible," some adjustments could be made. We think that this approach deserves further investigation.

In the case where there is a structure among the individual components of the vector θ, a hierarchical prior can be useful in practice. This, however, requires some prior knowledge about that structure and, therefore, might be better categorized as a model-building strategy. So, we will not consider this approach here.

2.3.1.3 Difficulties with objective Bayes

The motivation for the Bayesian approach comes from the classical setting where priors are meant to be subjective. Examples of such philosophical motivations include deductions based on rationality axioms, coherence, exchangeability and de Finetti's theorem, and Birnbaum's theorem on the likelihood principle, all reviewed in [111]. At best, these results suggest that *a* Bayesian analysis is appropriate, but they do not say anything about what prior to use. So, when there is no prior information available, these results are less than fully convincing.

Additional concerns, which have received less attention in the literature compared to those mentioned above, are about the interpretation and scaling of the posterior probabilities. We discussed in Chapter 1 that, if the prior has a frequency or belief probability interpretation, then the posterior inherits that. However, using a default (e.g., Jeffreys or reference) prior has neither interpretation, so then it is not clear in what sense the posterior probabilities should be interpreted. In regular problems, according to the Bernstein–von Mises theorem, the posterior will have a frequency interpretation only in the limit as sample size reaches infinity, but the interpretation will be vague up to that limit. The issue of scaling of posterior probabilities is also a concern. Mathematically, one can say that the posterior is absolutely continuous with respect to the prior, which means that any event with zero prior probability also has zero posterior probability. In other words, data cannot provide support for any assertion that corresponds to an event with prior probability zero, and this leads to some practical difficulties in constructing tests of point null hypotheses in the Bayesian framework; see Exercise 2.2. More generally, those events with small prior probability will also have small posterior probability, unless the information in data strongly contradicts that in the prior. From this point of view, it is clear that the posterior distribution is only as meaningful as the prior distribution, at least for moderately informative data. So, in this respect, we agree with Don Fraser in his assessment of Bayesian methods as just "quick and dirty confidence" [101].

2.3.2 Fiducial inference

2.3.2.1 A brief historical review

Motivated by his view of the limitations of Bayesian methods for scientific inference, Fisher [92, 94, 95], starting in the 1930s, began to develop a new framework of inference, which he called *fiducial inference*. He spent the rest of his life, about 30 years, working to develop his ideas and convince others of its soundness. Some of Fisher's ideas about fiducial inference were reinterpreted by Neyman [192] in the definition and construction of confidence intervals so, in a way, fiducial is connected to the modern-day frequentist framework. However, Fisher understood fiducial as something more than a tool for constructing an interval estimate, but his inability to clearly explain his fiducial ideas has been dubbed "Fisher's greatest failure" [266]. Fisher acknowledged the limited success of his efforts, but recognized that there was something valuable in his fiducial argument. He wrote:

> *I don't understand yet what fiducial probability does. We shall have to*

live with it a long time before we know what it's doing for us. But it should not be ignored just because we don't yet have a clear interpretation. [212, 266]

A wonderful survey of Fisher's thoughts on fiducial inference is given by Zabell [266]. Efforts continuing in this direction can be found in the later development of Fisher's fiducial inference and its variants, for example, Fraser's structural inference [98, 99, 56] and the Dempster–Shafer theory [62, 63, 64, 65, 66, 67, 221]; Hannig [126] gives a recent survey of work related to Fisher's fiducial argument, and a recent application of fiducial reasoning toward optimal estimation in a decision-theoretic setup is given in [234].

What exactly is fiducial inference? Somewhat trivially, fiducial inference is that which is obtained naturally, based on the rules of probability, after applying the "fiducial argument" about which Fisher wrote:

By contrast [to the Bayesian argument], the fiducial argument uses the observations only to change the logical status of the parameter from one in which nothing is known of it, and no probability statement about it can be made, to the status of a random variable having a well-defined distribution. [95], p. 54.

On Fisher's explanation of fiducial, Savage wrote:

The expressions "fiducial probability" and "fiducial argument" are Fisher's. Nobody knows just what they mean, because Fisher repudiated his most explicit, but definitely faulty, definition and ultimately replaced it with a few examples. [213]

This suggests that, at least in Savage's mind, there is no proper definition of fiducial inference. The word "fiducial," in English, is an adjective that means "based on trust," which, as we describe below in Section 2.3.2.2, helps to explain the construction of a fiducial distribution, but does not explain what exactly fiducial probability is. Despite the various controversies surrounding Fisher's ideas, e.g., [60, 61, 162], most would agree with Efron's [75] statement:

Here is a safe prediction for the 21st century: statisticians will be asked to solve bigger and more complicated problems. I believe there is a good chance... that something like fiducial inference will play an important role in this development. Maybe Fisher's biggest blunder will become a big hit in the 21st century!

The IM framework, which is the focus of this book, has some flavor of fiducial and the related Dempster–Shafer theory; see [186, 269] for some early IM developments, with connections to fiducial and Dempster–Shafer. It is debatable whether the IM framework may be this "something," but we leave this for the readers to consider.

2.3.2.2   *The approach and two examples*

The setup of fiducial inference is a special case of the association (A-step) in the basic IM framework; see Chapter 4. Start with a sufficient statistic $T = T(X)$ in the sampling model $X \sim \mathsf{P}_{X|\theta}$. The model for the sufficient statistic is assumed to be described as

$$T = a(\theta, U), \quad U \sim \mathsf{P}_U \tag{2.2}$$

where a is a known function and P_U is a known distribution. It is also assumed that the mapping a is such that, for any pair (t, u), there exists $\theta \in \Theta$ such that the relation $t = a(\theta, u)$ in (2.2) holds. Discussion of relaxing this latter assumption is provided in [126, 127]. In the fiducial framework, the variable U is called a *pivotal quantity*. Recall the English definition of "fiducial" discussed in Section 2.3.2.1—an adjective that means "based on trust." The logic of the fiducial argument is, in the words of Dempster [60], to "continue to regard" U as a quantity with distribution P_U after data X is observed. Here the conversion from frequency to belief probability is explicit, since, clearly, "$(U \mid X) \sim \mathsf{P}_U$" must be understood as a belief probability statement. This process is illustrated below with two scalar-parameter examples. There are widely recognized difficulties with the fiducial argument for the case of vector parameters [60, 61, 266]. Fiducial is also problematic in constrained parameter problems [81, 82].

Example 2.2. Consider the normal mean model $X \sim \mathsf{N}(\theta, 1)$, where X itself is the sufficient statistic. To obtain a fiducial distribution, write the model as

$$X = \theta + U, \quad U \sim \mathsf{N}(0, 1), \tag{2.3}$$

where U is the pivotal quantity. The familiar sampling model for X is obtained from (2.3) by assuming that $U \sim \mathsf{N}(0, 1)$ *a priori*, i.e., before seeing X. The fiducial arguments proceeds by "continuing to regard" U as a $\mathsf{N}(0, 1)$ random variable after seeing X. Then one can solve for θ in terms of X and U, i.e., $\theta = X - U$ and the (conditional) distribution of U, for given X, induces a (posterior) distribution for θ, namely, $(\theta \mid X) \sim \mathsf{N}(X, 1)$. This is the same as the Bayesian posterior distribution for θ based on the flat Jeffreys prior for θ, which is a special case of the general result in [162]. Consequently, the fiducial inference inherits both the desirable and undesirable properties of Bayesian inference, discussed above, in this case.

Example 2.3. Let $T = T(X)$ be a one-dimensional sufficient statistic, and let $F_\theta(t)$ denote the cdf of T, which depends on the scalar $\theta \in \Theta$. Assume that $1 - F_\theta(t)$, as a function of θ, is differentiable and satisfies

$$\inf_\theta [1 - F_\theta(t)] = 0 \quad \text{and} \quad \sup_\theta [1 - F_\theta(t)] = 1,$$

i.e., $G_t(\theta) = 1 - F_\theta(t)$, as a function of θ, has the properties of a cdf. Then the fiducial pdf is defined by differentiating this induced cdf:

$$g_T(\theta) = \left| \frac{\partial G_T(\theta)}{\partial \theta} \right|, \quad \theta \in \Theta. \tag{2.4}$$

Then the "continue to regard" operation follows by writing $U = F_\theta(T)$ and treating $U \sim \mathsf{Unif}(0, 1)$ as the pivotal quantity. One can produce fiducial (posterior) probabilities for events $A \subseteq \Theta$ by integrating g_T over A as usual.

2.3.2.3 Difficulties with fiducial

There are several difficulties with fiducial inference as envisaged by Fisher. A standard criticism is its non-uniqueness in multi-parameter problems, and we will not address this here. Another concern is that it is difficult to separate Fisher's ideas about fiducial from the more standard concepts of confidence intervals, etc. We mentioned previously that Neyman's development of confidence intervals was based on his attempt to understand Fisher's fiducial approach. Moreover, in his review of (an earlier edition of) Fisher [95], Pitman [199] says

> Up to this point the theory of fiducial probability and the theory of confidence intervals are essentially the same. The confidence interval terminology focuses attention on the real content of statements, and the fiducial distribution is an elegant way of exhibiting all confidence intervals.

We think this remark is representative of the typical statistician's understanding of fiducial, and explains the confounding of Fisher's ideas with Neyman's confidence, and the concept of *confidence distributions* to be discussed briefly in Section 2.3.4. Our belief is that Fisher had something different, and deeper, in mind in his thoughts about fiducial but, unfortunately, he was not able to make his points clearly.

In our estimation, the primary difficulty with the fiducial argument is its key "continue to regard" operation. For a simple illustration, reconsider the normal mean problem in Example 2.2. The point is that, when X is observed, the unobserved value of the pivotal quantity U is *exactly equal* to $X - \theta$, which is a fixed but unknown quantity. In other words, given X, U has a degenerate distribution at an unknown point, so the belief probabilities associated with the statement "$(U \mid X) \sim N(0,1)$" are not really believable. In a certain sense, this means that fiducial is not really "prior-free" as there is some outside information that is being used to justify the aforementioned probability statement. So, the fiducial probabilities have no meaningful interpretation, except for problems where there is an asymptotic frequency interpretation, such as in those problems, including the normal mean problem, where they agree with a standard objective Bayes posterior probabilities. In this sense, fiducial inference, even though it may give high quality answers in some examples, has difficulties similar to those outlined in Section 2.3.1.3 for the objective Bayes.

2.3.3 Dempster–Shafer theory

2.3.3.1 Brief history

The Dempster–Shafer (DS) theory is both an extension of Fisher's fiducial inference and a generalization of the Bayesian inference. The foundations of DS have been laid out by Dempster [62, 63, 64, 67] and Shafer [221, 222, 223, 224, 225]. The DS theory has been influential in many scientific areas, such as computer science and engineering. In particular, DS has played a major role in the theoretical and practical development of artificial intelligence. The 2008 edited volume [264] contains a selection of nearly 30 influential papers on DS theory and applications. For some recent statistical applications of DS theory, see [68, 151, 73].

2.3.3.2 DS models and combination operation

DS models are a flexible and powerful tool for representing uncertainty. In particular, a DS model on the parameter space Θ is specified by a conventional probability distribution, but not on Θ directly, but on the power set 2^{Θ}. That is, the DS model is described by the distribution of a random subset $\mathcal{S}$ of Θ. Those subsets of Θ which have positive probability, i.e., the support of $\mathcal{S}$, denoted by $\mathbb{S}$, are called *focal elements*. If we let $\mathsf{P}_{\mathcal{S}}$ denote the distribution of the random set $\mathcal{S}$, then we can write the belief and plausibility functions on Θ as

$$\text{bel}(A) = \mathsf{P}_{\mathcal{S}}(\mathcal{S} \subseteq A) \quad \text{and} \quad \text{pl}(A) = 1 - \mathsf{P}_{\mathcal{S}}(\mathcal{S} \subseteq A^c),$$

where $A \subseteq \Theta$ is any assertion concerning the whereabouts of θ. In some contexts, bel and pl are called the lower and upper probabilities, respectively. Dempster [67] suggested the use of a probability triple, written as (p, q, r), to summarize one's uncertainty about an assertion A. Here, p represents the probability of the truthfulness of A, q is the probability of the falsity of A, and r is the "don't know" probability. In particular,

$$p = \text{bel}(A), \quad q = \text{bel}(A^c), \quad \text{and} \quad r = \text{pl}(A).$$

It is easy to see that the three components are non-negative, and sum to unity. Therefore, as we mentioned in Chapter 1, the belief probabilities are subadditive in the sense that

$$\text{bel}(A) + \text{bel}(A^c) \leq 1.$$

Next is a simple but important example of a DS model. This example actually sheds important light on our views toward prior-free probabilistic inference.

Example 2.4. An important but somewhat unusual problem is to summarize mathematically the situation where one has no knowledge about the quantity of interest. We call this the knowledge of *total ignorance*, and we will discuss this further in Chapter 3. It turns out that this unusual situation can easily be described by a simple DS model. In particular, consider a DS model with just a single focal element, i.e., $\mathbb{S} = \{\Theta\}$. Then the distribution of the random set is degenerate, i.e., $\mathsf{P}_{\mathcal{S}}(\mathcal{S} = \Theta) = 1$, and it can be easily checked that, for any non-empty proper subset A of Θ,

$$\text{bel}(A) = 0 \quad \text{and} \quad \text{pl}(A) = 1.$$

That is, since we are ignorant about the whereabouts of the quantity of interest θ, we cannot assign any non-zero belief to a proper subset A of Θ. However, being that we are ignorant, we cannot rule out the possibility that $\theta \in A$, so any A is completely plausible. This is a perfectly satisfactory mathematical definition of ignorance.

In general, a DS analysis requires certain operations involving one or more DS models. Perhaps the most common operation is that of the DS combination, which follows Dempster's rule of combination. The idea is that one has several DS models, defined on a common space, say Θ, each representing uncertainty about the same quantity, say θ, coming from independent sources of information. The next subsection gives an example of inference on a Bernoulli parameter based on an iid sample.

The DS combination operation can be clearly defined mathematically using random sets. Let S_k, $k = 1, \ldots, K$ be random sets—all taking values in 2^Θ—encoding the information in K independent DS models concerning θ. The combination operation corresponds to considering a new random set

$$S = \bigcap_{k=1}^{K} S_k,$$

with the case where $S = \varnothing$, called a *conflict case*, ruled out in a certain way. That is, the belief for the combined DS models is

$$\text{bel}(A) = \frac{P_S(S \subseteq A, S \neq \varnothing)}{P_S(S \neq \varnothing)}, \quad A \subseteq \Theta.$$

While the above description of DS combination is quite simple, there is one subtle point that we glossed over. That is, in many cases, the individual DS models may not be defined on the same space. In such cases, one must augment the individual DS models with extra dimensions corresponding to the quantities present in at least one of the other DS models. With vacuous or ignorance DS models on the added dimensions, one obtains DS models on a common, enlarged state space. After the individual DS models are combined, one is usually interested in a marginal or projected DS model corresponding to some subset of all the involved quantities. We give an example below to show these steps for DS analysis in action.

2.3.3.3 DS inference on a Bernoulli parameter

Consider an iid sample $X = (X_1, \ldots, X_n)$ from a Bernoulli population with parameter $\theta \in \Theta = [0, 1]$. The goal is to make inference on θ. To carry out a DS analysis in this problem, we proceed as follows.

1. We construct a DS model on the individual stat space $\{0, 1\} \times [0, 1]$ for a single observation, say X_i, and the parameter θ. Care must be taken so that the DS model, when projected to the X_i margin, corresponds to the Bernoulli sampling model.

2. Given n independent DS models on $\{0, 1\} \times [0, 1]$, one for each i, the next step is to augment each individual DS model to create n new DS models defined on the "full" state space $\{0, 1\}^n \times [0, 1]$.

3. Combine the n DS models on the full space to a single DS model as above.

4. Finally, project the combined DS model down to the $[0, 1]$ margin for the quantity of interest, θ, and then compute relevant (p, q, r) summaries.

The individual DS models. Start by constructing a DS model for (X_i, θ) on $\{0, 1\} \times [0, 1]$ for a given $i = 1, \ldots, n$. A helpful device is to introduce an auxiliary variable U_i, similar to the pivotal quantity in the fiducial context as follows:

$$X_i = I_{[0, \theta]}(U_i), \quad U_i \sim \text{Unif}(0, 1),$$

where I_A is the indicator function, i.e., $I_A(x) = 1$ if $x \in A$ and $I_A(x) = 0$ otherwise. By introducing this auxiliary variable, it suggests focal elements—subsets of $\{0,1\} \times [0,1]$—given by

$$\{1\} \times [u,1] \cup \{0\} \times [0,u], \quad u \in (0,1).$$

Then we can write the random sets for the DS model for observation $i = 1, \ldots, n$ as

$$S_i = \{1\} \times [U_i, 1] \cup \{0\} \times [0, U_i], \quad U_i \sim \mathsf{Unif}(0,1).$$

As a quick sanity check, we want to make sure that the corresponding DS model, projected down to the space of X_i, corresponds to the posited Bernoulli sampling model. When considering the sampling model, θ is fixed at a particular value, and the simplest way to encode this information is to restrict the DS model for (X_i, θ) to the slice corresponding to that fixed value of θ. This corresponds to a separate DS model for θ with a single focal element $\{0,1\} \times \{\theta\}$. The combined DS model has a singleton random set

$$S_i^\theta = \{I_{[0,\theta]}(U_i)\} \times \{\theta\}, \quad U_i \sim \mathsf{Unif}(0,1),$$

and from this it is clear that the DS version of the sampling model is consistent with the Bernoulli model specified at the start.

The individual augmented DS models. Currently, we have n independent DS models, but even though the spaces $\{0,1\} \times [0,1]$ are the same for each, the first binary dimension corresponds to a different quantity, since X_i and X_j, for $j \neq i$, are different quantities living in their own copies of $\{0,1\}$. Therefore, we need to lift each individual DS model to an augmented DS model on $\{0,1\}^n \times [0,1]$. This should be done in a way that the jth margin of the ith DS model corresponds to an ignorance DS model. We can accomplish this by replacing the random set S_i above with

$$S_i^\star = \{0,1\}^{\times(i-1)} \times \{1\} \times \{0,1\}^{\times(n-i)} \times [U_i, 1]$$
$$\cup \{0,1\}^{\times(i-1)} \times \{0\} \times \{0,1\}^{\times(n-i)} \times [0, U_i], \quad U_i \sim \mathsf{Unif}(0,1).$$

Combined DS model for $(X_1, \ldots, X_n, \theta)$. As in the general discussion, this step is at least conceptually simple, as we only need to construct a new random set,

$$S = \bigcap_{i=1}^n S_i^\star,$$

based on an intersection, and then be sure to rule out conflict cases. Our interest is in θ, so we proceed to the final step without further distractions at this point.

Projected DS model for θ. We now have a single combined DS model on the $(n+1)$-dimensional space of $(X_1, \ldots, X_n, \theta)$, and the goal is to project this combined DS model down to the one-dimensional θ-space. Like in the sampling model calculation, the idea is to fix the $(X_1, \ldots, X_n)$ dimension at the observed values. A little reflection on the form $S_i^\star$ and some routine probability calculations reveals that the projected DS model, for given X, can be described by the random set

$$S^X = [U_{(T)}, U_{(T+1)}], \quad U_1, \ldots, U_n \overset{\mathrm{iid}}{\sim} \mathsf{Unif}(0,1),$$

where $T = T(X) = \sum_{i=1}^{n} X_i$ and $U_{(1)} < \cdots < U_{(n)}$ denote the order statistics. An interesting observation is that the projected DS model—which can be called a *DS posterior*—depends on data only through the minimal sufficient statistic T. Then we can compute the DS (p,q,r) for any assertion of interest using some basic probability calculations. For example, if $A = [0, \theta_0]$ for some fixed $\theta_0 \in (0,1)$, then

$$p = \text{bel}(A) = \mathsf{P}_{S^X}(S^X \subseteq A) = \mathsf{P}_U(U_{(T+1)} \leq \theta_0), \tag{2.5}$$

$$q = \text{bel}(A^c) = 1 - \mathsf{P}_{S^X}(S^X \subseteq A^c) = \mathsf{P}_U(U_{(T)} > \theta_0), \tag{2.6}$$

and, of course, $r = 1 - (p+q)$; see Exercise 2.7. Note that this DS posterior provides data-dependent probabilistic summaries of uncertainty about the whereabouts of θ, all without a prior distribution. More examples and discussion are given in [186].

2.3.3.4   DS as a generalization of Bayes

It was demonstrated by Dempster [61] that the DS framework provides an extension or generalization of the Bayesian methodology. For simplicity of presentation, the general remarks here will be given within the context of the Bernoulli example above.

Suppose that, in addition to the posited sampling model for $X = (X_1, \ldots, X_n)$, given θ, a prior $\pi(\theta)$ is available. Consider this prior and the prior-free DS posterior above as two independent sources of information about θ, or as independent DS models for θ. These two DS models can be combined using some routine calculations as follows. Start by computing the so-called *commonality function*, $c_X(\theta)$, for the DS posterior derived above. This is given by

$$\begin{aligned} c_X(\theta) &= \mathsf{P}_{S^X}(S^X \ni \theta) \\ &= \mathsf{P}_U(U_{(T)} \leq \theta \leq U_{(T+1)}) \\ &= \mathsf{P}_U(\text{exactly } T \text{ of } U_1, \ldots, U_n \text{ are less than } \theta) \\ &= \binom{n}{T} \theta^T (1-\theta)^{n-T}, \end{aligned}$$

where, recall, $T = T(X) = \sum_{i=1}^{n} X_i$. The key point is that the commonality function is proportional to the likelihood function $L_X(\theta)$ for the iid Bernoulli data X. As demonstrated in [221], combination of the two DS models, in this case, corresponds to treating the commonality function in the way the Bayes theorem treats the likelihood. Therefore, the DS posterior is exactly the usual Bayes posterior based on prior $\pi(\theta)$ and, in this sense, DS generalizes Bayes.

2.3.3.5   Difficulties with DS

As a successor of fiducial, DS includes fiducial as a special case and therefore suffers from the same difficulties as fiducial. First, the conditioning or projection-down operation shares with the fiducial argument the "continue to regard" operation which leads to problems in terms of the interpretation of the DS probability output. The weak belief method of [186] and [269] corrects this problem to some extent by loosening the connection between data X and auxiliary variable U, producing inferential results with desirable frequency properties. Second, computational methods are not yet available even for commonly used statistical models. This makes both applications and evaluations of DS analysis difficult.

It should also be noted that although DS is consistent with likelihood inference, the number of pivotal variables in the combined DS models can be larger than the number of sufficient statistics. For example, in the Bernoulli example, the final DS model for inference about θ is characterized by the pair of random variables $U_{(T)}$ and $U_{(T+1)}$, while the sufficient statistic $T = \sum_{i=1}^{n} X_i$ is a one-dimensional quantity. This observation helped to motivate the dimension-reduction techniques described in Chapters 6–7.

2.3.4   Other forms of distributional inference

By "distributional inference" we mean a method that, given data X, returns a X-dependent probability distribution on the parameter space. Examples of this include Bayes—both classical and objective—and fiducial as discussed. Other examples are *structural inference* [99], *generalized inference* [42, 255], *generalized fiducial inference* [71, 126, 127, 128, 129, 154, 246, 247, 248, 249] and *confidence distributions* [262, 263], and Bayesian inference with data-dependent priors [103, 185, 184], which have seen a lot of recent developments. Despite their success in applications, we want to differentiate this sort of distributional inference from our notion of probabilistic inference. The latter makes no claim that their "posterior probability" of an assertion $A \subseteq \Theta$ is a meaningful summary of the evidence in data supports the truthfulness of A. In fact, these methods are typically used to construct an interval estimate for θ by returning, say, the middle 95% of the posterior distribution, and relying on asymptotic theory to justify the claimed "95% confidence." So, while having a complete distribution provides more information than just a single point estimate or confidence interval, ultimately that distribution will be summarized by an interval and the interpretation of that interval is, in general, only meaningful in the asymptotic limit. Therefore, we would classify these distributional inference methods as just sophisticated tools for constructing frequentist procedures and, consequently, subject to Fraser's remarks in [101] concerning (objective) Bayes and confidence.

2.4   On the role of probability in statistical inference

Probability has played two roles in the statistical inference problem. The first is a natural one, as a model that relates the observable data to the unknown parameters. The second role, the one that is to be questioned here, is as a measure of evidence in data supporting the truthfulness or falsity of an assertion about the parameters. When a genuine prior probability distribution for θ is available, then summarizing uncertainty with a posterior probability is quite natural. However, when no prior information is available, there is no reason to assume that the posterior summaries should satisfy the properties of a probability measure. Indeed, as we have discussed above, existing methods which "force" the inferential output to be a posterior probability measure pay the price of having their posterior probabilities be meaningless except for some assertions in a large-sample limit. So, we think there is reason, as we describe in the next subsections, to consider looking beyond probability as a tool for measuring evidence. These comments will also provide motivation for our IM developments in later chapters.

2.4.1 Knowledge of total ignorance

As laid out in Chapter 1, we adopt the viewpoint, standard in applications, that we have no genuine prior information about the unknown parameter θ, except that it takes value in the known parameter space Θ. Of course, the problem is interesting only if Θ contains at least two points, so we make this assumption throughout.

Definition 2.3. A set function $b : 2^{\Theta} \to [0, 1]$ represents the knowledge of total ignorance about θ if $b(A) = 0$ for every proper subset A of Θ, and $b(\Theta) = 1$.

We gave a mathematical description of the state of ignorance using the Dempster–Shafer theory in Example 2.4, and the technical device was a random set with a degenerate distribution at Θ. The question here is whether there is a prior distribution which can also characterize the knowledge of ignorance. If so, then this should be the "non-informative" prior used by Bayesians; if not, then this casts doubt on the use of probability as a baseline tool for summarizing uncertainty.

Proposition 2.1. *There is no prior probability distribution that can represent the knowledge of ignorance about θ as in Definition 2.3.*

Proof. Suppose that there is such a probability P. Then, by Definition 2.3, $P(A)$ and $P(A^c)$ equal 0 for all events A. However, since P is a probability, it must satisfy $P(A^c) = 1 - P(A)$, which contradicts the supposition.   □

Hundreds of years ago, in the binomial model, both Bayes and Laplace had employed a uniform prior for θ based on a "principle of indifference," i.e., in a partition of $[0, 1]$ into intervals of equal length, all are equally likely to contain θ. However, as Proposition 2.1 shows, the uniform prior does not represent the state of ignorance. The point, we suppose, is that even "indifference" is knowledge, so a prior that encodes this cannot also encode ignorance.

Proposition 2.1 focuses only on the case of proper priors, but only because the definition of ignorance scales the function to $[0, 1]$. Obviously, any other finite scale would not affect the result. For the case of an improper prior, the result goes through basically as is. The point of the proof was that both A and A^c must be assigned value 0, which cannot be accomplished by a prior except one that has a density which takes value 0 almost everywhere. But such a prior cannot assign "full mass" to the parameter space Θ, which is sure to contain the true θ. See Exercise 2.9 for an alternative argument that improper priors cannot encode ignorance.

Our interpretation of Proposition 2.1 is as follows. If probability cannot adequately reflect our prior beliefs of ignorance, then we cannot trust the inferential output contained in the posterior probabilities obtained via the Bayes formula. We have previously raised the question of the meaningfulness of (objective) Bayesian posterior probabilities, except for possibly in a large-sample limit, and now Proposition 2.1 provides some explanation of what has gone wrong. In other words, the Bayes theorem is a perfectly valid tool for combining beliefs coming from different sources, but the belief of ignorance about θ, which we think is a typical one in applications, cannot be described by a (proper or improper) prior, and if one or both of the beliefs to be combined are not trustworthy, then neither is the output.

2.4.2   Additivity and evidence in data

Probability is predictive in nature, that is, it is designed to describe our uncertainties about a yet-to-be-observed outcome. In our particular statistical inference problem, however, the parameter of interest, θ, is a fixed but unknown quantity, not a random variable; there is no reason to require the summaries of our uncertainty—before or after seeing data—to satisfy the properties of a probability measure. We already saw in Section 2.4.1 that probability cannot properly describe the scientifically relevant position of having no genuine prior information. Here we want to address the question about the appropriateness of probability as a summary of evidence in data supporting truthfulness or falsity of an assertion about θ.

Let Π_X be a posterior distribution for θ given data X; in what follows, Π_X could also be a fiducial distribution, a generalized fiducial distribution, or a confidence distribution—anything that is a probability measure depending on X. For a fixed assertion $A \subseteq \Theta$, the posterior probability $\Pi_X(A)$ is to be interpreted as a measure of the evidence in X supporting the truthfulness of the claim "$\theta \in A$." Additivity of Π_X implies that the measure of evidence in X supporting the falsity of that same claim is $\Pi_X(A^c) = 1 - \Pi_X(A)$. The problem, which is a consequence of additivity, is clearly that there is no middle ground. In other words, evidence in X supporting the truthfulness of "$\theta \in A$" is necessarily evidence supporting the falsity of the negation. If θ were a yet-to-be-observed outcome from some experiment, then the dichotomy of "θ is in exactly one of A and A^c" is perfectly fine and natural. But our situation concerns a summary of evidence. In other situations, not so closely tied to our familiar world of probability, no such dichotomy exists. For example, consider a criminal court trial, where A is the assertion that the defendant is not guilty. Witness testimony that corroborates the defendant's alibi is evidence that supports the truthfulness of A but it does not provide direct evidence for the falsity of A^c. This suggests that there ought to be a third "don't know" category to which some evidence can be assigned, leading to a sub-additivity property, and we have already seen that belief functions are a suitable tool.

To further illustrate this point, let us take a simple statistical problem, $X \sim N(\theta, 1)$. Suppose we take the flat Jeffreys prior for θ, so that the posterior distribution for θ, given X, is $N(X, 1)$. Let $A = (-\infty, \theta_0]$, for fixed θ_0, be the assertion of interest. Consider the extreme case where $X \approx \theta_0$. In such a case, the posterior distribution assigns probability 0.5, roughly, to both A and A^c. Certainly, this data sitting on the boundary cannot distinguish between A and A^c, and the equally probable conclusion reached by the Bayesian posterior distribution is consistent with this. The question here is about the magnitude 0.5. As a measure of support, the value

0.5 suggests that there is some non-negligible evidence in the data supporting the truthfulness of both A and A^c, which is counterintuitive because a single observation at the boundary should not provide strong support for either conclusion. A more reasonable summary would be one that provides small support for both A and A^c, communicating the point that the single data point provides minimal support for both the assertion and its complement. Again, this suggests a sub-additivity property for the evidence measure, which a belief function would achieve. If the data were more informative, e.g., if the sample mean of 1000 observations is $\approx \theta_0$, then the posterior summary would assign values 1 and 0, roughly, to the two complementary assertions, and we agree that these would be meaningful and appropriately scaled summaries of the evidence in the data. So, in some sense, that third "don't know" category should disappear as the data becomes more informative, leaving a probability measure as a sort of "asymptotic" summary of evidence. This helps to further explain why those distributional inference methods, in general, provide valid summaries only asymptotically: they are large-sample approximations to a suitable belief function.

2.5 Our contribution in this book

In this chapter we have been a bit critical toward the existing methods for probabilistic inference, but we honestly think that these methods are important and useful. Indeed, the main point is given in the last line of the previous section: the "posterior distribution," the output of an existing probabilistic inferential method, provides a large-sample approximation to some other inferential output.

Following our criticism of existing methods, we think it is important again to highlight what specifically we are shooting for in the rest of this book. If one agrees with our characterization of existing methods as a sort of approximation to something else, then it is quite clear that we, as statisticians, should at least want to know what this something else is. There will be cases where the existing approximations to this "something else" will be fine and perhaps it is not worth the time and energy needed to scrap the existing methods and build up something completely new. However, there surely will be cases where the approximation is not completely satisfactory, and substantial improvements can be made by considering "something else." The new way of thinking, as presented in this book, has, in our opinion, already provided some new understanding which we hope to make clear in the subsequent chapters. Moreover, we expect further improvements in understanding, particularly as it relates to modern issues concerning the reproducibility of scientific discoveries.

We have been critical also of probability as it serves as a tool for summarizing evidence in data concerning assertions about the unknown parameter. This is not to say that probability is not useful. In fact, our view is that probability is the correct tool, but for describing uncertainty on the auxiliary variable-space, not on the θ-space. This makes some connections to the fiducial and Dempster–Shafer frameworks described above. A key difference, however, is that we will use probability for *predicting* the unobserved value of the auxiliary variable, which allows us to break the tight connection between the data and the auxiliary variable. To do a good job on prediction, it turns out that we need to use random sets. The distribution of these random sets must be tuned to the distribution of the auxiliary variable which, in turn, is connected to the sampling model for the data. It is this tuning of the random set distribution to the given sampling model which allows us to prove that the resulting inference is valid in the sense described in Chapter 1; see also, Chapter 3. Therefore, we will make heavy use of probability in our developments, just applied in a different way than what is familiar.

In conclusion, we want to reiterate that the main goal of this book is to provide some insight about this "something else" that existing methods are approximating. We believe that this other thing is the elusive "best possible inference," and in the remainder of the book we describe what we think this is and how we can get a handle on it. Since the classical methods that have stood the test of time must generally be good approximations, it remains to be seen the extent to which our new approach and understanding will change statistical practice. But we do not hesitate to say that progress toward understanding what is the "best possible inference" and when it can be reasonably approximated by existing tools is an important contribution.

2.6   Exercises

Exercise 2.1. Let $X = (X_1, \ldots, X_n)$ be iid $N(\mu, \sigma^2)$, where $\theta = (\mu, \sigma^2)$ is unknown.

(a) Consider the prior $\pi(\mu, \sigma^2) \propto (\sigma^2)^{-1}$ in (2.1), the reference prior and the right Haar prior. Find the posterior distribution for (μ, σ^2).

(b) Find the marginal posterior distribution for μ. That is, integrate out σ^2 from the joint posterior distribution for (μ, σ^2).

(c) Find a $100(1 - \alpha)\%$ credible upper bound for μ using the marginal posterior. That is, find $\bar{\mu}_\alpha = \bar{\mu}_\alpha(X)$ such that $\Pi_X(\mu \le \bar{\mu}_\alpha) = 1 - \alpha$.

(d) Think of $\bar{\mu}_\alpha = \bar{\mu}_\alpha(X)$ as a function of data. Show that the above credible upper bound is exactly probability matching. That is, show that

$$\mathsf{P}_{X|(\mu,\sigma^2)}\{\bar{\mu}_\alpha(X_1, \ldots, X_n) \ge \mu\} = 1 - \alpha, \quad \forall\, (\mu, \sigma^2).$$

Exercise 2.2. Recall the Bayes factor defined in Exercise 1.15. Consider a generalization where H_0 has a prior π_0 and likelihood $L_{n,0}$ for θ under H_0 and a prior π_1 and likelihood $L_{n,1}$ for θ under H_1. Then the Bayes factor is defined as

$$B_{01}(X) = \frac{\int L_{n,0}(\theta)\pi_0(\theta)\, d\theta}{\int L_{n,1}(\theta)\pi_1(\theta)\, d\theta}.$$

Show that the Bayes factor is not well-defined if one or both of the priors π_0 and π_1 are improper.

Exercise 2.3. Let $X \sim \mathsf{P}_{X|\theta}$ for a scalar parameter θ, and consider a collection of one-sided interval assertions

$$A_{\theta_0} = (-\infty, \theta_0], \quad \theta_0 \in \mathbb{R}.$$

The goal is to pick the prior such that the posterior Π_X satisfies

$$\Pi_X(A_{\theta_0}) \le_{\text{st}} \text{Unif}(0, 1) \quad \text{as a function of } X \sim \mathsf{P}_{X|\theta}, \ \theta > \theta_0$$

$$\Pi_X(A_{\theta_0}) \ge_{\text{st}} \text{Unif}(0, 1) \quad \text{as a function of } X \sim \mathsf{P}_{X|\theta}, \ \theta \le \theta_0,$$

where "$\leq_{st}$" and "$\geq_{st}$" denote stochastic order relations defined in Exercise 1.18. Then we say that the posterior is *valid*; this is related to the "probability matching" in Exercise 2.1 and the validity concept discussed in Chapter 1. Show that if $X \sim N(\theta, 1)$ and the prior is the flat Jeffreys prior, then the posterior Π_X is valid for one-sided assertions as defined above.

Exercise 2.4. Recall the notion of validity from Exercise 2.3. Let $X \sim \text{Bin}(n, \theta)$ with known n but unknown θ. Consider the Bayesian inference about θ from an observation from $\text{Bin}(n, \theta)$ with known n. Show that

(a) The posterior Π_X is valid for assertions $[0, \theta_0]$ with a $\text{Beta}(0, 1)$ prior.

(b) The posterior Π_X is valid for assertions $[\theta_0, 1]$ with a $\text{Beta}(1, 0)$ prior.

(This is an example where a class of priors is needed to be valid for all assertions. A general theory on the use of a class of priors and weakened posterior beliefs is called *imprecise Bayes*; see [244] for details and additional references.)

Exercise 2.5. The *likelihood principle* says that inference should depend only on the observed likelihood function, that is, other features concerning the data and model should be irrelevant. (Starting from Birnbaum [25], there has been lots of discussion, e.g., [17], and controversy surrounding the likelihood principle; see the recent papers [84] and [188].) The quantity of interest is the probability θ that a coin lands on heads, based on T heads in n total coin tosses.

(a) Fix a prior $\pi(\theta)$. Show that the posterior distribution for θ is the same whether sampling model for T is binomial or negative binomial. This suggests that Bayesian inference with a fixed prior $\pi(\theta)$ satisfies the likelihood principle.

(b) Argue that the Bayesian inference based on the Jeffreys prior does not satisfy the likelihood principle. [*Hint:* The prior depends on the sampling model.]

Exercise 2.6. Suppose the parameter θ can take only two values, say, θ_0 and θ_1. Then the Neyman–Pearson theorem tells us that there is a most powerful test of given size α for testing $H_0 : \theta = \theta_0$ versus $H_1 : \theta = \theta_1$, and that this test is based on the likelihood ratio. Show that, in this simple case, the likelihood ratio is a sufficient statistic. (See Theorem 6.12 in [158] for a generalization.)

Exercise 2.7. Complete the calculations in Equations (2.5) and (2.6) for the DS posterior for the Bernoulli parameter θ.

Exercise 2.8. Consider the trinomial model as the sampling model for the observed data $X \in \{1, 2, 3\}$ with probability mass function $\theta_i = P(X = i)$ for $i = 1, 2, 3$, where $\theta_i \geq 0$ and $\sum_{i=1}^{3} \theta_i = 1$.

(a) Suppose that $X = 1$ and θ_1 is known to equal t, a fixed value in the unit interval (e.g., $t = 0.5$). Take the DS model proposed in Dempster [62] and obtain the conditional DS model for inference about $\psi = \theta_2/\theta_3$ given $\theta_1 = t$. Show that this conditional DS model is not vacuous.

(b) Explain why this result is or is not counterintuitive.

Exercise 2.9. Let π be a non-negative continuous function playing the role of an improper prior "pdf" for θ in Θ. Suppose that extra information becomes available saying that θ is sure to reside in the compact subset Θ_c of Θ with a non-empty interior. Use the Bayes formula to find the conditional prior for θ, given that $\theta \in \Theta_c$, and argue that it is proper. Is it logical that an improper "ignorance" prior turns into a proper "non-ignorance" prior when combined with some incomplete information on the whereabouts of θ?

Chapter 3

Two Fundamental Principles

3.1 Introduction

Statistical inference is the process of converting experience, in the form of observed data, to knowledge about the underlying population in question, and is an essential part of the scientific method and of human discovery. The conclusion of Chapter 2 is that existing approaches provide only approximations to the "best possible inference." The goal of this chapter, and of the book more generally, is to describe our view of this elusive target. In particular, we provide two vague but hardly-disagreeable principles which we believe describe what this "best possible inference" should satisfy. The *validity principle* explains that probabilistic inference should be based on predicting an unobservable but predictable quantity, and that this prediction should be valid in the sense described in Chapter 1. That is, the inferential output ought to be calibrated to a meaningful scale for interpretation by intelligent minds. Since there are lots of ways to carry out a valid prediction of a predictable quantity, additional considerations are needed. The second principle, called the *efficiency principle*, says that the prediction should be made as efficient as possible, where efficiency is measured in terms of long-run frequencies. We explain why our calibration based on long-run frequencies is different from that for the more familiar frequentist procedures. These principles will be used to motivate the IM approach in Chapter 4.

To support our claims and to help those readers with a Bayesian background understand the subtle differences, we explain, in Section 3.5, that the general approach of conditioning (e.g., via Bayes formula) is meant primarily for improving inferential efficiency, not for any deeper philosophical reasons. A take-away message is that, whenever possible, conditioning should be carried out to improve efficiency.

3.2 Validity principle

As usual, we start with a sampling model $X \sim \mathsf{P}_{X|\theta}$ for the observable data, depending on a parameter θ known only to reside in Θ. As a starting point, recall our perspective on probabilistic inference, as in Definition 2.1. That is, we seek a set-function $b_X : 2^{\Theta} \to [0, 1]$, depending on X, such that $b_X(A)$ measures the evidence in data X supporting the truthfulness of the claim "$\theta \in A$" for any assertion $A \subseteq \Theta$. In order for $b_X(A)$ to be meaningful in the sense that it admits a common interpretation by intelligent minds, it must have a fixed objective scale. A natural choice of scale is the familiar uniform probability scale, so that, for example, if A is true, then $b_X(A)$, as a function of X, will rarely be small.

Our first claim is that the sampling model alone is insufficient to provide this common objective probabilistic scale. This is related to Fisher's view on the interpretation of the likelihood function not as a probability distribution for θ but as only providing a data-dependent partial order on the parameter space, i.e., the numerical values of the likelihood are meaningless; only the relative magnitudes are relevant. Therefore, a posterior probabilistic interpretation of the sampling model is possible only if something else is included, such as a prior distribution in the Bayesian setup. Here we explain an alternative way to add something to the sampling model in order to obtain (valid) probabilistic inference.

In order to introduce a probability scale, according to the basic setup of probability, there must be some quantity to be predicted having a known distribution, given certain features of the data X. This random element to be predicted will be called a predictable quantity. The goal is for the prediction of the predictable quantity to be valid in the sense that probabilities associated with its prediction are suitably calibrated relative to the familiar uniform scale.

Definition 3.1. Let U be an unobservable random variable with distribution $\mathsf{P}_{U|\theta}$ on $\mathbb{U}$, possibly depending on the parameter θ. Let $\mathcal{S}$ be a random set with distribution $\mathsf{P}_{\mathcal{S}|\theta}$, also possibly depending on θ, and focal elements $2^{\mathbb{U}}$. The prediction of the unobserved value of U with the random set $\mathcal{S}$ is called *valid* if

$$\gamma(U) \geq_{\mathrm{st}} \mathsf{Unif}(0,1),$$

where the *contour function* [226] γ is defined as

$$\gamma(u) = \gamma_{\mathcal{S}|\theta}(u) = \mathsf{P}_{\mathcal{S}|\theta}(\mathcal{S} \ni u). \tag{3.1}$$

In this case, we say that $\mathcal{S}$ is a valid *predictive random set*. In general, the unobserved value of U is called a *predictable quantity* if there exists a valid predictive random set for its prediction.

An important special case is that where the distribution of U and also of $\mathcal{S}$ do not depend on the parameter θ. This case is dealt with in Chapter 4. The general case is addressed in Chapter 7, specifically, in the famous Behrens–Fisher problem. Examples of valid predictive random sets are given in later chapters.

Now we are ready to explain what specifically we believe is needed for probabilistic inference beyond the sampling model. We state this formally as a principle.

Validity Principle. *Probabilistic inference requires associating an unobservable but predictable quantity to the observable data and unknown parameter. Then the probabilistic inferential output is obtained by predicting that predictable quantity with a valid predictive random set.*

With this new perspective, we can now take a different look at those existing methods for probabilistic inference.

- Bayesian inference based on a genuine prior satisfies the validity principle, since θ itself is a predictable quantity and the random sets $\mathcal{S} = \{\theta\}$, where $\theta \sim$

Π_X, is valid. However, validity of the aforementioned singleton random sets are not generally valid when the prior does not reflect real prior information.

- The fiducial (and Dempster–Shafer) approach can be understood within this framework by taking the pivotal quantity U to be the predictable quantity, and using the random set $\mathcal{S} = \{U\}$, $U \sim \mathsf{P}_U$. For singleton random sets, at least in the case of a continuously distributed pivotal quantity, the contour function (3.1) is identically zero (see Exercise 3.1), hence, strictly stochastically smaller than $\mathsf{Unif}(0, 1)$. In some examples, however (see Chapter 4.5), for certain assertions, one can construct valid fiducial predictive random sets, in which case valid probabilistic inference can be obtained; in fact, in those cases, the fiducial random set is a benchmark.

3.3   Efficiency principle

As we shall see in later chapters, for a given predictable quantity, there will be a variety of valid predictive random sets to choose from. Toward a resolution of this non-uniqueness, we appeal to the goals of statistical inference. For this, we make a simple analogy based on classical statistics. In a hypothesis testing problem, there will be many tests that control the Type I error probability, including the test that never rejects the null. Among these tests with suitable control on Type I errors, the goal is to pick the one with highest power, if it exists. We take an analogous approach here. Roughly speaking, in addition to validity, we want the inferential output b_X to be efficient in the sense that, if A is false, then $b_X(A)$ is stochastically as small as possible, as a function of X. (A definition of efficiency is given in Chapter 4.)

Efficiency Principle. *Subject to the validity constraint, probabilistic inference should be made as efficient as possible.*

The efficiency principle is purposefully vague, in particular, it allows for a variety of techniques to be employed to increase efficiency. Chapters 4 and 5 discuss the choice of good quality predictive random sets for efficient inference within the IM framework. In addition, Chapters 6 and 7 discuss general techniques for reducing the dimension of the predictable quantity for the purpose of improved efficiency.

3.4   Recap

The validity and efficiency principles suggest that the inferential output b_X, which must be a belief function determined by the distribution of an appropriate predictive random set, should have certain calibration properties relative to the sampling distribution of the data X. It is important for us to point out some important differences between our understanding of calibration and the kinds of frequency properties considered in existing approaches.

First, we insist that our inferential output be valid, and this is not an asymptotic statement, i.e., validity holds for all sample sizes. Compare this to existing methods, such as the objective Bayes, which provides valid inference asymptotically for some assertions. If we recall our characterization of existing attempts at probabilistic inference as approximations to the "best possible inference," then we can see that asymptotically valid inferential output must be an approximation to exactly valid inferential output. Therefore, we can see that the IM approach to be discussed in Chapter 4, motivated by the validity and efficiency principles, might very well be that elusive "best possible inference" that existing methods are approximating.

Second, our calibration is at the level of the inferential output, while classical frequentist methods focus on constructing procedures whose error rates are appropriately calibrated. The point is that the frequentist skips over the actual "inference" step, going directly for a procedure with desirable properties. The IM approach and, more generally, any approach that satisfies the validity and efficiency principles builds up meaningful and properly calibrated inferential output, which can then be used to construct decision procedures with desirable error rate control. In other words, by following the two principles discussed above, one goes beyond the scope of the classical frequentist methods to provide meaningful probabilistic inference from which exact and efficient decision procedures can be easily derived.

3.5   On conditioning for improved efficiency

A general approach that can be used to improve efficiency is conditioning, and this section provides a justification for this claim. More generally, reducing the dimension of the predictable quantity leads to improved efficiency.

Consider two random variables U_1 and U_2, taking values in $\mathbb{U}_1$ and $\mathbb{U}_2$, respectively with joint distribution characterized by the joint pdf $p(u_1, u_2)$. The marginal pdf of U_1 is given by

$$p_1(u_1) = \int_{\mathbb{U}_2} p(u_1, u_2)\, du_2, \quad u_1 \in \mathbb{U}_1.$$

Suppose that inference requires that we predict the unobserved value of U_1 only. Intuitively, the uncertainty encoded in p_1 determines the accuracy of the prediction of the value of U_1. A simple way to characterize this is with the (marginal) variance:

$$V(U_1) = E[\{U_1 - E(U_1)\}^2],$$

where we assume that the expectations exist.

Now suppose that the value u_2 of U_2 is observed, but inference still requires only accurate prediction of the value of U_1. In this new context, there are two possible choices of distribution of U_1 that can be used to develop a valid predictive random set. One is the marginal distribution with pdf $p_1(u_1)$ discussed above, and the other is the conditional distribution of U_1, given $U_2 = u_2$, with pdf given by

$$p_{1|2}(u_1 \mid u_2) = p(u_1, u_2)/p_2(u_2),$$

where $p_2(u_2)$ is the marginal pdf for U_2, just like that for U_1 above. It is commonly believed that the use of the conditional distribution would be better, or more efficient, than the use of the marginal distribution. Here we explain more precisely what is meant by "more efficient prediction." As before, measure the uncertainty in the conditional distribution via the conditional variance, i.e.,

$$V(U_1 \mid U_2 = u_2) = E[\{U_1 - E(U_1 \mid U_2 = u_2)\}^2 \mid U_2 = u_2].$$

One cannot make a direct comparison between $V(U_1)$ and $V(U_1 \mid U_2 = u_2)$, see Exercise 3.4, but a comparison can be made "on average," i.e.,

$$E\{V(U_1 \mid U_2)\} \leq V(U_1), \tag{3.2}$$

where expectation is with respect to the marginal distribution of U_2. Then we can say that prediction based on the conditional distribution of U_1, given $U_2 = u_2$, is more efficient in a long-run average sense.

Bayes's 1763 paper is an impeccable exercise in probability theory [78]. The above paragraph suggests that the use of a posterior distribution, a conditional distribution, can be motivated strictly from the probabilistic and efficiency point of view, without any other philosophical justification. That is, one carries out the conditioning on the observed data in Bayes formula only because the conditional distribution for θ, given X, provides more efficient inference, in the long-run average sense, than the marginal distribution for θ. We hope this view of the Bayesian update as motivated by some long-run frequency considerations will help the reader understand why the efficiency considerations in the IM framework make connection to long-run frequency properties.

3.6 Exercises

Exercise 3.1. Let $U \sim P_U$ be a continuous random variable, and consider a singleton random set $S = \{U\}$, $U \sim P_U$. Show that the contour function $\gamma(u) = P_S(S \ni u)$ in (3.1) satisfies $f(U) = 0$ with P_U-probability 1.

Exercise 3.2. Suppose that $U \sim N(\delta, 1)$ with δ known to be some value in $[-1, 1]$. Show that U is predictable.

Exercise 3.3. Suppose that Z_1 and Z_2 follow the bivariate normal distribution with mean vector $(0, 0)$ and covariance matrix $\Sigma > 0$. Show that

$$V(Z_1 \mid Z_2 = z_2) \geq V(Z_1), \quad \forall z_2 \in \mathbb{R}.$$

Under what condition does equality hold?

Exercise 3.4. Suppose that T_1 and T_2 follow the bivariate student-t distribution with degrees of freedom $v > 2$. Find all t_2 satisfying

$$V(T_1 \mid T_2 = t_2) > V(T_1).$$

Exercise 3.5. Suppose that a bivariate random variable (U_1, U_2) has a finite covariance matrix. Prove that

$$V(U_1) = E\{V(U_1 \mid U_2)\} + V\{E(U_1 \mid U_2)\}.$$

Use this result to show (3.2).

Exercise 3.6. Let $p(x,y)$ be the pdf of the joint distribution of two variables X and Y. Let $p_Y(y)$ be the marginal pdf of Y and $p_{Y|X}(y \mid x)$ the conditional pdf of Y, given $X = x$. The marginal entropy of Y is defined as

$$H(Y) = \int p_Y(y) \log p_Y(y) \, dy.$$

The expected conditional entropy is defined as

$$H(Y \mid X) = \int p_X(x) H(Y \mid X = x) \, dx,$$

where

$$H(Y \mid X = x) = \int p_{Y|X}(y \mid x) \log p_{Y|X}(y \mid x) \, dy.$$

(a) Suppose (X,Y) has a bivariate normal distribution, and derive $H(Y)$, $H(Y \mid X)$. Explain how entropy can be interpreted as a measure of uncertainty when the prediction of Y is of interest.

(b) Use Jensen's inequality to prove the *entropy inequality:* $H(Y \mid X) \leq H(Y)$.

(c) Explain what the entropy inequality says about the effect of conditioning in terms of reducing uncertainty.

Chapter 4

Inferential Models

Portions of the material in this chapter are from R. Martin and C. Liu, "Inferential models: A framework for prior-free posterior probabilistic inference," *Journal of the American Statistical Association* **108**, 301–313, 2013, reprinted by permission of the American Statistical Association, www.amstat.org.

4.1 Introduction

Posterior probabilistic statistical inference without priors is an important but so far elusive goal. Fisher's fiducial inference, the Dempster–Shafer theory of belief functions, and Bayesian inference with default priors are attempts to achieve this goal but, as we explained in Chapter 2, so far none has given a completely satisfactory solution. This chapter presents a new framework for prior-free probabilistic inference, called *inferential models* (IMs). This IM framework not only provides data-dependent probabilistic measures of uncertainty about the unknown parameter, but does so with an automatic long-run frequency calibration property. In other words, unlike existing approaches IMs produce exact prior-fee and prior-less probabilistic inference.

The key to this new approach is the specification of an unobservable auxiliary variable associated with observable data and an unknown parameter, and the prediction of this auxiliary variable with a random set before conditioning on data. Here we present a three-step IM construction, and prove a frequency-calibration property of the IM's belief function under mild conditions. A corresponding optimality theory is developed, which helps to resolve the non-uniqueness issue. Several examples are presented to illustrate the proposed approach.

This chapter is organized as follows. Section 4.2 introduces IMs at a somewhat non-technical level. Section 4.3 provides the details of the IM analysis, specifically the three-step construction, as well as a description of calculation and interpretation of the IM output: a posterior belief function. After arguing that the IM output provides a meaningful summary of one's uncertainty about unknown parameter after seeing the observed data, we prove a frequency calibration property of the posterior belief functions in Section 4.4 which establishes the meaningfulness of the posterior belief function across different users and experiments. As a consequence of this frequency-calibration property, we show in Section 4.4.4 that the IM output can easily be used to design new frequentist decision procedures having the desired control on error probabilities. Some basic but fundamental results on IM optimality are presented in Section 4.5. Section 4.6 gives IM-based solutions to two non-trivial examples, both involving some sort of marginalization. Nonetheless, these examples are relatively simple and they illustrate the advantages of the IM approach. Concluding remarks are given in Section 4.7.

4.2 Basic overview

As mentioned in previous chapters, in a statistical inference problem, one attempts to convert *experience*, in the form of observed data, to *knowledge* about the unknown parameter of interest. The fact that observed data is surely limited implies that there will be some uncertainty in this conversion, and probability is seemingly a natural tool to describe this uncertainty. But a statistical inference problem is different from the classical probability setting because everything—observed data and unknown parameter—is fixed, and so it is unclear where these probabilistic assessments of uncertainty should come from, and how they should be interpreted. In Chapter 2 we argued that a probability measure on the parameter space is not the appropriate tool for summarizing evidence in data supporting the truthfulness of falsity of an assertion about the parameter. In its place, we recommend the use of a belief function, the primary difference being that a belief function is sub-additive while probability measures are additive. Chapter 3 outlined the driving principles behind the "best possible inference." In particular, probabilistic inference requires identification of a predictable quantity and, according to the validity principle, we must find a valid predictive random set to predict the predictable quantity, which will generate the belief function to be used for inference.

The goal of this chapter is to develop the new IM framework for prior-free probabilistic inference. The seeds for this idea were first planted in [186] and [269]; here we formalize and extend these ideas toward a cohesive framework for statistical inference. Those two key principles presented in Chapter 3 will guide the technical development of IMs in the sense that we will consider the two important questions:

- *Is the inference valid?*

- *Can the inference be made more efficient without loss of validity?*

The jumping off point is a simple association of the observable data X and unknown parameter $\theta \in \Theta$ with an unobservable auxiliary variable U. In the language of Chapter 3, the unobserved value of U, which we denote here as $u^\star$, is the predictable quantity. For example, consider the simple signal plus noise model, $X = \theta + U$, where $U \sim N(0, 1)$. If $X = x$ is observed, then we know that $x = \theta + u^\star$. From this it is clear that knowing $u^\star$ is equivalent to knowing θ. So the IM approach attempts to accurately predict the value $u^\star$ before conditioning on $X = x$. The benefit of focusing on $u^\star$ rather than θ is that more information is available about $u^\star$: indeed, all that is known about θ is that it sits in Θ, while $u^\star$ is known to be a realization of a draw U from an *a priori* distribution, in this case $N(0, 1)$, that is fully specified by the postulated sampling model. However, this *a priori* distribution alone is insufficient for accurate prediction of $u^\star$. Therefore, we adopt a so-called *predictive random set* for predicting $u^\star$. When combined with the association between observed data, parameters, and auxiliary variables, these random sets produce prior-free, data-dependent probabilistic assessments of uncertainty about θ.

To summarize, an IM starts with an association between data, parameters, and auxiliary variables and a predictive random set, and produces prior-free, post-data probabilistic measures of uncertainty about the unknown parameter. The following associate-predict-combine steps provide a simple yet formal IM construction. The details of each of these three steps will be fleshed out in Section 4.3.

A-step. Associate the unknown parameter θ to each possible (x, u) pair to obtain a collection of sets $\Theta_x(u)$ of candidate parameter values.

P-step. Predict u^* with a valid predictive random set $\mathcal{S}$.

C-step. Combine $X = x$, $\Theta_x(u)$, and $\mathcal{S}$ to obtain a random set $\Theta_x(\mathcal{S}) = \bigcup_{u \in \mathcal{S}} \Theta_x(u)$. Then, for any assertion $A \subseteq \Theta$, compute the probability that the random set $\Theta_x(\mathcal{S})$ is a subset of A as a measure of the available evidence in x supporting A.

The A-step is meant to emphasize the use of unobservable but predictable auxiliary variables in the statistical modeling step. These auxiliary variables make it possible to introduce posterior probability-like quantities without a prior distribution for θ. The P-step is new and unique to the inferential model framework. The key is that $\Theta_x(\mathcal{S})$ contains the true θ if and only if $\mathcal{S}$ contains u^*. Then the validity condition in the P-step ensures that $\mathcal{S}$ will hit its target with large probability which, in turn, guarantees that probabilistic output from the C-step has a desirable frequency-calibration property. This, together with its dependence on the observed data x, makes the IM's probabilistic output meaningful both within and across experiments.

4.3   Inferential models

4.3.1   Auxiliary variable associations

If X denotes the observable sample data, then the sampling model is a probability distribution $\mathsf{P}_{X|\theta}$ on the sample space $\mathbb{X}$, indexed by a parameter $\theta \in \Theta$. The sampling model for X is induced by an auxiliary variable U, for given θ. Let $\mathbb{U}$ be an (arbitrary) auxiliary space, equipped with a probability measure P_U. In applications, $\mathbb{U}$ can often be a unit hyper-cube and P_U Lebesgue measure. The sampling model $\mathsf{P}_{X|\theta}$ shall be determined by the following "algorithm:"

$$\text{sample } U \sim \mathsf{P}_U \text{ and set } X = a(U, \theta), \tag{4.1}$$

for an appropriate mapping $a : \mathbb{U} \times \Theta \to \mathbb{X}$. The key is the association of the observable X, the unknown θ, and the auxiliary variable U through the relation $X = a(U, \theta)$. This particular formulation of the sampling model is not really a restriction. In fact, the two-step construction of the observable X in (4.1) is often consistent with scientific understanding of the underlying process under investigation; linear models form an interesting class of examples. As another example, suppose $X = (X_1, \ldots, X_n)$ consists of an independent sample from a continuous distribution. If the corresponding distribution function F_θ is invertible, then $a(\theta, U)$ may be written as

$$a(\theta, U) = \left(F_\theta^{-1}(U_1), \ldots, F_\theta^{-1}(U_n) \right), \tag{4.2}$$

where $U = (U_1, \ldots, U_n)$ is a set of independent $\mathsf{Unif}(0,1)$ random variables. Any model that can be simulated has the form (4.1) so, in our view, this particular formulation results in no loss of generality.

The notation $X = a(\theta, U)$ chosen to represent the association between (X, θ, U) is just for simplicity. In fact, this association need not be described by a formal equation. As the Poisson example below shows, all we need is a recipe, like that in (4.1), describing how to produce a sample X, for a given θ, based on a realization $U \sim \mathsf{P}_U$.

Example 4.1. Consider the problem of inference on the mean θ based on a single sample $X \sim \mathsf{N}(\theta, 1)$. In this case, the association linking X, θ, and an auxiliary variable U may be written as $U = \Phi(X - \theta)$ or, equivalently, $X = \theta + \Phi^{-1}(U)$, where $U \sim \mathsf{Unif}(0,1)$, and Φ is the standard Gaussian distribution function.

Example 4.2. Consider the problem of inference on the mean θ of a Poisson population based on a single observation X. For this discrete problem, the association for X, given θ, may be written as

$$F_\theta(X - 1) \leq 1 - U < F_\theta(X), \quad U \sim \mathsf{Unif}(0,1), \tag{4.3}$$

where F_θ denotes the $\mathsf{Pois}(\theta)$ distribution function. This representation is familiar for simulating $X \sim \mathsf{Pois}(\theta)$, i.e., one can first sample $U \sim \mathsf{Unif}(0,1)$ and then choose X so that the inequalities in (4.3) are satisfied. But here we also interpret (4.3) as a means to link data, parameter, and auxiliary variable.

It should not be surprising that, in general, there are many associations for a given sampling model. In fact, for a given sampling model $\mathsf{P}_{X|\theta}$, there are as many associations as there are triplets $(\mathbb{U}, \mathsf{P}_U, a)$ such that $\mathsf{P}_{X|\theta}$ equals the push-forward measure $\mathsf{P}_U a_\theta^{-1}$, with $a_\theta(\cdot) = a(\theta, \cdot)$. For example, if $X \sim \mathsf{N}(\theta, 1)$, then each of the following defines an association: $X = \theta + U$ with $U \sim \mathsf{N}(0,1)$, $X = \theta + \Phi^{-1}(U)$ with $U \sim \mathsf{Unif}(0,1)$, and

$$X = \begin{cases} \theta + U & \text{if } \theta \geq 0, \\ \theta - U & \text{if } \theta < 0, \end{cases} \quad \text{with } U \sim \mathsf{N}(0,1). \tag{4.4}$$

Presently, there appears to be no strong reason to choose one of these associations over the other. However, the optimality theory presented in Section 4.5 helps to resolve this non-uniqueness issue, that is, the optimal IM depends only on the sampling model, and not on the chosen association. From a practical point of view, we prefer, for continuous data problems, associations which are continuous in both θ and U, which rules out the latter of the three associations above. Also, we tend to prefer the representation with a uniform U, any other choice being viewed as just a reparametrization of this one. It will become evident that this view is without loss of generality for simple problems with a one-dimensional auxiliary variable. The case when U is moderate- to high-dimensional is more challenging and we defer its discussion to Section 4.7.

4.3.2   Three-step IM construction

4.3.2.1   Association step

The association (4.1) plays two distinct roles. Before the experiment, the association characterizes the predictive probabilities of the observable X. But once $X = x$ is observed, the role of the association changes. The key idea is that the observed x and the unknown θ must satisfy

$$x = a(u^\star, \theta) \qquad (4.5)$$

for some unobserved realization $u^\star$ of U. Although $u^\star$ is unobserved, there is information available about the nature of this quantity; in particular, we know exactly the distribution P_U from which it came.

Of course, the value of $u^\star$ can never be known, *but if it were*, the inference problem would be simple: given $X = x$, just solve the equation $x = a(u^\star, \theta)$ for θ. More generally, one could construct the set of solutions $\Theta_x(u^\star)$, where

$$\Theta_x(u) = \{\theta : x = a(u, \theta)\}, \quad x \in \mathbb{X}, \quad u \in \mathbb{U}. \qquad (4.6)$$

For continuous-data problems, $\Theta_x(u)$ is typically a singleton for each u; for other problems, it could be a set. In either case, given $X = x$, $\Theta_x(u^\star)$ represents the best possible inference in the sense that *the true θ is guaranteed to be in $\Theta_x(u^\star)$*.

Example 4.1 (cont). The Gaussian mean problem is continuous, so the association $x = \theta + \Phi^{-1}(u)$ identifies a single θ for each fixed (x, u) pair. Therefore, $\Theta_x(u) = \{x - \Phi^{-1}(u)\}$. In this case, clearly, if $u^\star$ were somehow observed, then the true θ could be determined with complete certainty.

Example 4.2 (cont). Integration-by-parts reveals that the $\mathrm{Pois}(\theta)$ distribution function F_θ satisfies $F_\theta(x) = 1 - G_{x+1}(\theta)$, where G_a is a $\mathrm{Gamma}(a, 1)$ distribution function; see Exercise 4.2. Therefore, from (4.3), we get the u-interval $G_{x+1}(\theta) < u \leq G_x(\theta)$. Inverting this u-interval produces the following θ-interval:

$$\Theta_x(u) = \left(G_x^{-1}(u), G_{x+1}^{-1}(u)\right]. \qquad (4.7)$$

If $u^\star$ was known, then $\Theta_x(u^\star)$ would provide the best possible inference in the sense that the true value of θ is guaranteed to sit inside this interval. But even in this ideal case it is not possible to identify the exact location of θ in $\Theta_x(u^\star)$.

4.3.2.2   Prediction step

The above discussion highlights the importance of the auxiliary variable for inference. According to the validity principle, it is, therefore, only natural that the inference problem should focus on accurately predicting the unobserved $u^\star$. To predict $u^\star$ with a certain desired accuracy, we employ a predictive random set. First we give the simplest description of a predictive random set and provide a useful example. More general descriptions will be given later.

Let $u \mapsto S(u)$ be a mapping from $\mathbb{U}$ to a collection of P_U-measurable subsets of $\mathbb{U}$; one decent example of such a mapping S is given in equation (4.8) below. Then the predictive random set $\mathcal{S}$ is obtained by applying the set-valued mapping S to a draw $U \sim P_U$, i.e., $S = S(U)$ with $U \sim P_U$. The intuition is that if a draw $U \sim P_U$ is a good prediction for the unobserved $u^\star$, then the random set $S = S(U)$ should be even better in the sense that there is high probability that $S \ni u^\star$.

Example 4.1 (cont). In this example we may predict the unobserved $u^\star$ with a predictive random set S defined by the set-valued mapping

$$S(u) = \{u' \in (0,1) : |u' - 0.5| < |u - 0.5|\}, \quad u \in (0,1). \tag{4.8}$$

As this predictive random set is designed to predict an unobserved uniform variate, we may also employ (4.8) in other problems, including the Poisson example.

There are, of course, other choices of $S(u)$, e.g., $[0,u)$, $(u,1]$, $(0.5u, 0.5 + 0.5u)$ and more. Although some other choice of $S = S(U)$ might perform slightly better depending on the assertion of interest, (4.8) seems to be a good default choice. See Sections 4.4 and 4.5 for more on the choice predictive random sets.

For the remainder of this chapter, we shall mostly omit the set-valued mapping S from the notation and speak directly about the predictive random set S. That is, the predictive random set S will be just a random subset of $\mathbb{U}$ with distribution P_S. In the above description, P_S is just the push-forward measure $P_U S^{-1}$. More details about random sets are given in Chapter 5.

4.3.2.3 Combination step

For the time being, let us assume that the predictive random set S is satisfactory for predicting the unobserved $u^\star$; this is actually easy to arrange, but we defer discussion until Section 4.4. To transfer the available information about $u^\star$ to the θ-space, our last step is to combine the information in the association, the observed $X = x$, and the predictive random set S. The intuition is that, if $u^\star \in S$, then the true θ must be in the set $\Theta_x(u)$, from (4.6), for at least one $u \in S$. So, logically, it makes sense to consider, for inference about θ, the expanded set

$$\Theta_x(S) = \bigcup_{u \in S} \Theta_x(u). \tag{4.9}$$

The set $\Theta_x(S)$ contains those values of θ which are consistent with the observed data and sampling model for at least one candidate $u^\star$ value $u \in S$. Since $\theta \in \Theta_x(S)$ if and only if the unobserved $u^\star \in S$, if we are willing to accept that S is satisfactory for predicting $u^\star$, then $\Theta_x(S)$ will do equally well at capturing θ.

Now consider an assertion A about the parameter of interest θ. To summarize the evidence in x that supports the truthfulness of the assertion A, we calculate the probability that $\Theta_x(S)$ is a subset of A, i.e.,

$$\mathrm{bel}_x(A) = P_S\{\Theta_x(S) \subseteq A \mid \Theta_x(S) \neq \varnothing\}. \tag{4.10}$$

We refer to $\mathrm{bel}_x(A)$ as the belief function at A. Naturally, bel_x also depends on the choice of association and predictive random set, but for now we suppress this dependence in the notation. As discussed previously, like in the Dempster–Shafer theory, bel_x is subadditive in the sense that if A is a non-trivial subset of Θ, then

$\mathrm{bel}_x(A) + \mathrm{bel}_x(A^c) \le 1$ with equality if and only if $\Theta_x(\mathcal{S})$ is a singleton with $\mathsf{P}_{\mathcal{S}}$-probability 1. However, our use of the predictive random set, and our emphasis on validity in Section 4.4, separates our approach from that of Dempster–Shafer.

Here we make two technical remarks about the belief function in (4.10). First, in the problems considered in this chapter, the case $\Theta_x(\mathcal{S}) = \varnothing$ is a $\mathsf{P}_{\mathcal{S}}$-null event, so the belief function can be simplified as $\mathrm{bel}_x(A) = \mathsf{P}_{\mathcal{S}}\{\Theta_x(\mathcal{S}) \subseteq A\}$, no conditioning. This simplification may not hold in problems where the observation $X = x$ can induce a constraint on the auxiliary variable u. For example, consider the Gaussian example from above, but suppose that the mean is known to satisfy $\theta \ge 0$. In this case, it is easy to check that $\Theta_x(\mathcal{S}) = \varnothing$ iff $\Phi^{-1}(\inf \mathcal{S}) > x$, an event which generally has positive $\mathsf{P}_{\mathcal{S}}$-probability. So, in general, we can ignore conditioning provided that

$$\Theta_x(u) \ne \varnothing \quad \text{for all } x \text{ and all } u. \tag{4.11}$$

The IM framework can be modified in cases where (4.11) fails, and we discuss this in Chapter 5. Second, measurability of $\mathrm{bel}_x(A)$, as a function of x for given A, which is important in what follows, is not immediately clear from the definition and should be assessed case-by-case. However, in our experience and in all the examples herein, $\mathrm{bel}_x(A)$ is a nice measurable function of x.

Unlike with an ordinary additive probability measure, to reach conclusions about A based on bel_x one must know *both* $\mathrm{bel}_x(A)$ and $\mathrm{bel}_x(A^c)$; for example, in the extreme case of "total ignorance" about A, one has $\mathrm{bel}_x(A) = \mathrm{bel}_x(A^c) = 0$. It is often more convenient to work with a different but related function

$$\mathrm{pl}_x(A) = 1 - \mathrm{bel}_x(A^c) = \mathsf{P}_{\mathcal{S}}\{\Theta_x(\mathcal{S}) \not\subseteq A^c \mid \Theta_x(\mathcal{S}) \ne \varnothing\}, \tag{4.12}$$

called the plausibility function at A; when $A = \{\theta\}$ is a singleton, we write $\mathrm{pl}_x(\theta)$ instead of $\mathrm{pl}_x(\{\theta\})$. From the subadditivity of the belief function, it follows that $\mathrm{bel}_x(A) \le \mathrm{pl}_x(A)$ for all A. In what follows, to summarize the evidence in x supporting A, we shall report the pair $\mathrm{bel}_x(A)$ and $\mathrm{pl}_x(A)$.

Example 4.1 (cont). With the predictive random set $\mathcal{S}$ in (4.8), the random set $\Theta_x(\mathcal{S})$ is given by

$$\begin{aligned}
\Theta_x(\mathcal{S}) &= \bigcup_{u \in \mathcal{S}}\{x - \Phi^{-1}(u)\} \\
&= \left(x - \Phi^{-1}(0.5 + |U - 0.5|), x - \Phi^{-1}(0.5 - |U - 0.5|)\right) \\
&= \left(\underline{\Theta}_x(U), \overline{\Theta}_x(U)\right), \quad \text{say,}
\end{aligned}$$

where $U \sim \mathrm{Unif}(0,1)$. For a singleton assertion $A = \{\theta\}$, it is easy to see that the belief function is zero. But the plausibility function is

$$\begin{aligned}
\mathrm{pl}_x(\theta) &= 1 - \mathsf{P}_U\{\underline{\Theta}_x(U) > \theta\} - \mathsf{P}_U\{\overline{\Theta}_x(U) < \theta\} \\
&= 1 - |2\Phi(x - \theta) - 1|.
\end{aligned} \tag{4.13}$$

A plot of $\mathrm{pl}_x(\theta)$, with $x = 5$, as a function of θ, is shown in Figure 4.1(a). The symmetry around the observed x is apparent, and all those θ values in a neighborhood of $x = 5$ are relatively plausible. See Section 4.4.4 for more applications of this graph.

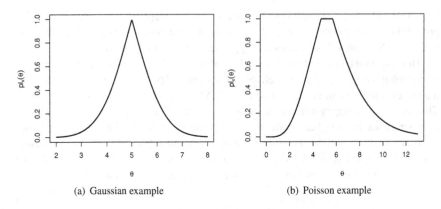

Figure 4.1 *Taken from [180]. Plot of the plausibility functions* $\mathsf{pl}_x(\theta)$, *as functions of* θ, *in (a) the Gaussian example and (b) the Poisson example. In both cases,* $X = 5$ *is observed.*

Example 4.2 (cont). With the same predictive random set as in the previous example, the random set $\Theta_x(\mathcal{S})$ is given by

$$
\begin{aligned}
\Theta_x(\mathcal{S}) &= \bigcup_{u \in \mathcal{S}} \left(G_x^{-1}(u), G_{x+1}^{-1}(u) \right] \\
&= \left(G_x^{-1}(0.5 - |U - 0.5|), G_{x+1}^{-1}(0.5 + |U - 0.5|) \right) \\
&= \left(\underline{\Theta}_x(U), \overline{\Theta}_x(U) \right), \quad \text{say,}
\end{aligned}
$$

where U is a random draw from $\mathsf{Unif}(0, 1)$. For a singleton assertion $A = \{\theta\}$, again the belief function is zero, but the plausibility function is

$$
\begin{aligned}
\mathsf{pl}_x(\theta) &= 1 - P_U\{\underline{\Theta}_x(U) > \theta\} - P_U\{\overline{\Theta}_x(U) < \theta\} \\
&= 1 - \max\{1 - 2G_x(\theta), 0\} - \max\{2G_{x+1}(\theta) - 1, 0\}. \quad (4.14)
\end{aligned}
$$

A graph of $\mathsf{pl}_x(\theta)$, with $x = 5$, as a function of θ is shown in Figure 4.1(b). The plateau indicates that no θ values in a neighborhood of 5 can be ruled out. As in the Gaussian example, θ values in an interval around 5 are all relatively plausible.

Dempster [67] gives a different analysis of this same Poisson problem. His plausibility function for the singleton assertion $A = \{\theta\}$ is $r_x(\theta) = e^{-\theta}\theta^x/x!$, which is the Poisson pmf treated as a function of θ. This function has a similar shape to that in Figure 4.1(b), but the scale is much smaller. For example, $r_5(5) = 0.175$, suggesting that the assertion $\{\theta = 5\}$ is relatively implausible, even though $X = 5$ was observed. Compare this to $\mathsf{pl}_5(5) = 1$. We would argue that, if $X = 5$ is observed, then no plausibility function threshold should be able to rule out $\{\theta = 5\}$; in that case, $\mathsf{pl}_5(5) = 1$ makes more sense. Furthermore, as Dempster's analysis is similar to ours but with an invalid predictive random set, namely, $\mathcal{S} = \{U\}$, with $U \sim \mathsf{Unif}(0, 1)$, the corresponding plausibility function is not properly calibrated for all assertions.

4.3.3 Interpretation of the belief function

It is clear that the belief function depends on the observed x and so must be meaningful within the problem at hand. But while it is data-dependent, $\text{bel}_x(A)$ is not a posterior probability for A in the familiar Bayesian sense. In fact, under our assumption that θ is fixed and non-random, there can be no non-trivial posterior distribution on Θ. The way around this limitation is to drop the requirement that posterior inference be based on a bona fide probability measure [134, 245, 254]. Therefore, we recommend interpreting $\text{bel}_x(A)$ and $\text{bel}_x(A^c)$ as degrees of belief, rather than ordinary probabilities, even though they manifest from P_U-probability calculations. More precisely, $\text{bel}_x(A)$ and $\text{bel}_x(A^c)$ represent the knowledge gained about the respective claims "$\theta \in A$" and "$\theta \notin A$" based on both the observed data x and prediction of the auxiliary variable.

4.3.4 Summary

The familiar sampling model appears in the A-step, but it is the corresponding association which is of primary importance. This association, in turn, determines the auxiliary variable which is to be the focus of the IM framework. Motived by the validity principle, we propose to predict the unobserved value of this auxiliary variable in the P-step with a predictive random set $\mathcal{S}$, which is chosen to have certain desirable properties (see Definition 4.1 below). This use of a predictive random set is likely the most unfamiliar aspect of the IM framework, but the intuition should be clear: one cannot hope to accurately predict a fixed value $u^\star$ by an ordinary continuous random variable. With the association, predictive random set, and observed $X = x$ in hand, one proceeds to the C-step where a random set $\Theta_x(\mathcal{S})$ on the parameter space is obtained. As this random set corresponds to a set of "reasonable" θ values, given x, it is natural to summarize the support of an assertion A by the probability that $\Theta_x(\mathcal{S})$ is a subset of A. This probability is exactly the belief function that characterizes the output of the IM and an argument is presented that justifies the meaningfulness of $\text{bel}_x(A)$ and $\text{pl}_x(A)$ as summaries of the evidence in favor of A.

Finally, we mention that the predictive random set $\mathcal{S}$ can depend on the assertion A of interest. That is, one might consider using one predictive random set, say $\mathcal{S}_A$, to evaluate $\text{bel}_x(A)$, and another predictive random set, say $\mathcal{S}_{A^c}$, to evaluate $\text{pl}_x(A) = 1 - \text{bel}_x(A^c)$. In Section 4.5 we show that this is actually a desirable strategy, in the sense that the optimal predictive random set depends on the assertion in question.

4.4 Theoretical validity of IMs

4.4.1 Intuition

In Section 4.3 we argued that $\text{bel}_x(A; \mathcal{S})$ and $\text{pl}_x(A; \mathcal{S})$ together provide a meaningful summary of evidence in favor of A for the given $X = x$; our notation now explicitly indicates the dependence of these function on the predictive random set $\mathcal{S}$. In this section we show that $\text{bel}_X(A; \mathcal{S})$ and $\text{pl}_X(A; \mathcal{S})$ are also meaningful as functions of the random variable $X \sim \mathsf{P}_{X|\theta}$ for a fixed assertion A. For example, we show that $\text{bel}_X(A)$

is frequency-calibrated in the following sense: if $\theta \notin A$, then $\mathsf{P}_{X|\theta}\{\mathrm{bel}_X(A;S) \geq 1 - \alpha\} \leq \alpha$ for each $\alpha \in [0,1]$. In other words, the amount of evidence in favor of a false A can be large with only small probability. This property means that the belief function is appropriately scaled for objective scientific inference. A similar property also holds for $\mathrm{pl}_X(A)$. We refer to this frequency-calibration property as *validity* (Definition 4.2).

4.4.2 Predictive random sets

We start with a few definitions. Recall the contour function defined in (3.1), i.e.,

$$\gamma_S(u) = \mathsf{P}_S(S \ni u), \quad u \in \mathbb{U}.$$

This represents the probability that the predictive random set S hits the specified target u. Ideally, S will be such that it hits most targets most of the time, which suggests a connection between the distributions P_S and P_U; see Theorem 4.1.

Definition 4.1. A predictive random set S is *valid* for predicting the unobserved auxiliary variable if $\gamma_S(U)$, as a function of $U \sim \mathsf{P}_U$, is stochastically no smaller than $\mathrm{Unif}(0,1)$. If $\gamma_S(U) \sim \mathrm{Unif}(0,1)$, then S is *efficient*.

In words, validity of S implies that the probability that it hits a target u is small for only a small P_U-proportion of possible u values. The predictive random set S defined by the mapping (4.8) is both valid and efficient. Indeed, it is easy to check that, in this case, $\gamma_S(u) = 1 - |2u - 1|$; see Exercise 4.3. Therefore, if $U \sim \mathrm{Unif}(0,1)$ then $\gamma_S(U) \sim \mathrm{Unif}(0,1)$ too. Corollary 4.1 below gives a simple and general recipe for constructing a valid and efficient S.

There is an important and apparently fundamental concept related to validity of predictive random sets, namely, *nesting*. We say that a collection of sets $\mathbb{S} \subseteq 2^{\mathbb{U}}$ is nested if, for any pair of sets S and S' in $\mathbb{S}$, we have $S \subseteq S'$ or $S' \subseteq S$. If the support $\mathbb{S}$ of a predictive random set S is nested, then we will call S a nested predictive random set. We need a few additional conditions on P_U and on the support S:

- For some $S \in \mathbb{S}$, $\mathsf{P}_U(S) > 0$;
- All closed subsets of $\mathbb{U}$ are P_U-measurable;
- S contains $\varnothing$, $\mathbb{U}$, and all of the other elements are closed sets.

The next theorem shows that if $\mathbb{S}$ is nested and the above regularity conditions hold, then there exists a nested predictive random set S, supported on $\mathbb{S}$, that is valid.

Theorem 4.1. *Let* $\mathbb{S}$ *and* P_U *satisfy the above conditions. Define a predictive random set* S, *with distribution* P_S, *supported on* $\mathbb{S}$, *such that*

$$\mathsf{P}_S\{S \subseteq K\} = \sup_{S \in \mathbb{S}:S \subseteq K} \mathsf{P}_U(S), \quad K \subseteq \mathbb{U}. \tag{4.15}$$

Then S *is valid in the sense of Definition 4.1.*

Proof. Set $Q_S(u) = 1 - \gamma_S(u) = \mathsf{P}_S\{S \not\ni u\}$. For any $\alpha \in (0,1)$, let S_α be the smallest $S \in \mathbb{S}$ such that $\mathsf{P}_S\{S \subseteq S\} \equiv \mathsf{P}_U(S) \geq 1 - \alpha$. In particular, $S_\alpha = \bigcap\{S \in \mathbb{S} : \mathsf{P}_U(S) \geq$

$1 - \alpha$}. Since each S is closed, so is S_α; it is also measurable by our assumptions about the richness of the σ-algebra on $\mathbb{U}$. The key observation is that $Q(u) > 1 - \alpha$ iff $u \in S_\alpha^c$. Therefore, by continuity of P_U from above, we get

$$\mathsf{P}_U\{Q_S(U) > 1 - \alpha\} = \mathsf{P}_U(S_\alpha^c) = 1 - \mathsf{P}_U(S_\alpha) = 1 - \lim \mathsf{P}_U(S),$$

where the limit is over all S decreasing to S_α. By construction, each such S satisfies $\mathsf{P}_U(S) \geq 1 - \alpha$. So, finally, we get $\mathsf{P}_U\{Q_S(U) > 1 - \alpha\} \leq \alpha$ and, since α is arbitrary, we get $Q_S(U) \leq_{\text{st}} \text{Unif}(0, 1)$, proving validity. $\qquad \square$

Remark 4.1. The particular random set constructed (4.15) in Theorem 4.1, whose distribution we call the *natural measure*, is an essential part of the construction of optimal IMs. We will have more to say specifically about this predictive random set in Chapter 5, and applications in later chapters.

It is clear that, if P_U is absolutely continuous and the nested support $\mathbb{S}$ is sufficiently rich, then the predictive random set defined above is also efficient. Specifically, if $\mathbb{U} \in \mathbb{S}$ and, for S_α defined in the proof above, $\mathsf{P}_U(S_\alpha) = 1 - \alpha$ for every $\alpha \in (0, 1)$. This vague argument for efficiency is made more precise in the next important special case.

Corollary 4.1. *Suppose the P_U is non-atomic, and let h be a real-valued function on $\mathbb{U}$. Then the predictive random set $S = \{u \in \mathbb{U} : h(u) < h(U)\}$, with $U \sim \mathsf{P}_U$, is valid. If h is continuous and constant only on P_U-null sets, then it is also efficient.*

Proof. Validity is a consequence of Theorem 4.1 and the fact that this S is nested. To prove the efficiency claim, let H be the distribution function of $h(U)$ when $U \sim \mathsf{P}_U$. Then, for $u \in \mathbb{U}$, $Q_S(u) = \mathsf{P}_U\{h(U) \leq h(u)\} = H(h(u))$. If h satisfies the stated conditions, then $h(U)$ is a continuous random variable. Therefore, if $U \sim \mathsf{P}_U$, then $Q_S(U) = H(h(U)) \sim \text{Unif}(0, 1)$, so efficiency follows. $\qquad \square$

The above results demonstrate that nesting is a sufficient condition for predictive random set validity. But nesting is not a necessary condition [186]. The real issue, however, is the performance of the corresponding IM. We show in Section 4.5 that for any non-nested predictive random set S, there is a nested predictive random set S' such that the IM based on S' is "at least as good" as that based on S.

4.4.3 IM validity

Validity of the underlying predictive random set S is essentially all that is needed to prove the meaningfulness of the corresponding IM/belief function. Here meaningfulness refers to a calibration property of the belief function.

Definition 4.2. *Suppose $X \sim \mathsf{P}_{X|\theta}$ and let A be an assertion of interest. Then the IM with belief function bel_x is valid for A if, for each $\alpha \in (0, 1)$,*

$$\sup_{\theta \notin A} \mathsf{P}_{X|\theta}\{\text{bel}_x(A; S) \geq 1 - \alpha\} \leq \alpha. \qquad (4.16)$$

The IM is valid if it is valid for all A.

By (4.12), the validity property can also be stated in terms of the plausibility function. That is, the IM is valid if, for all assertions A and for any $\alpha \in (0,1)$,

$$\sup_{\theta \in A} \mathsf{P}_{X|\theta}\{\mathsf{pl}_X(A;\mathcal{S}) \leq \alpha\} \leq \alpha. \tag{4.17}$$

Theorem 4.2. *Suppose the predictive random set $\mathcal{S}$ is valid, and $\Theta_x(\mathcal{S}) \neq \varnothing$ with $\mathsf{P}_\mathcal{S}$-probability 1 for all x. Then the IM is valid.*

Proof. For any A, take (x,u,θ) such that $\theta \notin A$ and $x = a(\theta,u)$. Since $A \subseteq \{\theta\}^c$, $\mathsf{bel}_x(A;\mathcal{S}) \leq \mathsf{bel}_x(\{\theta\}^c;\mathcal{S}) = \mathsf{P}_\mathcal{S}\{\Theta_x(\mathcal{S}) \not\ni \theta\} = \mathsf{P}_\mathcal{S}\{\mathcal{S} \not\ni u\}$ by monotonicity. Validity of $\mathcal{S}$ implies that the right-hand side, as a function of $U \sim \mathsf{P}_U$, is stochastically smaller than $\mathsf{Unif}(0,1)$. This, in turn, implies the same of $\mathsf{bel}_x(A;\mathcal{S})$ as a function of $X \sim \mathsf{P}_{X|\theta}$. Therefore, $\mathsf{P}_{X|\theta}\{\mathsf{bel}_x(A;\mathcal{S}) \geq 1 - \alpha\} \leq \mathsf{P}\{\mathsf{Unif}(0,1) \geq 1 - \alpha\} = \alpha$. Taking a supremum over $\theta \notin A$ on the left-hand side completes the proof. $\qquad\square$

A key feature of the validity theorem above is that it holds under minimal conditions on the predictive random set. Validity of the IM does not depend on the particular form of predictive random set, only that it is valid. Recall that the condition "$\Theta_x(\mathcal{S}) \neq \varnothing$ with $\mathsf{P}_\mathcal{S}$-probability 1" holds whenever (4.11) holds. See, also, [82].

The following corollary states that the validity theorem remains true even after a suitable—possibly θ-dependent—change of auxiliary variable. In other words, the validity property is independent of the choice of auxiliary variable parametrization. This reparametrization comes in handy in examples, including those in Section 4.6.

Corollary 4.2. *Consider a one-to-one transformation $v = \phi_\theta(u)$, possibly depending on θ, but only to the extent that the push-forward measure $\mathsf{P}_V = \mathsf{P}_U \phi_\theta^{-1}$ on $\mathbb{V} = \phi_\theta(\mathbb{U})$ does not depend on θ. Suppose $\mathcal{S}$ is valid for predicting $v^\star = \phi_\theta(u^\star)$, and $\Theta_x(\mathcal{S}) \neq \varnothing$ with $\mathsf{P}_\mathcal{S}$-probability 1 for all x. Then the corresponding belief function satisfies (4.16) and the transformed IM is valid.*

Proof. See Exercise 4.4. $\qquad\square$

4.4.4 Application: IM-based frequentist procedures

In addition to providing problem-specific measures of certainty about various assertions of interest, the belief/plausibility functions can easily be used to create frequentist procedures. First consider testing $H_0 : \theta \in A$ versus $H_1 : \theta \in A^c$. Then an IM-based counterpart to a frequentist testing rule is of the following form:

$$\text{Reject } H_0 \text{ if } \mathsf{pl}_x(A) \leq \alpha, \text{ for a specified } \alpha \in (0,1). \tag{4.18}$$

According to (4.17) and Theorem 4.2, if the predictive random set $\mathcal{S}$ is valid, then the probability of a Type I error for such a rejection rule is $\sup_{\theta \in A} \mathsf{P}_{X|\theta}\{\mathsf{pl}_x(A) \leq \alpha\} \leq \alpha$. Therefore, the test (4.18) controls the Type I error probability at level α.

Next consider the class of singleton assertions $\{\theta\}$, with $\theta \in \Theta$. As a counterpart to a frequentist confidence region, define the $100(1 - \alpha)\%$ *plausibility region*

$$\Pi_x(\alpha) = \{\theta : \mathsf{pl}_x(\theta) > \alpha\}. \tag{4.19}$$

Now the coverage probability of the plausibility region (4.19) is

$$P_{X|\theta}\{\Pi_X(\alpha) \ni \theta\} = P_{X|\theta}\{\mathsf{pl}_X(\theta) > \alpha\} = 1 - P_{X|\theta}\{\mathsf{pl}_X(\theta) \le \alpha\} \ge 1 - \alpha,$$

where the last inequality follows from Theorem 4.2. Therefore, this plausibility region has at least the nominal coverage probability.

Example 4.1 (cont). Suppose $X = 5$. Then, using the predictive random set $\mathcal{S}$ in (4.8), the plausibility function is $\mathsf{pl}_5(\theta) = 1 - |2\Phi(5 - \theta) - 1|$. The 90% plausibility interval for θ, determined by the inequality $\mathsf{pl}_5(\theta) > 0.10$, is $5 \pm \Phi^{-1}(0.05)$, the same as the classical 90% z-interval for θ given in standard textbooks.

Example 4.2 (cont). For the predictive random set determined by $\mathcal{S}$ in (4.8), the plausibility function $\mathsf{pl}_x(\theta)$ is displayed in (4.14). For observed $X = 5$, a 90% plausibility interval for θ, characterized by the inequality $\mathsf{pl}_5(\theta) > 0.10$, is $(1.97, 10.51)$. This interval is not the best possible; in fact, the one presented in Section 4.5.3.2 is better. But these plausibility intervals have exact coverage properties, which means that they may be too conservative at certain θ values for practical use. This is the case for all exact intervals in discrete data problems [34, 36].

4.5 Theoretical optimality of IMs

4.5.1 Intuition

Martin et al. [186] showed that fiducial inference and Dempster–Shafer theory are special cases of the IM framework corresponding to a singleton predictive random set. But it is easy to show that, for some assertions A, the fiducial probability is not valid in the sense of Definition 4.2. To correct for this bias, we propose to replace the singleton with some larger $\mathcal{S}$. But taking $\mathcal{S}$ to be too large will lead to inefficient inference. So the goal is to take $\mathcal{S}$ just large enough that validity is achieved.

4.5.2 Preliminaries

Throughout the subsequent discussion, we shall assume (4.11), i.e., $\Theta_x(u) \ne \varnothing$ for all x and u. This allows us to ignore conditioning in the definition of belief functions.

For the predictive random set $\mathcal{S}_0 = \{U\}$, with $U \sim \text{Unif}(0, 1)$, the belief function at A is $\mathsf{bel}_x(A; \mathcal{S}_0) = P_U\{\Theta_x(U) \subseteq A\}$, where $\Theta_x(u) = \{\theta : x = a(\theta, u)\}$ as in (4.6). This is exactly the fiducial probability for A given $X = x$. For a general predictive random set $\mathcal{S}$, we have $\mathsf{bel}_x(A; \mathcal{S}) = P_{\mathcal{S}}\{\Theta_x(\mathcal{S}) \subseteq A\}$, where $\Theta_x(\mathcal{S}) = \bigcup_{u \in \mathcal{S}} \Theta_x(u)$ is defined in (4.9). In light of the discussion in Section 4.5.1, we shall compare the two belief functions $\mathsf{bel}_x(A; \mathcal{S})$ and $\mathsf{bel}_x(A; \mathcal{S}_0)$. Toward this, we have the following result which says that the fiducial probability is an upper bound for the belief.

Proposition 4.1. *If* (4.11) *holds and the predictive random set $\mathcal{S}$ is valid in the sense of Definition 4.1, then* $\mathsf{bel}_x(A; \mathcal{S}) \le \mathsf{bel}_x(A; \mathcal{S}_0)$ *for each fixed x.*

Proof. Let $\mathbb{U}_x(A) = \{u : \Theta_x(u) \subseteq A\}$; note that $\mathcal{S} \subseteq \mathbb{U}_x(A)$ iff $\Theta_x(\mathcal{S}) \subseteq A$. Also, put $b = \mathsf{bel}_x(A; \mathcal{S})$ and $b_0 = \mathsf{bel}_x(A; \mathcal{S}_0) \equiv P_U\{\mathbb{U}_x(A)\}$. If $u \notin \mathbb{U}_x(A)$, then

$$Q_{\mathcal{S}}(u) \equiv P_{\mathcal{S}}\{\mathcal{S} \not\ni u\} \ge P_{\mathcal{S}}\{\mathcal{S} \subseteq \mathbb{U}_x(A)\} = P_{\mathcal{S}}\{\Theta_x(\mathcal{S}) \subseteq A\} = b.$$

Therefore, $P_U\{Q_S(U) \geq b\} \geq P_U\{\mathbb{U}_x(A)^c\} = 1 - b_0$. Also, validity of S implies $P_U\{Q_S(U) \geq b\} \leq 1 - b$. Then $1 - b_0 \leq 1 - b$, i.e., $\text{bel}_x(A; S) \leq \text{bel}_x(A; S_0)$. $\quad\square$

For given assertion A and predictive random set S, consider the ratio

$$R_A(x; S) = \text{bel}_x(A; S)/\text{bel}_x(A; S_0), \quad x \in \mathbb{X}. \tag{4.20}$$

We call this the relative efficiency of the IM based on S. Proposition 4.1 guarantees that this ratio is bounded by unity, provided that the denominator $\text{bel}_x(A; S_0)$ is non-zero. Our main goal is to choose S to make this ratio large in some sense. Toward this goal, we have the following "complete-class theorem" which says that nested predictive random sets—which, by Theorems 4.1 and 4.2, produce valid IMs—are the only kind of predictive random sets that deserve consideration.

Before we get to the complete-class theorem, we need one technical result.

Lemma 4.1. *On a space $\mathbb{U}$ equipped with probability P_U, let S be a valid predictive random set for $U \sim P_U$. Choose a collection of P_U-measurable subsets $\{\mathbb{U}_x : x \in \mathbb{X}\}$ of $\mathbb{U}$, and set $\eta(x) = P_S\{S \subseteq \mathbb{U}_x\}$. Then*

$$\inf_{x \in \mathbb{X}_0} \eta(x) \leq P_U\left\{\bigcap_{x \in \mathbb{X}_0} \mathbb{U}_x\right\}$$

for any subset $\mathbb{X}_0$ of $\mathbb{X}$ such that $\bigcap_{x \in \mathbb{X}_0} \mathbb{U}_x$ is P_U-measurable.

Proof. First, note that if $u \in \mathbb{U}_x^c$, then $Q(u) \equiv P_S\{S \not\ni u\} \geq \eta(x)$. Therefore, if $u \in \bigcup_{x \in \mathbb{X}_0} \mathbb{U}_x^c$, then $Q(u) \geq \inf_{x \in \mathbb{X}_0} \eta(x)$. This argument implies

$$P_U\left\{Q(U) \geq \inf_{x \in \mathbb{X}_0} \eta(x)\right\} \geq P_U\left\{\bigcup_{x \in \mathbb{X}_0} \mathbb{U}_x^c\right\} = 1 - P_U\left\{\bigcap_{x \in \mathbb{X}_0} \mathbb{U}_x\right\}.$$

Since S is valid, we have

$$P_U\left\{Q(U) \geq \inf_{x \in \mathbb{X}_0} \eta(x)\right\} \leq 1 - \inf_{x \in \mathbb{X}_0} \eta(x);$$

Combining this with the inequality in the previous display, we get

$$1 - \inf_{x \in \mathbb{X}_0} \eta(x) \geq 1 - P_U\left\{\bigcap_{x \in \mathbb{X}_0} \mathbb{U}_x\right\},$$

which implies $\inf_{x \in \mathbb{X}_0} \eta(x) \leq P_U\{\bigcap_{x \in \mathbb{X}_0} \mathbb{U}_x\}$. $\quad\square$

To avoid measurability difficulties in what follows, we introduce some regularity conditions. First, note that if the sampling model $P_{X|\theta}$ is discrete, then none of these additional conditions are needed. In the proof of Proposition 4.1 we introduced the *a-event* $\mathbb{U}_x(A) = \{u \in \mathbb{U} : \Theta_x(u) \subseteq A\}$. In the proof of Theorem 4.3 below, we will need that arbitrary intersections of a-events are P_U-measurable. To ensure this, we will replace each a-event by its closure. Since all closed sets are P_U-measurable by assumption, and arbitrary intersections of closed sets are closed (see Exercise 4.5), measurability issues are avoided. Replacing a-events by their closure does not affect any properties of the resulting belief function when P_U is non-atomic. In all the examples we have considered, P_U can be taken as continuous.

Theorem 4.3. *Suppose that either* $\mathbb{X}$ *is a discrete space, or that the assumptions in the previous paragraph hold. Fix* $A \subseteq \Theta$ *and assume condition* (4.11). *Given any valid predictive random set* $\mathcal{S}$, *there exists a nested and valid predictive random set* $\mathcal{S}'$ *such that* $\mathrm{bel}_x(A; \mathcal{S}') \geq \mathrm{bel}_x(A; \mathcal{S})$ *for each* $x \in \mathbb{X}$.

Proof. For the given A and $\mathcal{S}$, set $b(x) \equiv \mathrm{bel}_x(A; \mathcal{S})$. Define a collection $\mathbb{S}' = \{S'_x : x \in \mathbb{X}\}$ of subsets of $\mathbb{U}$ as follows:

$$S'_x = \bigcap_{y \in \mathbb{X}: b(y) \geq b(x)} \mathbb{U}_y(A), \tag{4.21}$$

where $\mathbb{U}_x(A)$ is the closure of the basic a-event. If necessary, add $\varnothing$ and $\mathbb{U}$ to $\mathbb{S}'$ to satisfy the requirement in Theorem 1′. This collection $\mathbb{S}'$ will serve as the support for the new predictive random set $\mathcal{S}'$. First, we can see that $\mathbb{S}'$ is nested: if $b(y) \geq b(x)$, then $S_y \supseteq S_x$. Second, since the new a-events are closed, each S'_x in (4.21) is closed and, hence, $\mathsf{P}_\mathbb{U}$-measurable. Third, define the measure $\mathsf{P}_{\mathcal{S}'}$ for $\mathcal{S}'$ to satisfy

$$\mathsf{P}_{\mathcal{S}'}\{\mathcal{S}' \subseteq K\} = \sup_{x: S'_x \subseteq K} \mathsf{P}_\mathbb{U}(S'_x).$$

According to Theorem 1′, the new $\mathcal{S}'$ is valid. Moreover, by the lemma and the definition of S'_x, we have

$$\mathsf{P}_\mathbb{U}(S'_x) \geq \inf_{y \in \mathbb{X}: b(y) \geq b(x)} b(y) = b(x) \equiv \mathrm{bel}_x(A; \mathcal{S}). \tag{4.22}$$

Finally, we have a comparison of the belief functions corresponding to $\mathcal{S}$ and $\mathcal{S}'$:

$$\mathrm{bel}_x(A; \mathcal{S}') = \mathsf{P}_{\mathcal{S}'}\{\mathcal{S}' \subseteq \mathbb{U}_x(A)\} \geq \mathsf{P}_{\mathcal{S}'}\{\mathcal{S}' \subseteq S'_x\} = \mathsf{P}_\mathbb{U}(S'_x) \geq \mathrm{bel}_x(A; \mathcal{S});$$

the first inequality follows from monotonicity of $\mathsf{P}_{\mathcal{S}'}\{\mathcal{S}' \subseteq \cdot\}$ and the fact that $S'_x \subseteq \mathbb{U}_x(A)$ for each x, and the second inequality follows from (4.22). $\qquad\square$

In the special case where the collection $\{\mathbb{U}_x(A) : x \in \mathbb{X}\}$ is itself nested, then, in general, one can construct an "optimal" $\mathcal{S}$ such that $R_A(X; \mathcal{S}) \equiv 1$. This is done explicitly for the special case in Section 4.5.3.1 below.

4.5.3 Optimality in special cases

Throughout this section, we will focus on scalar X and θ. However, this is just for simplicity, and not a limitation of the method; see Section 4.6. Indeed, special dimension-reduction techniques, akin to Fisher's theory of sufficient statistics, are available to reduce the dimension of observed X to that of θ within the IM framework, which are considered in Chapters 6–7. Also, there is no conceptual difference between scalar and vector θ problems so, since the ideas are new, we prefer to keep the presentation as simple as possible.

4.5.3.1   One-sided assertions

Here we consider a one-sided assertion, e.g., $A = \{\theta \in \Theta : \theta < \theta_0\}$, where θ_0 is fixed. This "left-sided" assertion is the kind we shall focus on, but other one-sided assertions can be handled similarly. In this context, we can consider a very strong definition of optimality.

Definition 4.1. Fix a left-sided assertion A. For two nested predictive random sets $\mathcal{S}$ and $\mathcal{S}'$, the IM based on $\mathcal{S}$ is said to be more efficient than that based on $\mathcal{S}'$ if, as functions of $X \sim \mathsf{P}_{X|\theta}$ for any $\theta \in A$, $R_A(X;\mathcal{S})$ is stochastically larger than $R_A(X;\mathcal{S}')$. The IM based on $\mathcal{S}^\star$ is optimal, or most efficient, if $R_A(X;\mathcal{S}^\star)$ is stochastically largest, in the sense above, among all nested predictive random sets.

That optimality here is described via a stochastic ordering property is natural in light of the notion of validity used throughout. This definition is particularly strong because it concerns the full distribution of $R_A(X, \mathcal{S})$ as a function of $X \sim \mathsf{P}_{X|\theta}$, not just a functional thereof. Next we establish a strong optimality result for one-sided assertions; when the assertion is not one-sided, it may not be possible to establish such a strong result.

Theorem 4.4. *Let* $A = \{\theta \in \Theta : \theta < \theta_0\}$ *be a left-sided assertion. Suppose that* $\Theta_x(u)$, *defined in (4.6), is such that, for each x, the right endpoint* $\sup \Theta_x(u)$ *is a non-decreasing (resp. non-increasing) function of u. Then, for the given A, the optimal predictive random set is* $\mathcal{S}^\star = [0, U]$ *(resp.* $\mathcal{S}^\star = [U, 1]$*), where* $U \sim \mathsf{Unif}(0, 1)$.

Proof. First observe that both forms of $\mathcal{S}^\star$ are nested. We shall focus on the non-decreasing case only; the other case is similar. Since $\sup \Theta_x(u)$ is non-decreasing in u, it follows that $\sup \Theta_x([0, U]) = \sup \Theta_x(U)$. Therefore,

$$\mathrm{bel}_x(A;\mathcal{S}^\star) = \mathsf{P}_U\{\sup \Theta_x([0, U]) < \theta_0\} = \mathsf{P}_U\{\sup \Theta_x(U) < \theta_0\} = \mathrm{bel}_x(A;\mathcal{S}_0).$$

This holds for all x, so $R_A(\cdot;\mathcal{S}^\star) \equiv 1$, its upper bound. Consequently, $R_A(X;\mathcal{S}^\star)$ is stochastically larger than $R_A(X;\mathcal{S})$ for any other $\mathcal{S}$, so optimality of $\mathcal{S}^\star$ obtains. $\square$

Example 4.1 (cont). We showed previously that $\Theta_x(u) = \{x - \Phi^{-1}(u)\}$. If we treat this as a degenerate interval, then we see that the right endpoint $x - \Phi^{-1}(u)$ is a strictly decreasing function of u. Therefore, by Theorem 4.4, the optimal predictive random set for a left-sided assertion is $\mathcal{S}^\star = [U, 1]$, $U \sim \mathsf{Unif}(0, 1)$.

As an application, consider the testing problem $H_0 : \theta \geq \theta_0$ versus $H_1 : \theta < \theta_0$. If we take $A = (-\infty, \theta_0)$, then the IM-based rule (4.18) rejects H_0 iff $1 - \mathrm{bel}_x(A;\mathcal{S}^\star) \leq \alpha$. With the optimal $\mathcal{S}^\star = [U, 1]$ as above, we get $\mathrm{bel}_x(A;\mathcal{S}^\star) = \Phi(\theta_0 - x)$. So the IM-based testing rule rejects H_0 iff $\Phi(\theta_0 - x) \geq 1 - \alpha$ or, equivalently, iff $x \leq \theta_0 - \Phi^{-1}(1 - \alpha)$. The reader will recognize this as the uniformly most powerful size-α test based on the classical Neyman–Pearson theory.

Example 4.2 (cont). In this case, $\Theta_x(u) = (G_x^{-1}(u), G_{x+1}^{-1}(u)]$; see (4.7). The right endpoint $G_{x+1}^{-1}(u)$ is strictly increasing in u. So Theorem 4.4 states that, for left-sided assertions, the optimal predictive random set is $\mathcal{S}^\star = [0, U]$, $U \sim \mathsf{Unif}(0, 1)$. The same connection with the Neyman–Pearson uniformly most powerful test in the Gaussian example holds here as well, but we omit the details.

4.5.3.2 Two-sided assertions

Consider the case where $A = \{\theta_0\}^c$ is the two-sided assertion of interest, with θ_0 a fixed interior point of $\Theta \subseteq \mathbb{R}$. This is an important case, which we have already considered in Section 4.3, just in a different form. These problems are apparently more difficult than their one-sided counterparts, just as in the classical hypothesis testing context. Here we present some basic results and intuitions on IM optimality for two-sided assertions.

Assume $P_{X|\theta}$ is a continuous distribution. Then the fiducial probability $\text{bel}_X(\{\theta_0\}^c; S_0)$ for the two-sided assertion is unity, and so the relative efficiency (4.20) is simply $\text{bel}_x(\{\theta_0\}^c; S)$. Here we focus on predictive random sets S with the property that $\text{bel}_X(\{\theta_0\}^c; S) \sim \text{Unif}(0, 1)$ under $P_{X|\theta_0}$; see Corollary 4.1. Based on the intuition developed in Section 4.3, $\text{bel}_X(\{\theta_0\}^c; S)$ should be smallest (probabilistically) under $P_{X|\theta}$ for $\theta = \theta_0$. We shall, therefore, impose the following condition on the predictive random set S:

$$P_{X|\theta}\{\text{bel}_X(\{\theta_0\}^c; S) \leq \alpha\} < \alpha, \quad \forall \theta \neq \theta_0, \quad \forall \alpha \in (0, 1). \qquad (4.23)$$

Roughly speaking, condition (4.23) states that the belief function at $\{\theta_0\}^c$ is stochastically larger under $P_{X|\theta}$ than under $P_{X|\theta_0}$. There is also a loose connection between (4.23) and the classical unbiasedness condition imposed to construct optimal tests when the alternative hypothesis is two-sided [159]. Our goal in what follows is to find a "best" predictive random set that satisfies (4.23).

To make things formal, suppose that both $\mathbb{X}$ and Θ are one-dimensional, that $P_{X|\theta}$ is continuous with distribution function $F_\theta(x)$ and density function $f_\theta(x)$, and that the usual regularity conditions hold; in particular, we assume that the expectation with respect to $P_{X|\theta}$ and differentiation with respect to θ can be interchanged. Let

$$T_\theta(x) = (\partial/\partial\theta)\log f_\theta(x),$$
$$V_\theta(x) = T_\theta(x)^2 + (\partial/\partial\theta)T_\theta(x).$$

The quantity $T_\theta(x)$ is called the *score function*. Then, under the usual regularity conditions, we have $E_{X|\theta}\{T_\theta(X)\} = 0$ and $E_{X|\theta}\{V_\theta(X)\} = 0$ for all θ.

If we assume that $\text{bel}_X(\{\theta_0\}^c; S) \sim \text{Unif}(0, 1)$ under $P_{X|\theta_0}$, then there exists a collection of measurable subsets $\mathbb{X}(\alpha) \subseteq \mathbb{X}$, depending implicitly on θ_0 and S, such that, for each α, $P_{X|\theta_0}\{\mathbb{X}(\alpha)\} = \alpha$, and $\text{bel}_x(\{\theta_0\}^c; S) \leq \alpha$ iff $x \in \mathbb{X}(\alpha)$. It follows that, for any θ,

$$P_{X|\theta}\{\text{bel}_X(\{\theta_0\}^c; S) \leq \alpha\} = \psi_\alpha(\theta) := \int_{\mathbb{X}(\alpha)} f_\theta(x)\,dx.$$

By definition, $\psi_\alpha(\theta_0) = \alpha$. Now, (4.23) is equivalent to $\psi_\alpha(\theta) < \psi_\alpha(\theta_0)$ for all α, or, to put it another way, $\psi_\alpha(\theta)$ is maximized at $\theta = \theta_0$ for all α. Under the stated regularity conditions, this maximization is equivalent to the claim that, for all

$\alpha \in (0,1)$, the first and second derivatives of $\psi_\alpha(\theta)$ at $\theta = \theta_0$ satisfy

$$\psi'_\alpha(\theta_0) = \int_{\mathbb{X}(\alpha)} T_{\theta_0}(x) f_{\theta_0}(x)\, dx = 0, \qquad (4.24)$$

$$\psi''_\alpha(\theta_0) = \int_{\mathbb{X}(\alpha)} V_{\theta_0}(x) f_{\theta_0}(x)\, dx < 0. \qquad (4.25)$$

Since $T_{\theta_0}(X)$ has mean zero under $\mathsf{P}_{X|\theta_0}$, we can see that (4.24) requires $\mathbb{X}(\alpha)$ to be somehow symmetric, or balanced, with respect to the distribution of $T_{\theta_0}(X)$. We, therefore, refer to (4.24) as the *score-balance* condition. This condition, expressed in terms of $\mathbb{X}(\alpha)$ in (4.24), can be traced back to a corresponding condition on the predictive random set.

Let us now assume that $\mathcal{S}_B$ is such that $\mathrm{bel}_X(\{\theta_0\}^c; \mathcal{S}_B) \sim \mathrm{Unif}(0,1)$ under $\mathsf{P}_{X|\theta_0}$; in the main text we construct a particular score-balanced predictive random set and show that that this assumption holds. Then, as we argued above, for any $\alpha \in (0,1)$, there exists $t(\alpha) \in \mathbb{T}$ such that $\mathrm{bel}_x(\{\theta_0\}^c; \mathcal{S}_B) \leq \alpha$ iff $T_{\theta_0}(x) \in B_{t(\alpha)}$. In this case, for any θ,

$$\mathsf{P}_{X|\theta}\{\mathrm{bel}_X(\{\theta_0\}^c; \mathcal{S}_B) \leq \alpha\} = \int_{T_{\theta_0}(x) \in B_{t(\alpha)}} f_\theta(x)\, dx,$$

and the right-hand side is $\psi_\alpha(\theta)$ as defined previously. From the definition of B, differentiating under the integral sign reveals that (4.24) holds. We can now prove

Proposition 4.2. *Focus on predictive random sets $\mathcal{S}$ such that $\mathrm{bel}_X(\{\theta_0\}^c; \mathcal{S}) \sim \mathrm{Unif}(0,1)$ under $\mathsf{P}_{X|\theta_0}$. Then condition (4.23) holds for all θ in a neighborhood of θ_0 iff the predictive random set $\mathcal{S} = \mathcal{S}_B$ is score-balanced and*

$$\int_{T_{\theta_0}(x) \in B_t} V_{\theta_0}(x) f_{\theta_0}(x)\, dx < 0, \qquad \forall\, t \in \mathbb{T}. \qquad (4.26)$$

Proof. Take θ close enough to θ_0 such that the remainder terms in a second-order Taylor approximation of $\psi_\alpha(\theta)$ about $\theta = \theta_0$ can be ignored. That is, for any α,

$$\psi_\alpha(\theta) - \psi_\alpha(\theta_0) = \int_{T_{\theta_0}(x) \in B_{t(\alpha)}} T_{\theta_0}(x) f_{\theta_0}(x)\, dx \cdot (\theta - \theta_0)$$
$$+ \frac{1}{2} \int_{T_{\theta_0}(x) \in B_{t(\alpha)}} V_{\theta_0}(x) f_{\theta_0}(x)\, dx \cdot (\theta - \theta_0)^2.$$

The first term vanishes and the second term is negative by (4.26). Therefore $\psi_\alpha(\theta) < \psi_\alpha(\theta_0)$ for all α and, hence, (4.23) holds for all θ in a neighborhood of θ_0. $\qquad \square$

The above arguments suggest that a good predictive random set $\mathcal{S}$ must have a support with certain symmetry or balance properties with respect to the sampling distribution of $T_{\theta_0}(X)$. In particular, let $B = \{B_t : t \in \mathbb{T}\}$ be a generic collection of nested measurable subsets of $\mathbb{T} = T_{\theta_0}(\mathbb{X})$. The collection B shall be called score-balanced if

$$\mathsf{E}_{X|\theta_0}\{T_{\theta_0}(X) I_{B_t}(T_{\theta_0}(X))\} = 0, \qquad \forall\, t \in \mathbb{T}. \qquad (4.27)$$

For a score-balanced collection $B = \{B_t\}$ satisfying (4.27) we can define a corresponding *score-balanced predictive random set* $\mathcal{S} = \mathcal{S}_B$ as follows. Define the class $\mathbb{S} = \{S_t : t \in \mathbb{T}\}$ of subsets of $\mathbb{U} = [0, 1]$ given by

$$S_t = F_{\theta_0}(\{x : T_{\theta_0}(x) \in B_t\}).$$

For simplicity, and without loss of generality, assume $\mathbb{S}$ contains $\varnothing$ and $\mathbb{U}$. Now take a predictive random set $\mathcal{S}_B$, supported on $\mathbb{S}$, such that its measure $\mathsf{P}_{\mathcal{S}_B}$ satisfies

$$\mathsf{P}_{\mathcal{S}_B}\{\mathcal{S}_B \subseteq K\} = \sup_{t : S_t \subseteq K} \mathsf{P}_U(S_t), \quad K \subseteq [0, 1],$$

where P_U is the $\mathsf{Unif}(0,1)$ measure. (The set S_t is P_U-measurable for all t by the assumed measurability of B_t, T_{θ_0}, and F_{θ_0}.) The corresponding score-balanced belief function is

$$\begin{aligned}
\mathrm{bel}_x(\{\theta_0\}^c; \mathcal{S}_B) &= \mathsf{P}_{\mathcal{S}_B}\{\mathcal{S}_B \not\ni F_{\theta_0}(x)\} \\
&= \mathsf{P}_{X|\theta_0}\{B_{T_{\theta_0}(X)} \not\ni T_{\theta_0}(x)\} \\
&= \mathsf{P}_{X|\theta_0}\{T_{\theta_0}(X) \in B_{T_{\theta_0}(x)}\},
\end{aligned}$$

where the last equality follows from the assumed nesting of $\{B_t\}$.

But there are many such $\mathcal{S}_B$ to choose from, so we now consider finding a "best" one. A reasonable definition of optimal score-balanced predictive random set is one that makes the difference between the right- and left-hand sides of (4.23) as large as possible for each θ in a neighborhood of θ_0. Then, for two-sided assertions, we have

Definition 4.3. Let $B^\star = \{B_t^\star : t \in \mathbb{T}\}$ be such that, for each t,

$$\int_{T_{\theta_0}(x) \in B_t^\star} V_{\theta_0}(x) f_{\theta_0}(x) \, dx \tag{4.28}$$

is minimized subject to the score-balance constraint (4.27). Then $\mathcal{S}^\star = \mathcal{S}_{B^\star}$ is the optimal score-balanced predictive random set.

Here we give a general construction of an an optimal score-balanced predictive random sets. Proving that the predictive random sets satisfy the conditions of Definition 4.3 will require assumptions about the model. Start with the following class of intervals:

$$B_t^\star = \big(\xi_-(t), \xi_+(t)\big), \quad t \in T_{\theta_0}(\mathbb{X}), \tag{4.29}$$

where the functions ξ_-, ξ_+ (which depend implicitly on θ_0) are such that (4.27) holds. In addition, we shall assume these functions are continuous and satisfy

- $\xi_-(t)$ is non-positive, $\xi_-(t) = t$ for $t \in (-\infty, 0)$ and is decreasing for $t \in [0, \infty)$;
- $\xi_+(t)$ is non-negative, $\xi_+(t) = t$ for $t \in [0, \infty)$ and is increasing for $t \in (-\infty, 0)$.

The functions ξ_-, ξ_+ describe a sort of symmetry/balance in the distribution of $T_{\theta_0}(X)$: they satisfy $\xi_+(\xi_-(t)) = t$ and $\xi_-(\xi_+(-t)) = -t$ for all $t \geq 0$. In some cases, for given t, expressions for $\xi_-(t)$ and $\xi_+(t)$ can be found analytically, but typically numerical solutions are required. Set $\mathcal{S}^\star = \mathcal{S}_{B^\star}$. We claim that, under certain conditions on $V_{\theta_0}(x)$, $\mathcal{S}^\star$ is optimal in the sense of Definition 4.3.

Before we get to the optimality considerations, we first verify the assumption that $\mathrm{bel}_X(\{\theta_0\}^c; \mathcal{S}^\star) \sim \mathrm{Unif}(0,1)$ under $\mathsf{P}_{X|\theta_0}$. From the definition of $B_t^\star$, it is clear that

$$T \in B_t^\star \iff \xi_-(t) < T < \xi_+(t)$$
$$\iff \xi_-(t) < \xi_-(T) < \xi_+(T) < \xi_+(t)$$
$$\iff \xi_+(T) - \xi_-(T) < \xi_+(t) - \xi_-(t).$$

Consequently, if $D_{\theta_0}(X) = \xi_+(T_{\theta_0}(X)) - \xi_-(T_{\theta_0}(X))$, then

$$\mathrm{bel}_x(\{\theta_0\}^c; \mathcal{S}^\star) = \mathsf{P}_{X|\theta_0}\{T_{\theta_0}(X) \in B_{T_{\theta_0}}^\star(x)\} = \mathsf{P}_{X|\theta_0}\{D_{\theta_0}(X) < D_{\theta_0}(x)\}.$$

Therefore, since $D_{\theta_0}(X)$ is a continuous random variable, an argument like that in Corollary 4.1 shows that $\mathrm{bel}_X(\{\theta_0\}^c; \mathcal{S}^\star) \sim \mathrm{Unif}(0,1)$ under $\mathsf{P}_{X|\theta_0}$.

We are now ready for the optimality of $\mathcal{S}^\star$. Write $V(t)$ for $V_{\theta_0}(x)$, when treated as a function of $t = T_{\theta_0}(x)$. The condition to be imposed is:

$$V(t) \text{ is uniquely minimized at } t = 0, \text{ and } V(0) < 0. \tag{4.30}$$

This condition holds, e.g., for all exponential families with θ the natural parameter.

Proposition 4.3. *Under condition* (4.30), *the score-balanced predictive random set* $\mathcal{S}^\star = \mathcal{S}_{B^\star}$, *with $B^\star$ described above, is optimal in the sense of Definition 4.3.*

Proof. The proof is simple but tedious, so here we just sketch the main idea. Under (4.30), the intervals $B_t^\star$ which are "balanced" around $T_{\theta_0}(x) = 0$, make most efficient use of the space where $V_{\theta_0}(x)$ is smallest in the following sense. They are exactly the right size to make $\mathcal{S}_{B^\star}$ efficient, so any other efficient score-balanced predictive random set $\mathcal{S}_B$ must be determined by sets $B = \{B_t\}$ other than intervals concentrated around $T_{\theta_0}(x) = 0$. Since such sets are where $V_{\theta_0}(x)$ is smallest, the integral in (4.28) corresponding to B_t must be larger than that corresponding to $B_t^\star$. Therefore, $\mathcal{S}^\star$ satisfies the conditions of Definition 4.3 and, hence, is optimal. $\qquad\square$

Unfortunately, (4.30) is not always satisfied. For example, it can fail for exponential families not in natural form. But we claim that (4.30) is not absolutely essential. Assume $V(t)$ is convex and $V(0) < 0$. This relaxed assumption holds, e.g., for all exponential families. To keep things simple, suppose that $V(t)$ is minimized at $\hat{t} > 0$. Although the argument to be given is general, Figure 4.2(a) illustrates the phenomenon for the exponential distribution with mean $\theta_0 = 1$. The heavy line there represents $V(t)$, and the thin lines represent $th(t)$ (black) and $V(t)h(t)$ (gray), where $h(t)$ is the density of T. The horizontal lines represent the intervals $B_t^\star$ in (4.29) for select t. By convexity of $V(t)$, there exists t_0 such that $\hat{t} \in (0, t_0)$ and $V(t) < V(0)$ for each $t \in (0, t_0)$; this is $(0, 0.5)$ in the figure. For $t \in (0, t_0)$, the intervals $B_t^\star$ do not contain $(0, t_0)$; these intervals are shown in black. In such cases, the integral (4.28) can be reduced by breaking B_t^* into two parts: one part takes more of $(0, t_0)$, where $V(t)$ is smallest, and the other part is chosen to satisfy the score-balance condition (4.27). But when $t \geq t_0$, no improvement can be made by changing B_t^*; these cases are shown in gray. So, in this sense, the intervals B_t^* in (4.29) are not too bad even if (4.30) fails.

On the other hand, violations of (4.30) are due to the choice of the parametrization. Indeed, under mild assumptions, there exists a transformation $\eta = \eta(\theta)$ such that the corresponding $V(t)$ function for η satisfies (4.30). Then the predictive random set S^* in Proposition 4.3 is the optimal for this transformed problem.

Example 4.1 (cont). This is a natural exponential family distribution, so Proposition 4.3 holds, and S^* is the optimal score-balanced predictive random set. Here the score function is $T_\theta(x) = x - \theta$. Under $X \sim N(\theta, 1)$, the distribution of $T_\theta(X)$ is symmetric about 0. Therefore, $B_t^* = (-|t|, |t|)$, and the corresponding predictive random set is supported on subsets S_t given by

$$S_t = F_{\theta_0}(\{x : |x - \theta_0| \leq |t|\}) = (\Phi(-|t|), \Phi(|t|)),$$

with belief function $\mathrm{bel}_x(\{\theta_0\}^c; S^*) = 2\Phi(|x - \theta_0|) - 1$. This is exactly one minus the plausibility function in (4.13) based on the default predictive random set (4.8). Therefore, we conclude that the (4.8) is, in fact, the optimal score-balanced predictive random set in the Gaussian problem. This is consistent with our intuition, given that the results based on this default choice in the Gaussian example match up with good classical results.

Example 4.2 (cont). Although the theory above holds only for continuous models, the score-balanced predictive random set performs well in discrete problems too. The interested reader is referred to [178] for the details.

Example 4.3. Suppose X is an exponential random variable with mean θ, as discussed above. Unlike the Gaussian, this distribution is asymmetric, so, for the optimal score-balanced IM, a numerical method is needed to identify the set $B_{T_{\theta_0}(x)}$ for each observed x. Plots of the corresponding plausibility functions $\mathrm{pl}_x(\theta; S) = 1 - \mathrm{bel}_x(\{\theta\}^c; S)$ for two different predictive random sets based on $X = 5$ are shown in Figure 4.2(b). The black line is based on the optimal score-balanced predictive random set, and the gray line is based on the default predictive random set in (4.8). 90% plausibility intervals, determined by the horizontal line at $\alpha = 0.1$, are much shorter for the score-balanced IM compared to the default in this case. For comparison, one might consider a crude nominal 90% confidence interval for θ, namely, $(Xe^{-1.65}, Xe^{1.65})$, based on a variance-stabilizing transformation and normal approximation. These intervals tend to be shorter than both plausibility intervals, but their coverage probability (≈ 0.82) is too small.

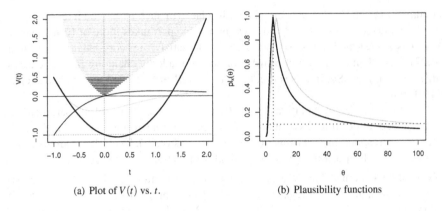

(a) Plot of $V(t)$ vs. t.   (b) Plausibility functions

Figure 4.2 *Taken from [180]. Specifics of Panel (a) are discussed in the text. Panel (b) shows* $\mathrm{pl}_x(\theta; \mathcal{S})$, *as a function of the scale parameter* θ, *for two predictive random sets* $\mathcal{S}$: *optimal score-balanced (black) and default (gray). The vertical line marks the observed* $X = 5$.

4.6 Two more examples

4.6.1 A standardized mean problem

Suppose that $X_1, \ldots, X_n$ are independent $N(\mu, \sigma^2)$ observations. The goal is to make inference on $\psi = \mu/\sigma$, the standardized mean, or signal-to-noise ratio. Following [60], we start with a reduction of the full data to the sufficient statistics for $\theta = (\mu, \sigma^2)$, namely $(\bar{X}, S^2)$, the sample mean and variance. Formal IM-based justification for this reduction is given in Chapter 6.

For the A-step, we take the association to be

$$\bar{X} = \mu + n^{-1/2}\sigma U_1 \quad \text{and} \quad S = \sigma U_2, \tag{4.31}$$

where $U = (U_1, U_2) \sim P_U = N(0, 1) \times \{\text{ChiSq}(n-1)/(n-1)\}^{1/2}$. After replacing σ in the left-most identity in (4.31) with S/U_2, a bit of algebra reveals that

$$n^{1/2}\bar{X}/S = (n^{1/2}\psi + U_1)/U_2 \quad \text{and} \quad S = \sigma U_2.$$

For $\theta = (\psi, \sigma)$, make a change of auxiliary variable $v = \phi_\theta(u)$, given by

$$v_1 = F_\psi\left(\frac{n^{1/2}\psi + u_1}{u_2}\right) \quad \text{and} \quad v_2 = \frac{\exp(u_2)}{1 + \exp(u_2)},$$

where F_ψ is the distribution function for $t_{n-1}(n^{1/2}\psi)$, a non-central Student-t distribution with $n-1$ degrees of freedom and non-centrality parameter $n^{1/2}\psi$. Note that the full generality of the parameter-dependent change-of-variables in Corollary 4.2 is needed here. Then the transformed association is

$$n^{1/2}\bar{X}/S = F_\psi^{-1}(V_1) \quad \text{and} \quad S = \sigma \log\{V_2/(1 - V_2)\},$$

and the measure P_V on the space of $V = (V_1, V_2)$ has a $\mathsf{Unif}(0,1)$ marginal on the V_1-space; the distribution on V_1-slices of the V_2 space can be worked out, but it is not needed in what follows. For the P-step, we predict $v^\star = \phi_\theta(u^\star)$ with a rectangle predictive random set $\mathcal{S}$ defined by the following set-valued mapping:

$$v = (v_1, v_2) \mapsto \{v_1' : |v_1' - 0.5| < |v_1 - 0.5|\} \times [0, 1]. \tag{4.32}$$

Optimality considerations along the lines in Section 4.5.3.2 could be pursued here, but we choose to keep things simple since analysis of the non-central Student-t distribution is non-trivial. An important direction of future research is to develop numerical methods for evaluating optimal IMs. Using a predictive random set that spans the entire v_2-space for each v has the effect of "integrating out" the nuisance parameter σ. For the predictive random set $\mathcal{S}$ in (4.32), if $z = n^{1/2}\bar{x}/s$, then the C-step gives the following set $\Theta_x(\mathcal{S}) = \Psi_x(\mathcal{S}) \times \Sigma_x(\mathcal{S})$ of candidate (ψ, σ) pairs:

$$\{\psi : |F_\psi(z) - 0.5| < |V_1 - 0.5|\} \times \{\sigma : \sigma > 0\}, \quad V \sim \mathsf{P}_V. \tag{4.33}$$

For assertions $A = \{(\psi, \sigma) : \sigma > 0\}$ the plausibility function is given by

$$\mathsf{pl}_x(A) = \mathsf{P}_{\mathcal{S}}\{\Theta_x(\mathcal{S}) \not\subseteq A^c\} = \mathsf{P}_{\mathcal{S}}\{\Psi_x(\mathcal{S}) \ni \psi\} = 1 - |2F_\psi(z) - 1|.$$

In this case, the $100(1 - \alpha)\%$ plausibility interval $\Pi_x(\alpha)$ for ψ is obtained by inverting the inequality $1 - |2F_\psi(z) - 1| > \alpha$, i.e., $\Pi_x(\alpha) = \{\psi : \alpha/2 < F_\psi(z) < 1 - \alpha/2\}$.

This is exactly the usual frequentist confidence interval based on the sampling distribution of the standardized sample mean; it also agrees with the fiducial intervals obtained by [56, 60]. The standard frequentist approach relies on an informal "plug-in style" marginalization, whereas the IM approach above shows exactly how σ is ignored via cylinder assertions. More sophisticated IM marginalization techniques are available, see Chapter 7.

4.6.2 A many-exponential-rates problem

For our last example, we consider a high-dimensional problem. Suppose that $X = (X_1, \ldots, X_n)$ consists of independent observations $X_i \sim \mathsf{Exp}(\theta_i)$, $i = 1, \ldots, n$, with unknown rates $\theta_1, \ldots, \theta_n$. The goal is to give a probabilistic measure of the support in $X = x$ for the assertion $A = \{\theta_1 = \cdots = \theta_n\}$ that the rates are equal. A version of this problem was also discussed in [186], but here we simplify the presentation, emphasize the three-step IM construction, and produce much better results.

Start, in the A-step, with the association $X_i = U_i/\theta_i$, $i = 1, \ldots, n$, where P_U is the product measure $\mathsf{Exp}(1)^{\times n}$. Make a change of auxiliary variables $v = \phi(u)$:

$$v_0 = \sum_{i=1}^n u_i \quad \text{and} \quad v_i = u_i/v_0, \quad i = 1, \ldots, n.$$

The new vector $v = (v_0, v_1, \ldots, v_n)$ takes values in $\mathbb{V} = (0, \infty) \times \mathbb{P}_{n-1}$, where $\mathbb{P}_{n-1}$ is the $(n-1)$-dimensional probability simplex in $\mathbb{R}^n$, and $\mathsf{P}_V = \mathsf{P}_U \phi^{-1}$ is the product measure $\mathsf{Gamma}(n, 1) \times \mathsf{Dir}_n(1_n)$. Then the modified association is

$$X_i = V_0 V_i/\theta_i, \quad i = 1, \ldots, n, \quad \text{where} \quad V = (V_0, V_1, \ldots, V_n) \sim \mathsf{P}_V. \tag{4.34}$$

For the P-step, we shall consider the following predictive random set $\mathcal{S}$ characterized by $V \sim \mathsf{P}_V$ and the set-valued mapping $v \mapsto \{v' : h(v') < h(v)\}$. In this case, we take

$$h(v) = - \sum_{i=1}^{n-1} \left[a_i \log t_i(v) + b_i \log\{1 - t_i(v)\} \right],$$

with $t_i(v) = \sum_{j=1}^{i} v_i$, $a_i = 1/(n - i - 0.3)$, and $b_i = 1/(i - 0.3)$. A few remarks on this choice of $\mathcal{S}$ are in order. First, it follows from Corollary 4.1 that $\mathcal{S}$ is efficient. Second, the random vector $(t_1(V), \ldots, t_{n-1}(V))$, for $V \sim \mathsf{P}_V$, has the distribution of a vector of $n - 1$ sorted $\mathrm{Unif}(0,1)$ random variables, and [268] shows that $\mathcal{S}$ provides an easy-to-compute alternative to the well-performing hierarchical predictive random set for predicting sorted uniforms used in [186]. Finally, that the first component v_0 of v is essentially ignored in $\mathcal{S}$ is partly for convenience, and partly because v_0 is related to the overall scale of the problem which is irrelevant to the assertion A.

For the C-step, combining the observed data, the association model (4.34), and the predictive random set $\mathcal{S}$ above, we get the following random set for θ:

$$\Theta_x(\mathcal{S}) = \{\theta : h(v(x, \theta)) < h(V)\}, \quad V \sim \mathsf{P}_V,$$

where $v(x, \theta) = (\theta_1 x_1, \ldots, \theta_n x_n) / \sum_{j=1}^{n} \theta_j x_j$. Since the assertion $A = \{\theta_1 = \cdots = \theta_n\}$ is a one-dimensional subset of Θ, the belief function is zero. It is also important to note that when θ is a constant vector, $v(x, \theta)$ is independent of that constant, i.e., $v(x, \theta) = v(x, 1_n)$, which greatly simplifies computation of the plausibility function at A. Indeed,

$$\mathsf{pl}_x(A) = \mathsf{P}_V\{h(V) > h(v(x, 1))\},$$

which can easily be evaluated using Monte Carlo. As described in Section 4.4.4, the level α IM-based tests rejects the assertion A if and only if $\mathsf{pl}_x(A) \leq \alpha$.

For illustration, we compare our results with those of [186]. They consider the basic likelihood ratio test, which is based on the test statistic $\left\{ \left(\prod_{i=1}^{n} x_i \right)^{1/n} / \bar{x} \right\}^n$. They also consider a different sort of IM solution, based on thresholding the plausibility function, but with a default type of predictive random set that uses a Kullback–Leibler neighborhood for predicting the component $(V_1, \ldots, V_n)$ of V. We compare the power of these three tests in several different cases. In each setup, $n = n_1 + n_2 = 100$ observations are available, but the first n_1 exponential rates equal 1 while the last n_2 equal θ. Figure 4.3 shows the power functions over a range of θ values for two configurations of (n_1, n_2). Here we see that, in both cases, the likelihood ratio and old IM tests have similar power, possibly because of the common connection to the Kullback–Leibler divergence. On the other hand, the new IM-based test presented above has strikingly larger power than the other two. This substantial improvement in power is likely due to the close relationship between our choice of $\mathcal{S}$ and the assertion of interest. So while the comparison between the new IM results and those of the other "default" methods is not entirely fair, it is interesting to see that an assertion-specific choice of predictive random set can lead to drastically improved performance.

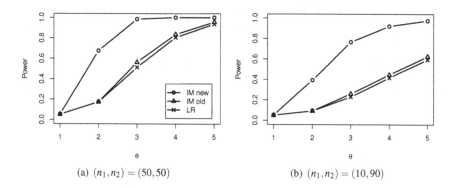

Figure 4.3 *Taken from [180]. Estimated powers of the likelihood ratio and two IM-based tests for the simulation described in Section 4.6.2. Here θ is the ratio of the rate of the last n_2 observations to that of the first n_1.*

4.7 Concluding remarks

The conversion of experience to knowledge is fundamental to the advancement of science, and statistical inference plays a crucial role. Here we have described a three-step procedure to construct IMs for prior-free, post-data probabilistic inference, and proved that IMs yield frequency-calibrated probabilities under very general conditions. The point is that the values of the corresponding belief/plausibility function are meaningful both within and across experiments, accomplishing both the frequentist and Bayesian goals simultaneously.

The proposed IM approach is surely new, but since new is not always better, it is natural to ask what is the benefit of using IMs. Our response is that, although it will take time for users to familiarize themselves with the thought process, the IM framework is logical, intuitive, and able to produce meaningful and frequency-calibrated probabilistic measures of uncertainty about θ without a prior distribution. The latter property is something that no other inferential framework is able to achieve.

Admittedly, the final IM depends on the user's choice of association and predictive random set, but we do not believe that this is particularly damning. Section 4.5 laid the foundation for a theory of optimal predictive random sets, and further efforts to develop "default" predictive random sets are ongoing, particularly for multi-parameter problems. But a case can be made to prefer the ambiguity of the choice of predictive random set over that of a frequentist's choice of statistic or a Bayesian's choice of prior. The point is that neither a frequentist sampling distribution nor a Bayesian prior distribution adequately describes the source of uncertainty about θ. As we argued above, this uncertainty is fully characterized by the fact that, whatever the association, the value of $u^\star$ is missing. Therefore, it seems only natural to prefer the IM framework that features a direct attack on the source of uncertainty over another that attacks the problem indirectly. Moreover, as was demonstrated in Section 4.6.2, choosing the predictive random set that depends on the problem and/or assertion of interest can lead to drastically improved results.

We note that differences between IM outputs from different predictive random sets are slight for assertions involving one-dimensional quantities. However, for high-dimensional auxiliary variables, the choice of predictive random set deserves special attention. In such cases, our approach is to construct predictive random sets for functions of auxiliary variables that are most relevant to the assertions of interest. This leads to a practically useful auxiliary variable dimension reduction. It is interesting that this approach has some close connections to Fisher's theory of sufficient statistics [182]. For nuisance parameter problems, like those in Section 4.6, there is a different form of dimension reduction required [183].

To this end, it is worth mentioning that IMs can help see when fiducial and DS generate valid or exact inference and when they do not. Simply put, perhaps it can be said that their results are valid if and only if they can be obtained from an IM formulation, For assertions that correspond to pre-specified nested predictive random sets in the auxiliary space, fiducial and DS produce meaningful or valid probabilities. This is typically true for one-sided assertions with simple unconstrained single parameter problems. Otherwise, care must be taken in scientific inference.

IMs can also be used to justify certain frequentist calculations. For example, p-values are a mainstay in conventional statistics but are often misinterpreted. Martin and Liu [181] propose a new interpretation of the p-value as a meaningful plausibility, where this is to be interpreted formally within the inferential model framework. They show that, for most practical hypothesis testing problems, there exists an inferential model such that the corresponding plausibility function, evaluated at the null hypothesis, is exactly the p-value. The advantages of this representation are that the notion of plausibility is consistent with the way practitioners use and interpret p-values, and the plausibility calculation avoids the troublesome conditioning on the truthfulness of the null. This connection with plausibilities also reveals a shortcoming of standard p-values in problems with non-trivial parameter constraints. An IM approach to problems with parameter constraints is straightforward via the use of *elastic* predictive random sets. This is discussed in the next chapter. It may also be helpful to take a fresh look at p-values from an optimal IM perspective. That is, the plausibility function of an optimal IM yields a p-value with the desirable interpretation but perhaps with different numerical values and properties.

4.8 Exercises

Exercise 4.1. Show that X defined in (4.4) satisfies $X \sim N(\theta, 1)$.

Exercise 4.2. Show that if F_θ is the Pois(θ) distribution function and G_a is the Gamma($a, 1$) distribution function, then $F_\theta(x) = 1 - G_{x+1}(\theta)$ for all x and all θ. (*Hint:* Use integration-by-parts.)

Exercise 4.3. Show that, for the predictive random set S defined by the mapping (4.8), the contour function γ_S satisfies $\gamma_S(u) = 1 - |2u - 1|$ for $u \in (0, 1)$.

Exercise 4.4. Prove Corollary 4.2.

Exercise 4.5. Let $\{C_t : t \in T\}$ be an arbitrary collection of closed sets. Show that $\bigcap_{t\in T} C_t$ is also closed.

Exercise 4.6. Recall the beta distribution $\text{Beta}(\alpha, \beta)$ with shape parameters $\alpha \geq 0$ and $\beta \geq 0$. The pdf is

$$f_{\alpha,\beta}(x) = \frac{\Gamma(\alpha+\beta)}{\Gamma(\alpha)\Gamma(\beta)} x^{\alpha-1}(1-x)^{\beta-1} \quad x \in [0,1]).$$

Let $F_{\alpha,\beta}$ be the corresponding cdf. Fix a positive integer n and a number $\theta \in [0,1]$. For $U \sim \text{Unif}(0,1)$, define a random variable X, taking values in $\{0,1,\ldots,n\}$, that satisfies

$$F_{X+1,n-X}(\theta) < U \leq F_{X,n-X+1}(\theta). \tag{4.35}$$

Show that $X \sim \text{Bin}(n,\theta)$. (*Hint:* Use integration-by-parts.)

Exercise 4.7. Consider inference about the point assertion $A_{\theta_0} = \{\theta_0\}$ about the binomial proportion θ in the binomial distribution from the observed count X.

(a) Take Equation (4.35) in Exercise 4.6 above as an association and the default predictive random set $\mathcal{S}$ for predicting $U \sim \text{Unif}(0,1)$. Show that $\text{bel}_X(A_{\theta_0}) = 0$ and

$$\text{pl}_X(A_{\theta_0}) = 2\min\{F_{x,n-x+1}(\theta_0), 1 - F_{x+1,n-x}(\theta_0), 1/2\}.$$

(b) Develop the optimal IMs for inference about $A_{\theta_0} = \{\theta_0\}$. (*Hint:* See [181].)

Exercise 4.8. Use the result in the previous exercise to calculate the 95% plausibility intervals for θ given

(a) $n = 5$ and $X = 1$.

(b) $n = 50$ and $X = 0$.

(c) $n = 100$ and $X = 10$.

Chapter 5

Predictive Random Sets

Portions of the material in this chapter are from D. Ermini Leaf and C. Liu, "Inference about constrained parameters using the elastic belief method," *International Journal of Approximate Reasoning* **53**, 709–727, 2012, reprinted with permission by Elsevier.

5.1 Introduction

As we have argued in the previous chapters, for the purpose of prior-free probabilistic inference, it is natural to specify the sampling distribution of the observable data, given the parameter, with unobservable auxiliary variables. Existing approaches, such as fiducial and Dempster–Shafer, adopt this strategy. However, as discussed in detail in Chapter 2, these approaches (implicitly) make an artificial change in the auxiliary variable distribution when data are fixed, making the interpretation of their inferential output questionable.

To avoid this, we have proposed to predict the unobserved value of the auxiliary variable, rather than making an artificial update of its *a priori* distribution. By doing so, we can keep the same interpretation of probability throughout the entire statistical problem, from modeling, to estimation and inference. This is the chief novelty of the proposed IM framework. From a practical point of view, to implement the IM framework, one must introduce a valid predictive random set for predicting that unobserved value of the auxiliary variable.

We encountered the concept of predictive random set in Chapter 4. Here, in this chapter, we provide some further details about these predictive random sets. Section 5.2, based in part on [177], provides some background about random sets in general and how these new predictive random sets fit within this general theory. The remaining sections in this chapter, based in part on [82], focus on a particularly challenging problem in statistics, namely, when the parameter has some non-trivial constraints. For valid prior-free probabilistic inference in constrained parameter problems, we propose the use of an *elastic* predictive random set. The basic idea is to start with a valid predictive random set for the unconstrained problem, but then "stretch" it out just enough so that it is consistent with at least one parameter value in the constraint set. This is accomplished by introducing an elasticity parameter, and we demonstrate that the corresponding elastic predictive random set provides valid and efficient probabilistic inference for constrained parameters. Moreover, we show that this approach is more efficient than the Dempster-like rule of combination.

5.2 Random sets

5.2.1 Some history and a general definition

Random sets are random elements taking values as subsets of some space, serving as general mathematical models for set-valued observations and irregular geometrical patterns, and they generalize the traditional concept of random variables and vectors. Random sets appeared in sampling design theory [124], in stochastic geometry [146], and in statistics [204]. The first systematic treatment of random (closed) sets as bona fide random elements on Hausdorff, locally compact and second countable topological spaces is [187]. A comprehensive treatment of the theory of random sets is given by Molchanov [189], upon which our summary is based.

Our goal here is to present the minimal amount of technical details necessary to understand the analysis in this book involving predictive random sets. To start, the standard theory of random sets, as in [189], focuses on the case of closed random sets, i.e., random sets whose values are closed sets (with probability 1). Write $\mathbb{S}$ for the support of our random set $\mathcal{S}$, a collection of subsets of the space $\mathbb{U}$; we assume that $\mathbb{S}$ contains both $\varnothing$ and $\mathbb{U}$. Let $\mathbb{U}$ be a separable metric space and take $\mathscr{U}$ to be a σ-algebra of subsets that contains all the closed subsets of $\mathbb{U}$. Assume that each set, or focal element, $S \in \mathbb{S}$ is closed, relative to the topology on $\mathbb{U}$.

To define the random set $\mathcal{S}$, consider a probability space $(\Omega, \mathscr{A}, P)$, and let $\mathcal{S}$ be mapping from Ω to $\mathbb{S}$ which is measurable in the sense that

$$\{\omega : \mathcal{S}(\omega) \cap K \neq \varnothing\} \in \mathscr{A}, \quad \text{for all compact } K \subseteq \mathbb{U}.$$

We define the distribution $\mathsf{P}_{\mathcal{S}}$ of $\mathcal{S}$ as the push-forward measure $P\mathcal{S}^{-1}$. Also, for compact K, the event $\{\mathcal{S} \subset K\}$ is measurable, so the probability $\mathsf{P}_{\mathcal{S}}\{\mathcal{S} \subset K\}$ discussed below makes sense. Moreover, since $\mathbb{U}$ is separable and $\mathcal{S}$ is closed, the indicator stochastic process $\{I_{\mathcal{S}}(u) : u \in \mathbb{U}\}$ is separable and, by Proposition 4.10 in [189], the distribution $\mathsf{P}_{\mathcal{S}}$ is determined by probabilities assigned to the events $\{\mathcal{S} \subset K\}$, and can be extended to include arbitrary sets K.

An important class of examples are the predictive random sets described in Corollary 4.1 of Chapter 4. In particular, start by taking the probability space $(\Omega, \mathscr{A}, P)$ to be $(\mathbb{U}, \mathscr{U}, \mathsf{P}_U)$. Next, for a continuous function $h : \mathbb{U} \to \mathbb{R}$, define the set-valued mapping ψ_h on $\mathbb{U}$ as

$$\psi_h(u) = \{u' \in \mathbb{U} : h(u') \leq h(u)\}, \quad u \in \mathbb{U}.$$

Then the focal elements $S \in \mathbb{S}$ are closed level sets of the function h, and if $\mathscr{U}$ is rich enough to contain all the closed sets, then ψ_h is a measurable function and, consequently, $\mathcal{S} = \psi_h(U)$, for $U \sim \mathsf{P}_U$, is a closed random set.

There are several ways to describe the distribution $\mathsf{P}_{\mathcal{S}}$ of the random set. One approach is via the the plausibility function $\mathrm{pl}(K) = \mathsf{P}_{\mathcal{S}}\{\mathcal{S} \cap K \neq \varnothing\}$, $K \subseteq \mathbb{U}$, also called the capacity functional [189]. An equivalent quantity, often used in artificial intelligence applications, is the belief function, $\mathrm{bel}(K) = 1 - \mathrm{pl}(K^c)$. The belief function is a formal analogue to the distribution function of a random variable. One key difference when dealing with random sets, compared to random variables, is that the complementation law generally fails, i.e., $\mathrm{pl}(K) + \mathrm{pl}(K^c) \geq 1$, with equality for all K if and only if pl is a probability measure if and only if S is a singleton set with P_S-probability 1. One can discuss belief and plausibility functions without explicitly talking about random sets [223], though we shall not do so here. There is now a substantial literature on belief and plausibility functions; see [264].

5.2.2 Nested predictive random sets and validity

In the IM context, for reasons discussed in Chapter 4 and below, we focus only on predictive random sets with nested support $\mathbb{S}$, where nested means:

$$\text{for any } S, S' \in \mathbb{S}, \text{ either } S \subseteq S' \text{ or } S \supseteq S'. \tag{5.1}$$

Such supports can be, and usually are, constructed as in the small example above, but require that the function h has a unique minimizer and that it is non-decreasing (or, for simplicity, increasing) as it moves further away from that minimizer. The default predictive random set, for the case $\mathbb{U} = [0,1]$, takes $h(u) = |u - 0.5|$, which is of the form just described. Using the terminology in [226], we could call a predictive random set with nested support consonant; see, also, [4, 5, 221]. Consonant random sets have the simplest distributional properties. Indeed, it is easy to check (Exercise 5.1) that the plausibility function corresponding to a consonant random set satisfies

$$\mathrm{pl}(K_1 \cup K_2) = \max\{\mathrm{pl}(K_1), \mathrm{pl}(K_2)\}, \quad K_1, K_2 \subseteq \mathbb{U}. \tag{5.2}$$

This immediately extends to finite unions and, with some standard continuity assumptions on pl, to countable unions. Here we consider a stronger version of continuity, called *condensability* [193, 221, 226]. That is, the plausibility function is condensable if, for any upward net $\mathcal{K}$ of subsets in $\mathbb{U}$,

$$\mathrm{pl}\left(\bigcup_{K \in \mathcal{K}} K\right) = \sup_{K \in \mathcal{K}} \mathrm{pl}(K); \tag{5.3}$$

by an upward net we mean a collection of sets with the property that, for any K_1 and K_2 in $\mathcal{K}$, there exists K_3 in $\mathcal{K}$ such that $K_3 \supset K_1 \cup K_2$. That this generalizes the usual notion of continuity of set functions is clear.

Section V.G of [226] shows that consonance and condensability together imply that the plausibility function is fully characterized by the contour function, given by

$$\gamma_S(u) = \mathrm{P}_S\{S \ni u\}, \quad u \in \mathbb{U}, \tag{5.4}$$

i.e., the probability that the random set S catches the fixed point $u \in \mathbb{U}$. The contour function resembles the "commonality function" that appears occasionally in the Dempster–Shafer theory literature [221]. Since γ_S is an ordinary function, not a set function, it will be easier to work with than the full plausibility function. That this function captures the entire plausibility function can be seen by the formula $\mathrm{pl}(K) = \sup_{u \in K} \gamma_S(u)$, a special case of (5.3).

Given the auxiliary variable space $(\mathbb{U}, \mathscr{U}, \mathrm{P}_U)$, let $\mathbb{S}$ be a collection of closed measurable subsets of $\mathbb{U}$, assumed to contain $\varnothing$ and $\mathbb{U}$. To avoid measurability issues, assume that $\mathcal{U}$ contains all closed sets. Write $\mathcal{S}$ for the predictive random set supported on $\mathbb{S}$, $\mathsf{P}_{\mathcal{S}}$ for its distribution, and $\gamma_{\mathcal{S}}(u)$ for the corresponding contour function (5.4). We say the predictive random set $\mathcal{S}$ is *valid* if

$$\gamma_{\mathcal{S}}(U) \geq_{\text{st}} \text{Unif}(0,1) \quad \text{when } U \sim \mathsf{P}_U. \tag{5.5}$$

We have already seen that validity is a core concept in the IM framework and in scientific inference, more generally. How to construct a valid predictive random set? To start, take $\mathcal{S}$ to be nested, so that its support $\mathbb{S}$ satisfies (5.1), and assume its plausibility function is condensable. As discussed in the previous subsection, this implies that the contour function $\gamma_{\mathcal{S}}$ fully characterizes the distribution $\mathsf{P}_{\mathcal{S}}$. Validity implicitly requires some connection between $\mathsf{P}_{\mathcal{S}}$ and P_U, so we will also assume that $\mathcal{S}$, with contour function, $\gamma_{\mathcal{S}}$, satisfies

$$\gamma_{\mathcal{S}}(u) = 1 - \sup_{S \in \mathbb{S}: S \not\ni u} \mathsf{P}_U(S), \quad u \in \mathbb{U}. \tag{5.6}$$

The following result says that these conditions are sufficient for validity.

Proposition 5.1. *Suppose $\mathcal{S}$ is nested with condensable plausibility function. If its contour function satisfies (5.6), then it is valid.*

Proof. See Exercise 5.2.   □

A good example of a nested predictive random set is the "default" discussed in Chapter 4. That is, suppose P_U is a $\text{Unif}(0,1)$ distribution and define $\mathcal{S}$ as

$$\mathcal{S} = \{u : |u - \tfrac{1}{2}| \leq |U - \tfrac{1}{2}|\}, \quad \text{with} \quad U \sim \mathsf{P}_U. \tag{5.7}$$

Then the support $\mathbb{S}$ of $\mathcal{S}$ contains all symmetric intervals centered at $\frac{1}{2}$ of width less than or equal to 1, which is clearly a nested collection. Also, note that the random set $\mathcal{S}$ is determined by a typical probability space $(\mathbb{U}, \mathcal{U}, \mathsf{P}_U)$ and a set-valued mapping $u \mapsto G(u) = \{u' : |u' - \tfrac{1}{2}| \leq |u - \tfrac{1}{2}|\}$. This matches the setup in [63], so condensability of the corresponding plausibility function follows from general theory; see Section V.F in [226]. For the contour function,

$$\gamma_{\mathcal{S}}(u) = \mathsf{P}_{\mathcal{S}}\{\mathcal{S} \ni u\} = \mathsf{P}_U\{|U - \tfrac{1}{2}| \geq |u - \tfrac{1}{2}|\} = 1 - |2u - 1|.$$

Then condition (5.8) clearly holds, since the supremum is attained at $G(u)$, which has P_U-probability $|2u - 1|$. Therefore, $\mathcal{S}$ satisfies the conditions of Proposition 5.1 and, hence, is valid. However, the validity of $\mathcal{S}$ can be shown directly by checking that its contour function above satisfies $\gamma_{\mathcal{S}}(U) \sim \text{Unif}(0,1)$ for $U \sim \mathsf{P}_U$. These arguments can be easily generalized to those discussed above determined by level sets of a function h, which come in handy when the auxiliary variable space $\mathbb{U}$ is of dimension two or more.

5.2.3 Admissible predictive random sets

Theorem 4.1 in Chapter 4 is a fundamental result which says that, given a predictive random set, there is a nested predictive random set which is at least as efficient. As mentioned in Remark 4.1, that particular nested predictive random set constructed in the proof of that theorem plays an important role in subsequent developments concerning efficiency and optimality. This motivates the following

Definition 5.1. A predictive random set $\mathcal{S}$, supported on $\mathbb{S}$ is called *admissible* if

- $\mathbb{S}$ is nested in the sense of (5.1);
- Its contour function satisfies (5.6) or, equivalently, if

$$\mathsf{P}_{\mathcal{S}}\{\mathcal{S} \subset K\} = \sup_{S \in \mathcal{S}: S \subset K} \mathsf{P}_U(S). \tag{5.8}$$

As in Remark 4.1, we call the distribution $\mathsf{P}_{\mathcal{S}}$ in (5.8) the *natural measure* relative to the given support $\mathbb{S}$ and the distribution P_U; see Exercise 5.3. This natural measure will be used extensively in Chapters 6–7 and 10.

Finally, since we now have a general mathematical concept of a random set, we can ask the question if, for a given $\mathbb{S}$ and P_U, there exists a random set $\mathcal{S}$ with distribution given by the natural measure (5.8). In other words, although the right-hand side in (5.8) is well-defined, is there random set $\mathcal{S}$ with distribution $\mathsf{P}_{\mathcal{S}}$ so that the left-hand side equals the right-hand side? Fortunately, the famous Choquet capacity theorem, Theorem 1.13 in [189], can be applied to show existence of such an $\mathcal{S}$. The argument is relatively simple, given the close connection to the probability measure P_U, but we omit the unnecessary details here.

5.3 Predictive random sets for constrained problems

5.3.1 Motivating examples

The parameter space Θ of a probability model $\mathsf{P}_{X|\theta}$ may extend beyond what is consistent with the physical world. It is clear that ignoring such constraints can result in inference which is, at best, difficult to interpret and, at worst, misleading; see Exercises 5.4–5.5. Currently, there is no widely agreed upon approach for incorporating such constraints into the statistical analysis. Perhaps the simplest strategy is, within a Bayesian framework, to restrict the prior distribution to the constrained parameter space. Such an approach makes sense when the prior distribution is meaningful. However, the advertised properties of the posterior distribution based on a default non-informative prior generally will not hold when that same prior is restricted to a subset of the full parameter space; see Exercise 5.5.

Here we consider an IM approach to this problem. For motivation, we use two examples of particular interest to high energy physicists during the past twenty years, namely, inference about a bounded quantity measured with Gaussian error, and inference about the Poisson rate from a contaminated observed count.

Suppose X is the measurement of a nonnegative quantity, μ, with Gaussian error distribution. Choosing the variance, $\sigma^2 = 1$, for simplicity, this can be represented by the probability model $X \sim N(\mu, 1)$ and the constraint $\mu \geq 0$. The Gaussian model for X allows any real-valued μ. For this unrestricted case, many inference methods have proven to be simple and produce practically the same results for μ. Somewhat surprisingly, when μ is known to belong to a restricted interval, the same problem becomes challenging; see, for example, [81]. As discussed in [174], this problem arises when measuring particle masses, which must be non-negative and are expected to be relatively small, if nonzero.

In the Poisson example, the observed count, Y, is known to be comprised of signal and background events, each coming from their own independent Poisson distributions. Suppose the background rate, b, is known, but the signal rate, λ, is unknown. Let $S \sim \text{Pois}(\lambda)$ be the number of signal events and $B \sim \text{Pois}(b)$ be the number of background events. Both S and B are unobserved, but the observed count, $Y = S + B$, comes from a $\text{Pois}(\theta)$ distribution with $\theta = \lambda + b$. The Poisson model for Y only requires that θ be nonnegative or, equivalently, $\lambda \geq -b$. However, negative values of λ are not valid and so the constraint $\theta \geq b$ is required. This model is used in experiments measuring a number of events caused by neutrino oscillations. Some of the observed events are due to random background sources, but are indistinguishable from the signal events of interest. Detailed discussion and references to experimental results can be found in [174].

Much existing work on the two example problems was aimed at developing confidence intervals that involve the constraints. Methods were developed within both the Bayesian and frequentist frameworks; see [174]. In particular, Gleser's comment focuses on how the likelihood function can quantify uncertainty about the unknown mean in the Gaussian example and, later, [105] argued in favor of reporting the likelihood and one-tailed p-value as a function of hypothetical parameter values. This allows each individual to make their own judgment about the strength of evidence required for rejection. However, inference using likelihood and p-values generally lacks the predictive interpretation sought by practitioners. As articulated in [95] care must be taken in making a probabilistic interpretation of p-values; see, also, [181].

5.3.2   IMs for constrained parameter problems

Recall that an IM is created in the C-step where the predictive random set is combined with the association and given data. Here, as is often the case in practice, we assume that the predictive random set $\mathcal{S}$ is the result of a set-valued mapping $\mathcal{S} : \mathbb{U} \to 2^{\mathbb{U}}$. Then the C-step defines a collection of subsets of Θ, indexed by $\mathbb{U}$:

$$\Theta_x(\mathcal{S}) = \Theta_x(\mathcal{S}(u)) = \bigcup_{u' \in \mathcal{S}(u)} \{\theta : \theta \in \Theta, x = a(\theta, u')\}, \quad u \in \mathbb{U}. \qquad (5.9)$$

In this case, the belief function, bel_x supporting the assertion A is computed as follows, conditioned on the focal elements being nonempty:

$$\text{bel}_x(A) = \mathsf{P}_U\{\Theta_x(\mathcal{S}(U)) \subseteq A \mid \Theta_x(\mathcal{S}(U)) \neq \varnothing\}. \qquad (5.10)$$

We now consider how to incorporate non-trivial parameter space constraints into the IM (5.9). Throughout this section assume θ is known to be in some constraint set $C \subset \Theta$, e.g., $C = \{\theta : \theta \geq \theta_0\}$.

As described in Chapter 4, a predictive random set S is designed to be valid and efficient for predicting the auxiliary variable U in $\mathbb{U}$. After $X = x$ is observed, C can be mapped to a subset of $\mathbb{U}$ by inverting the sampling model association:

$$\mathbb{U}_{C,x} = \bigcup_{\theta \in C} \{u : x = a(\theta, u)\}.$$

We call $\mathbb{U}_{C,x}$ the a-constraint set. Let θ^* be the true, unobserved value of the parameter. Then, there must exist $u^* \in \mathbb{U}_{C,x}$ such that $x = a(\theta^*, u^*)$. If $u^* \notin \mathbb{U}_{C,x}$, then the corresponding θ^* is not in the constraint set C, which is impossible. Thus, when $S(u) \cap \mathbb{U}_{C,x} = \varnothing$, the focal element of the IM contains only values of θ that are not in the constraint set, i.e., $\Theta_x(S) \cap C = \varnothing$. Such a focal element is called a conflict case, and these are indexed by the set

$$\mathbb{U}_{\varnothing,x} = \{u : S(u) \cap \mathbb{U}_{C,x} = \varnothing\}.$$

The problem of incorporating parameter constraints into an IM can, therefore, be framed in terms of handling conflict cases.

The probability on the set $\mathbb{U}_{\varnothing,x}$ can been seen as measuring discord between the observed value of x, its probability model, and the parameter constraint set. If the probability on conflict cases is very large, one should question whether the probability model is appropriate for the observed data or whether the constraint is correct. However, the presence of conflict alone does not justify modifying the model. Section 5.3.3 introduces a new method that modifies the predictive random set in a data-dependent way while preserving validity and striving for high efficiency. The result is that conflict cases become evidence for certain values of θ. In Section 5.4.1, the new method is compared to an existing conditioning method, which can use the probability on conflict cases to represent an additional layer of uncertainty about the model assumptions.

5.3.3   Elastic predictive random sets

The predictive random set, $S(u)$, and the P_U distribution on $\mathbb{U}$ represent a set of predictions and a measure of uncertainty about those predictions. Intuitively, a conflict case results from $S(u)$ being too small. If the probability model for the observed x and the parameter constraint are not in doubt, then $S(u)$ should be enlarged in an adaptive fashion. The basic idea is to eliminate conflict cases by allowing the predictive random set to stretch until it includes at least one member of $\mathbb{U}_{C,x}$ while retaining the same P_U distribution. The resulting IM method is called *elastic belief* [82], which is used here in this chapter to distinguish it from the alternative IM method using a type of Dempster's rule of combination; see Section 5.4.1). Technically, the elastic belief method equips the predictive random set with an elasticity parameter, $e \in [0, 1]$, thus forming a predictive random set collection, $S = \{S_e : e \in [0, 1]\}$, called an *elastic predictive random set*.

Definition 5.2. A collection of predictive random sets, $\mathcal{S}$, indexed by $e \in [0,1]$, is called *elastic* if,

(a) For any $e \in [0,1]$, $\mathcal{S}_e$ is valid, i.e., $\mathcal{S}_e$ satisfies Definition 4.1;

(b) For any $e_1 \le e_2$, $\mathcal{S}_{e_1}(U) \subseteq \mathcal{S}_{e_2}(U)$ with probability one, as $U \sim \mathsf{P}_U$; and

(c) For any $u \in \mathbb{U}$ and any $(x, \theta) \in \mathbb{X} \times \Theta$, there exists an $e \in [0,1]$ and $u' \in \mathcal{S}_e(u)$ such that $x = a(\theta, u')$.

These three properties are referred to as (a) *validity*, (b) *monotonicity*, and (c) *completeness*.

Example 5.1. For an auxiliary variable $U \sim \mathsf{N}(0,1)$, consider the predictive random set $\mathcal{S}(u) = \{u' : |u'| \le |u|\}$. One way to make S elastic is

$$\mathcal{S}_e(u) = \begin{cases} \{u' : |u'| \le \frac{1}{1-e}|u|\}, & \text{if } e \in [0,1), \\ \mathbb{R} & \text{if } e = 1. \end{cases}$$

Using the Gaussian association, $X = \theta + U$, it is easy to verify that $\mathcal{S} = \{\mathcal{S}_e : e \in [0,1]\}$ satisfies Definition 5.2.

The existence of an elastic predictive random set is ensured by the nature of the association. Let $\mathcal{S}_0$ be a valid predictive random set, i.e., satisfying Definition 4.1 and let $\mathcal{S}_1(u) \equiv \mathbb{U}$. For $e \in (0,1)$, an arbitrary $\mathcal{S}_e$ increasing in e from $\mathcal{S}_0$ to $\mathbb{U}$ will satisfy (a) and (b) in Definition 5.2. Since $a(\theta, U)$ has the same distribution as X for fixed $\theta \in \Theta$, then $\mathbb{X} = \bigcup_{u \in \mathbb{U}} a(\theta, u)$. Thus, for all $(x, \theta) \in \mathbb{X} \times \Theta$, there must exist $u' \in \mathcal{S}_1(u)$ such that $x = a(\theta, u')$, which satisfies (c).

To use the elastic belief method, each focal element $\Theta_x(\mathcal{S})$ in the IM (5.9)) is simply replaced with

$$\widetilde{\Theta}_x(\mathcal{S}) = \mathcal{C} \cap \Theta_x(\mathcal{S}_{\hat{e}}),$$

where

$$\hat{e} = \min\{e : \mathcal{S}_e(u) \cap \mathbb{U}_{\mathcal{C},x} \ne \varnothing\} = \min\{e : \mathcal{C} \cap \Theta_x(\mathcal{S}_e) \ne \varnothing\}.$$

In effect, the elastic belief method stretches the IM focal element until it is just large enough to intersect with $\mathcal{C}$. The amount of stretching is characterized by $\hat{e}$. The completeness property (c) in Definition 5.2 ensures that $\{e : \mathcal{S}_e(u) \cap \mathbb{U}_{\mathcal{C},x} \ne \varnothing\}$ is not empty for any u. Therefore, if $\hat{e}$ exists, then $\widetilde{\Theta}_x(\mathcal{S})$ is also not empty. Finally, if applying the elastic belief method results in any duplicate focal elements (i.e., $\widetilde{\Theta}_x(\mathcal{S}(u)) = \widetilde{\Theta}_x(\mathcal{S}(u'))$ for $u \ne u'$), they can be considered as a single element with mass aggregated from the duplicate elements. After building an IM with the elastic belief method, the belief for any assertion, $A \subseteq \mathcal{C}$ can be computed as similar to that in the IM of Chapter 4:

$$\mathrm{bel}_x(A) = \mathsf{P}_U\{\widetilde{\Theta}_x(\mathcal{S}(U)) \subseteq A\}. \tag{5.11}$$

When using elastic predictive random sets, all the IM focal elements are non-empty. The conditioning on non-empty focal elements encountered in Chapter 4 is omitted in (5.11) because $\mathsf{P}_U\{\widetilde{\Theta}_x(\mathcal{S}(U)) \ne \varnothing\} = 1$.

Example 5.2. Suppose it is known that the Gaussian mean, μ, lies in the range $[a,b]$ for some known constants $a < b$. The elastic belief method can be applied with the association and the elastic predictive random set from Example 5.1. This gives

$$\Theta_x(S_e) = \left[x - \tfrac{1}{1-e}|u|, x + \tfrac{1}{1-e}|u|\right].$$

To handle conflict cases, the elastic predictive random set is expanded with

$$\hat{e} = \min\{e : [a,b] \cap M_x(u, S_e) \neq \varnothing\}$$

$$= \begin{cases} 1 + \frac{|u|}{x-a} & \text{if } x + |u| < a; \\ 1 - \frac{|u|}{x-b} & \text{if } x - |u| > b; \\ 0 & \text{otherwise.} \end{cases}$$

This yields an IM with the following focal elements,

$$\widetilde{\Theta}_x(S) = \begin{cases} \{a\}, & \text{if } |u| < a - x; \\ \left[\max\{a, x - |u|\}, \min\{b, x + |u|\}\right], & \text{if } |u| \geq \max\{a - x, x - b\}; \\ \{b\}, & \text{if } |u| < x - b. \end{cases}$$

Now, suppose $a = -1/4$, $b = 1/4$, and $x = -1$ is observed. Then, for the assertion $A = \{\mu : \mu \geq 0\}$, the elastic belief method gives

$$\text{bel}_x(A) = P_U\{u : -1 - |u| \geq 0\} = 0$$
$$\text{bel}_x(A^c) = P_U\{u : -1 + |u| < 0\} = 1 - 2\Phi(-1) \approx 0.68,$$

which is the same result as in the case where the constraint was not part of the IM.

5.4   Theoretical results on elastic predictive random sets

The following theorem shows how $\widetilde{\Theta}_x(S)$ can be used for valid inference in the sense of Definition 4.2.

Theorem 5.1. *Let $S = \{S_e : e \in [0,1]\}$ satisfy properties (b) and (c) of Definition 5.2. For convenience, let $\text{bel}_{x,S}(A)$ and $\text{bel}_x^*(A)$ be defined as in (5.10) and (5.11), respectively. Then, for any $x \in \mathbb{X}$, the following are true:*

(i) For any assertion $A \subset C$,

$$\text{bel}_{x,S_0}(A) \leq \text{bel}_x^*(A) \leq \text{bel}_{x,S_0}(A \cup C^c).$$

(ii) If $\text{bel}_{x,S_0}(A \cup C^c)$ and $\text{bel}_{x,S_0}(A^c \cup C^c)$ satisfy

$$P_{X|\theta}\{x : \text{bel}_{x,S}(A) \geq 1 - \alpha\} \leq \alpha \quad \text{for all } \theta \in A^c \qquad (5.12)$$

for some assertion $A \subset C$, then bel_x^* is valid for inference about A.

(iii) If S also satisfies property (a) of Definition 5.2, then bel_x^* is valid for inference about every assertion $A \subset C$.

Proof.   (i) The relationships between bel_x^* and bel_x can be shown by expanding
the definition of $\text{bel}_x^*(A)$:

$$\text{bel}_x^*(A) = P_U\{\widetilde{\Theta}_x(\mathcal{S}(U)) \subseteq A\}$$
$$= P_U\{\Theta_x(\mathcal{S}_0(U)) \cap \mathcal{C} \subseteq A, \Theta_x(\mathcal{S}_0(U)) \cap \mathcal{C} \neq \varnothing\}$$
$$+ P_U\{\Theta_x(\mathcal{S}_{\hat{e}}(U)) \cap \mathcal{C} \subseteq A, \Theta_x(\mathcal{S}_0(U)) \cap \mathcal{C} = \varnothing\}.$$

The inequality, $\text{bel}_{x,\mathcal{S}_0}(A) \leq \text{bel}_x^*(A)$, follows from the fact that

$$P_U\{\Theta_x(\mathcal{S}_0(U)) \cap \mathcal{C} \subseteq A, \Theta_x(\mathcal{S}_0(U)) \cap \mathcal{C} \neq \varnothing\} \geq P_U\{\Theta_x(\mathcal{S}_0(U)) \subseteq A\}$$
$$= \text{bel}_{x,\mathcal{S}_0}(A).$$

The inequality,
$$\text{bel}_x^*(A) \leq \text{bel}_{x,\mathcal{S}_0}(A \cup \mathcal{C}^c), \tag{5.13}$$

is determined by the mass of conflict cases that support A in the IM resulting
from the elastic belief method:

$$P_U\{\Theta_x(\mathcal{S}_{\hat{e}}(U)) \cap \mathcal{C} \subseteq A, \Theta_x(\mathcal{S}_0(U)) \cap \mathcal{C} = \varnothing\} \leq P_U\{\Theta_x(\mathcal{S}_0(U)) \cap \mathcal{C} = \varnothing\}$$
$$= K_x,$$

with equality when all of the conflict cases support A after using the elastic
belief method. It follows that

$$\text{bel}_x^*(A) \leq P_U\{\Theta_x(\mathcal{S}_0(U)) \cap \mathcal{C} \subseteq A, \Theta_x(\mathcal{S}_0(U)) \cap \mathcal{C} \neq \varnothing\} + K_x$$
$$= P_U\{\Theta_x(\mathcal{S}_0(U)) \subseteq A \cup \mathcal{C}^c\} = \text{bel}_{x,\mathcal{S}_0}(A \cup \mathcal{C}^c).$$

(ii) The validity of $\text{bel}_x^*(A)$ follows by considering the random variables, $\text{bel}_X^*(A)$
and $\text{bel}_{X,\mathcal{S}_0}(A \cup \mathcal{C}^c)$ as functions of the random variable, X. By satisfying
(5.12),
$$P_{X|\theta}\{\text{bel}_{X,\mathcal{S}_0}(A \cup \mathcal{C}^c) \geq 1 - \alpha\} \leq \alpha$$

for any $\theta \in (A \cup \mathcal{C}^c)^c = A^c \cap \mathcal{C}$. The inequality (5.13) implies $\text{bel}_X^*(A)$ is
stochastically smaller than $\text{bel}_{X,\mathcal{S}_0}(A \cup \mathcal{C}^c)$. Thus,

$$P_{X|\theta}\{\text{bel}_X^*(A) \geq 1 - \alpha\} \leq P_{X|\theta}\{\text{bel}_{X,\mathcal{S}_0}(A \cup \mathcal{C}^c) \geq 1 - \alpha\} \leq \alpha.$$

for any $\theta \in A^c \cap \mathcal{C}$. Also, by the same argument when $\text{bel}_{X,\mathcal{S}_0}(A^c \cup \mathcal{C}^c)$ satisfies
(5.12),

$$P_{X|\theta}\{\text{bel}_X^*(A^c) \geq 1 - \alpha\} \leq P_{X|\theta}\{\text{bel}_{X,\mathcal{S}_0}(A^c \cup \mathcal{C}^c) \geq 1 - \alpha\} \leq \alpha.$$

for any $\theta \in A \cap \mathcal{C} = A$.

(iii) If $\mathcal{S}$ satisfies property (a) of Definition 5.2, then $\mathcal{S}_0$ is valid, *i.e.*, satisfies Definition 4.1. By Theorem 1 of [186], for any $A \subset \mathcal{C}$, bel_x is valid for inference about $A \cup \mathcal{C}^c$ and $A^c \cup \mathcal{C}^c$. The result follows from applying part (ii).

$\square$

An important application of Theorem 5.1 is when an IM has been created using a predictive random set satisfying the efficiency criteria in Chapter 4 without considering constraints. That predictive random set can be used for S_0 when creating an elastic predictive random set. Thus, when constraints are incorporated using the elastic method, bel_x will be valid for inference about any $A \subset C$.

5.4.1 Dempster's rule of combination

Another method for incorporating parameter constraints into an IM is the conditioning rule described in [63] known as "Dempster's rule of combination" or "Dempster's conditioning rule" [221]. In effect, this method uses $S(u) \cap \mathbb{U}_{C,x}$ as the predictive random set and conditions the distribution, P_U, on the event, $\mathbb{U}^c_{\varnothing,x}$. Define $K_x = P_U\{\Theta_x(S) \cap C = \varnothing\}$ as a function of the observed data, x. For any practical assertion, $A \subset C$, we have

$$\text{bel}_x(A \mid C) = \frac{P_U\{\Theta(S(U)) \cap C \subseteq A, \Theta_x(S(U)) \cap C \neq \varnothing\}}{1 - K_x}$$

and

$$\text{bel}_x(A^c \mid C) = \frac{P_U\{\Theta_x(S(U)) \cap C \subseteq A^c, \Theta_x(S(U)) \cap C \neq \varnothing\}}{1 - K_x}.$$

The following theorem, an extension of Theorem 1 in [186], states that over repeated observations of X, the conditional $\text{bel}_x(A \mid C)$ and $\text{bel}_x(A^c \mid C)$ will be valid for A.

Theorem 5.2. *For a given value of $\alpha \in (0,1)$, suppose S is valid by Definition 4.1 and that $P_U\{\Theta_x(S(U)) = \varnothing\} > 0$ for some $x \in \mathbb{X}$. Then, $\text{bel}_x(A \mid C)$ and $\text{bel}_x(A^c \mid C)$ will be valid according to Definition 4.2 for any $A \subset C$.*

Proof. For the sample space, $\mathcal{X}$, and parameter space, Θ, let U be defined in the a-space, $\mathbb{U}$ with distribution P_U. Then, let $S(U) \subseteq \mathbb{U}$ be the predictive random set and obtain the IM (5.9) for Θ with focal elements $\{\Theta_x(S(u))\}_{u \in \mathbb{U}}$. Assume $\Theta_x(S(u))$ was designed so that $P_U\{\Theta_x(S(U)) = \varnothing\} = 0$ for every $x \in \mathcal{X}$. This gives

$$\text{bel}_x(A^c) = P_U\{\Theta_x(S(U)) \subseteq A^c\} \tag{5.14}$$

as the evidence *against* the assertion, A. Further, suppose $\Theta_x(S)$ is valid for inference about A as in Definition 4.2 so that,

$$P_{X|\theta}\{\text{bel}_x(A^c) \geq 1 - \alpha\} \leq \alpha,$$

for $\alpha \in (0,1)$ and every $\theta \in A$.

Next, suppose a constraint on the parameter space, $C \subset \Theta$, is introduced such that θ is known to lie inside C. The evidence against an assertion, $A \subseteq C$, is defined by the conditioning rule as the conditional probability,

$$\text{bel}_x(A^c \mid C) = P_U\{\Theta_x(S(U)) \cap C \subseteq A^c \mid \Theta_x(S) \cap C \neq \varnothing\}.$$

This can be written as:

$$\text{bel}_x(A^c \mid C) = \frac{\text{bel}_x(A^c \cup C^c) - \text{bel}_x(C^c)}{1 - \text{bel}_x(C^c)}.$$

The proof follows from the fact that $A \subseteq C$ implies $C^c \subseteq A^c$. Thus,

$$\text{bel}_x(A^c \mid C) = \frac{\text{bel}_x(A^c) - \text{bel}_x(C^c)}{1 - \text{bel}_x(C^c)}.$$

The proof is completed by noting that $\frac{a-b}{1-b} \leq a$ when $a, b \in [0, 1)$ and $a \geq b$. Here,

$$a = \text{bel}_x(A^c) \quad \text{and} \quad b = \text{bel}_x(C^c).$$

Thus, for every $x \in \mathbb{X}$, we have $\text{bel}_x(A^c \mid C) \leq \text{bel}_x(A^c)$, with equality when $P_U\{\Theta_x(\mathcal{S}(U)) \cap C = \varnothing\} = 0$. Therefore,

$$P_{X\mid\theta}\{\text{bel}_X(A^c \mid C) \geq 1 - \alpha\} \leq P_{X\mid\theta}\{\text{bel}_X(A^c) \geq 1 - \alpha\} \leq \alpha$$

for every $\theta \in A$. The same argument can be repeated with $\text{bel}_X(A \mid C) = \text{bel}_X((A^c)^c \mid C)$ to show that

$$P_{X\mid\theta}\{\text{bel}_X(A \mid C) \geq 1 - \alpha\} \leq P_{X\mid\theta}\{\text{bel}_X(A) \geq 1 - \alpha\} \leq \alpha$$

for every $\theta \in A^c$. This shows that the conditioning rule preserves the IM's validity for an assertion in the presence of a constraint on Θ.   □

Although not required for the proof, it is worthwhile to consider conditions under which $P_U\{\Theta_x(\mathcal{S}(U)) \cap C = \varnothing\} = 0$. If S has a neutral point, u_0, such that $u_0 \in \mathcal{S}(u)$ for every $u \in \mathbb{U}$, then we can partition $\mathbb{X}$ by the impossibility of conflict cases. Let $\Theta_x(u)$ be the basic IM obtained with the singleton predictive random set, $\mathcal{S}(u) = \{u\}$ and define

$$\mathbb{X}_{NC} = \{x : \Theta_x(u_0) \cap C \neq \varnothing\}.$$

Then, for any $x \in \mathbb{X}_{NC}$ and any $u \in \mathbb{U}$, we have $\Theta_x(u_0) \subseteq \Theta_x(\mathcal{S})$ and so $\Theta_x(\mathcal{S}) \cap C \neq \varnothing$. Therefore, on $\mathbb{X}_{NC}$,

$$\{u : \Theta_x(\mathcal{S}) \cap C = \varnothing\} = \varnothing,$$

which implies $\text{bel}_x(A^c \mid C) = \text{bel}_x(A^c)$ for any assertion, A.

Although beliefs resulting from conditioning are valid, the following example illustrates that they may be less likely to suggest the truth of A or A^c than beliefs computed with the unconstrained IM (5.9).

Example 5.3. For the constraint set $C = [a, b]$, the unconstrained Gaussian IM

$$\Theta_x(\mathcal{S}(U)) = [x - |U|, x + |U|], \quad U \sim N(0, 1),$$

will have conflict cases for focal elements indexed by $\{u : |u| < \max(x - b, a - x)\}$. If $a = -1/4$, $b = 1/4$, and $x = -1$, as in Example 5.2, then for $A = \{\mu : \mu \geq 0\}$, the conditioning rule gives

$$\text{bel}_x(A \mid C) = 0,$$

which is the same as usual IM without constraints, but

$$\text{bel}_x(A^c \mid \mathcal{C}) = \frac{\Phi(1) - \Phi(3/4)}{1 - \Phi(3/4)} \approx 0.30,$$

which is less than half of what was found in the case where no constraint on μ was known.

As shown by the above examples, introducing the constraint on μ values leads to weaker indications of whether or not $\mu \geq 0$, given the same evidence: $x = -1$. This is due to the large mass of conflict cases, $2\Phi(3/4) - 1 \approx 0.55$. The conditioning rule effectively ignores all these cases and distributes their mass over the non-conflict set. While both $\text{bel}_x(A^c)$ in the usual case without constraints and $\text{bel}_x(A^c \mid \mathcal{C})$ in Example 5.3 represent uncertainty about whether or not A^c is true, the reduction in $\text{bel}_x(A^c \mid \mathcal{C})$ could be attributed to additional uncertainty about the data and the model assumptions that led to conflict. However, for any $\mu \in [-1/4, 1/4]$ the probability of observing $x \leq -1$ is greater than 0.10. Under the model assumptions, it is not a rare event to observe $x = -1$ or something more extreme. If there is no other reason to doubt the validity of the data or the model, then it seems paradoxical that introducing more information about possible μ values in the form of a constraint leads to weaker indications of whether or not $\mu \geq 0$. Conflict cases become subsets of $\mathcal{C}$ when using elastic belief and therefore may become evidence for an assertion, $A \subseteq \mathcal{C}$. Thus, when more information is known about a parameter via constraints, and there is no reason to doubt the data and model assumptions, the elastic belief method may find stronger evidence for an assertion where the conditioning rule would find weaker evidence.

In some sense, the conditioning rule can also be understood as a different way of stretching $\mathcal{S}(u)$ by replacing it with a larger one, especially when the predictive random set $\{\mathcal{S}(u)\}_{u \in \mathbb{U}}$ forms a nested sequence. In that case, only those $\mathcal{S}(u)$ large enough to intersect with the a-constraint set, $\mathbb{U}_{\mathcal{C},x}$, will be considered. For $u \in \mathbb{U}_{\varnothing,x}$, the set, $\mathcal{S}(u)$, is too small and will be thrown away. Compared to elastic belief, the conditioning rule stretches stochastically more than necessary. This explains intuitively why the conditioning rule is valid but sometimes inefficient.

5.5   Two examples of the elastic belief method

5.5.1   Bounded normal mean

Consider computing the bel_x for the nonnegative mean example using the IM obtained with the elastic belief method in Example 5.2. In this case $a = 0$ and $b \to \infty$. Let $A = \{\mu : \mu = \mu_0\}$ be the assertion of interest. Then,

$$\text{bel}_x(\{\mu_0\}) = \begin{cases} 1 - 2\Phi(x), & \text{if } \mu_0 = 0 \text{ and } x < 0; \\ 0, & \text{otherwise}; \end{cases}$$

$$\text{bel}_x(\{\mu_0\}^c) = \begin{cases} 2\Phi(|x - \mu_0|) - 1, & \text{if } \mu_0 > 0; \\ 2\Phi(x) - 1, & \text{if } \mu_0 = 0 \text{ and } x > 0; \\ 0, & \text{otherwise}; \end{cases}$$

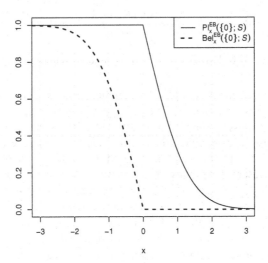

Figure 5.1 *Taken from [82]. Belief and plausibility functions* bel_x *and* pl_x *for the assertion* $A_0 = \{\mu : \mu = 0\}$ *with* $\sigma^2 = 1$ *in the Gaussian example of Section 5.5.1.*

for $\mu_0 \in [0, \infty)$. In many situations, the goal is to infer the presence or absence of a signal and so $\mu = 0$ is the assertion of interest. Figure 5.1 illustrates bel_x and pl_x for $A_0 = \{\mu : \mu = 0\}$. For negative values of x, $\mathrm{bel}_x(A_0)$ is large. For $x \leq -2$, $\mathrm{bel}_x(A_0) > 0.95$, suggesting that A_0 is true. For positive x values, $\mathrm{bel}_x(A_0)$ drops to zero and $\mathrm{pl}_x(A_0)$ becomes small. For $x \geq 2$, $\mathrm{pl}_x(A_0) < 0.05$, suggesting that A_0 is false. When x is close to zero, $\mathrm{bel}_x(A_0)$ is small while $\mathrm{pl}_x(A_0)$ is large. In these cases, it may be difficult to make any conclusion about A_0.

Applying the conditioning rule to this problem, one obtains:

$$\mathrm{bel}_x(\{\mu_0\} \mid \mathcal{C}) = 0$$

for any μ_0 and

$$\mathrm{bel}_x(\{\mu_0\}^c \mid \mathcal{C}; S) = \begin{cases} \frac{\Phi(\mu_0 - x) - \Phi(-x)}{1 - \Phi(-x)}, & \text{if } x < 0; \\ 2\Phi(|x - \mu_0|) - 1, & \text{if } x \geq 0. \end{cases}$$

For the assertion $A_0 = \{\mu : \mu = 0\}$, both the conditioning rule and elastic belief methods give the same plausibility. No matter what value of x is observed, $\mathrm{pl}_x(A_0) = \mathrm{pl}_x(A_0 \mid \mathcal{C})$. However, $\mathrm{bel}_x(A_0 \mid \mathcal{C}) = 0$, and so, unlike with the elastic belief method, no x observation ever supports the assertion.

The IM obtained with the elastic belief method in Example 5.2 can also be used to create a plausibility interval for μ based on the observed x. For a level $\gamma \in (0, 1)$, let $z_\gamma = \Phi^{-1}(\frac{1+\gamma}{2})$. Then,

$$\{\mu_0 : \mathrm{pl}_x(\{\mu_0\}) \geq 1 - \gamma\} = [\max\{0, x - z_\gamma\}, \max\{0, x + z_\gamma\}].$$

A level γ plausibility interval has coverage probability of at least γ over repeated experiments. Similarly, a level γ plausibility interval for μ can be created using the IM formed with the conditioning rule:

$$[\max\{0, x - z_\gamma\}, x + z_\gamma\}], \quad x \geq 0,$$
$$[0, x + \Phi^{-1}(\gamma + (1 - \gamma)\Phi(-x))], \quad x < 0.$$

Both of the above intervals are illustrated in Figure 5.2 for $\gamma = 0.9$. The shaded region is the plausibility interval obtained by the elastic belief method while the dashed line marks the boundary of the plausibility interval found with the conditioning rule. Their lower boundaries coincide for every x and their upper boundaries coincide when x is non-negative. When $x \in (-z_\gamma, \infty)$, for any specific μ_0 in the elastic belief-based interval interior $\mathrm{bel}_x(\{\mu_0\}) < \gamma$ and $\mathrm{bel}_x(\{\mu_0\}^c) < \gamma$. So there is not enough evidence to either support or deny μ_0 at level γ. When $x \in (-\infty, -z_\gamma]$, the elastic belief-based interval collapses to a single point where one concludes that $\mu = 0$ with $\mathrm{bel}_x(\{0\}) \geq \gamma$. If there is no reason to doubt the data and model assumptions, then the elastic method says that these improbable observations are consistent with $\mu = 0$ far more than any other value of μ, and in fact these improbable x values support the hypothesis that $\mu = 0$. As expressed in [39], an interval construction that collapses to a point for improbable observations is a reflection of the strength of evidence. Using the elastic belief method, this is explicitly quantified by computing $\mathrm{bel}_x(\{\mu_0\})$ and $\mathrm{pl}_x(\{\mu_0\})$ for hypothetical μ_0 values. The interval found with the conditioning rule has strictly positive length for any observed x. All μ_0 values within the interval are plausible with respect to level γ. However, since $\mathrm{bel}_x(\{\mu_0\} \mid C) < \gamma$ in the interval, no specific μ_0 value is supported, no matter how improbable the observed x is.

5.5.2 Poisson with known background rate

Consider inference on the signal rate, λ, from a Poisson count, Y, with known background rate, b. With b known, the overall rate is $\theta = \lambda + b$. For inference about λ, it is sufficient to perform inference about θ with constraint set $C = \{\theta : \theta \geq b\}$.

The A-step relies on a relationship between the Poisson and gamma distributions. Let G_y be the cdf for the gamma distribution with shape y and scale 1. Also, let F_θ be the cdf for the Poisson distribution with rate θ. Then, for non-negative integer y,

$$G_{y+1}(\theta) = 1 - F_\theta(y).$$

Let $G_0(\theta) = 1 - F_\theta(-1) = 1$. The auxiliary variable, U, has a uniform distribution over $\mathbb{U} = [0, 1]$. Because Y is discrete, the association is a many-to-one mapping:

$$\begin{aligned} a(\theta, u) &= \{Y : F_\theta(Y - 1) \leq u \leq F_\theta(Y)\} \\ &= \{Y : G_{Y+1}(\theta) \leq 1 - u \leq G_Y(\theta)\} \\ &= \{Y : G_Y^{-1}(1 - u) \leq \theta \leq G_{Y+1}^{-1}(1 - u)\}, \end{aligned} \quad (5.15)$$

where G_y^{-1} is the quantile function for the gamma distribution with shape y and scale

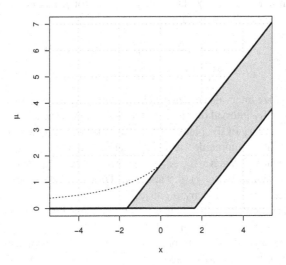

Figure 5.2 *Taken from [82]. Level $\gamma = 0.9$ plausibility intervals for μ with $\sigma^2 = 1$ in the Gaussian example of Section 5.5.1. The shaded region is the plausibility interval found with the elastic belief method, which collapses to the point $\mu = 0$ for $x < \Phi^{-1}(0.05)$. The dotted lines marks the boundary of the plausibility interval obtained from the conditioning rule. Both methods have the same lower boundary.*

1, and $G_0^{-1}(\theta) \equiv 0$. An alternative association can be derived from the waiting times in a Poisson process [67]; see Exercise 5.6. That association introduces additional challenges because the auxiliary variable has more than one dimension and its distribution depends on x.

For an unconstrained model,

$$\mathcal{S}(u) = \left[\tfrac{1}{2} - |u - \tfrac{1}{2}|, \tfrac{1}{2} + |u + \tfrac{1}{2}|\right] \tag{5.16}$$

is an efficient predictive random set for predicting U from a uniform distribution on $[0, 1]$. This can be adapted for an elastic predictive random set in the P-step:

$$\mathcal{S}_e(u) = \left[(1 - e)\left(\tfrac{1}{2} - |u - \tfrac{1}{2}|\right), (1 - e)\left(\tfrac{1}{2} + |u - \tfrac{1}{2}|\right) + e\right], \qquad e \in [0, 1].$$

which is (5.16) when $e = 0$ and increases to $\mathbb{U} = [0, 1]$ as $e \to 1$. This gives

$$\Theta_y(\mathcal{S}_e) = \left[G_y^{-1}\left((1 - e)(\tfrac{1}{2} - |u - \tfrac{1}{2}|)\right), G_{y+1}^{-1}\left((1 - e)(\tfrac{1}{2} + |u - \tfrac{1}{2}|) + e\right)\right].$$

To handle conflict cases, the elastic predictive random set is expanded with

$$\hat{e} = \min\{e : [b, \infty) \cap \Theta_y(S_e) \neq \varnothing\}$$

$$= \begin{cases} \dfrac{G_{y+1}(b) - \frac{1}{2} - |u - \frac{1}{2}|}{\frac{1}{2} - |u - \frac{1}{2}|} & \text{if } F_b(y) < \frac{1}{2} - |u - \frac{1}{2}|; \\ 0 & \text{otherwise.} \end{cases}$$

The resulting IM focal elements are:

$$\tilde{\Theta}_y(S) = \left[\max\left\{b, G_y^{-1}\left(\tfrac{1}{2} - |u - \tfrac{1}{2}|\right)\right\}, \max\left\{b, G_{y+1}^{-1}\left(\tfrac{1}{2} + |u - \tfrac{1}{2}|\right)\right\}\right].$$

For point assertions of the form $A = \{\theta : \theta = \theta_0\}$ we have the following bel_y when $\theta_0 = b$,

$$\text{bel}_y(\{b\}) = \begin{cases} 2G_{y+1}(b) - 1, & \text{if } F_b(y) \leq 1/2; \\ 0, & \text{otherwise;} \end{cases}$$

$$\text{bel}_y(\{b\}^c) = \begin{cases} 1 - 2G_y(b), & \text{if } F_b(y-1) > 1/2; \\ 0, & \text{otherwise;} \end{cases}$$

and for $\theta_0 > b$:

$$\text{bel}_y(\{\theta_0\}) = 0$$

$$\text{bel}_y(\{\theta_0\}^c) = \begin{cases} 2G_{y+1}(\theta_0) - 1, & \text{if } F_{\theta_0}(y) < 1/2; \\ 1 - 2G_y(\theta_0), & \text{if } F_{\theta_0}(y-1) > 1/2; \\ 0, & \text{otherwise.} \end{cases}$$

Just as in the constrained Gaussian example, we can test for the absence of a signal. This is represented by the assertion $A_b = \{\theta : \theta = b\}$. Figure 5.3 illustrates bel_y and pl_y for this assertion when $b = 15$.

A plausibility interval can also be created for the unknown θ. For $\gamma \in (0, 1)$,

$$\{\theta_0 : \text{pl}_y(\{\theta_0\}) \geq 1 - \gamma\} = [\max\{b, G_y^{-1}(\tfrac{1-\gamma}{2})\}, \max\{b, G_{y+1}^{-1}(\tfrac{1+\gamma}{2})\}].$$

The interval behaves similarly to the Gaussian interval: when $F_b(y) \leq \frac{1-\gamma}{2}$, then $\text{bel}_y(\{b\}) \geq \gamma$, but for $F_b(y) > \frac{1-\gamma}{2}$, any θ_0 on the interval interior has $\text{bel}_y(\{\theta_0\}) < \gamma$ and $\text{bel}_y(\{\theta_0\}^c) < \gamma$. Fig. 5.4 illustrates the level 0.9 plausibility interval for $b = 15$.

The level γ plausibility interval coverage probability is at least γ in repeated experiments. The following methods were also designed to achieve proper coverage probability. Numerical results illustrate the relative performance of the new Poisson plausibility interval compared to the existing methods.

Feldman and Cousins [85] constructed confidence bounds with proper coverage by filling acceptance intervals with points ordered according to a likelihood ratio. Giunti [115] argued that it is undesirable for the upper confidence bound to decrease in b when small values of Y are observed and proposed a modification to the ranking method that lessens the rate of decrease.

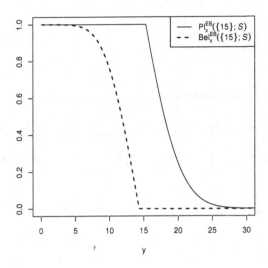

Figure 5.3 *Taken from [82]. Belief and plausibility functions* bel_y *and* pl_y *for the assertion* $\mathcal{A}_b = \{\theta : \theta = b\}$ *with* $b = 15$ *in the Poisson example of Section 5.5.2.*

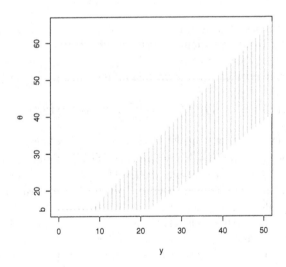

Figure 5.4 *Taken from [82]. Level* $\gamma = 0.9$ *plausibility interval for* θ *with* $b = 15$ *in the Poisson example of Section 5.5.2.*

Roe and Woodroofe [208] noted that observing $Y = 0$ is equivalent to observing $S = 0$ and $B = 0$. When the number of signal events is known, the interval bounds for λ should not depend on b. This issue is addressed in [208] by forming an interval conditioned on the fact that $B \leq y$ when $Y = y$ is observed. This method may undercover over all repeated experiments. Mandelkern and Schultz [172] provided an "ad hoc" remedy [174] by shifting the upper bound of each acceptance interval until proper unconditional coverage was achieved.

The conditional probability used to form intervals in [208] has the same form as the posterior density for λ when given a uniform prior over $[0, \infty)$. Roe and Woodroofe [209] developed this into a procedure for constructing a Bayesian credible interval. While this method has appropriate conditional coverage probability, they employed an "ad hoc" adjustment of the bounds to obtain appropriate unconditional coverage.

Confidence intervals derived from maximum likelihood estimators [173] differ from other methods in that the interval bounds remain constant for all observations outside of the constrained parameter space. Constructing the interval from the sampling distribution of the estimator ensures proper coverage.

For $\gamma = 0.9$, $b = 3$, and λ ranging from 0 to 4, Figure 5.5 shows the plausibility interval coverage probability compared to the existing methods. The Feldman and Cousins [85] and Roe and Woodroofe [209] methods had coverage probability at least as large the plausibility interval for most values of λ. The "ad hoc" adjustment of [172] to the conditional intervals in [208] tended to have coverage closer to 0.9 than the plausibility interval. However, the construction of the plausibility interval guarantees proper coverage so that ad hoc adjustments are not necessary. Furthermore, to our knowledge, there is no analytical expression for the [208] interval nor an expression that includes the [172] adjustment. For this Poisson example, the plausibility interval expression requires less computation to produce numerical values. The intervals of [115] and [173] provided coverage closer to 0.9 than the plausibility interval for most values of λ, with the former providing the best coverage of all.

Table 5.5.2 lists the level 0.9 interval bounds obtained from the elastic belief-based plausibility interval and the other methods for several values of y when $b = 3$. The interval widths for the different methods are plotted in Fig. 5.6. For $y < b$, the plausibility interval is narrower than those produced by most of the other methods. When $y \geq b$, the plausibility interval becomes wider than the others. This greater width causes the peaks in coverage probability seen in Figure 5.5. For example, λ values in $[1.70, 1.74]$ are covered by the plausibility interval when $y \in [1, 9]$. Hence, the coverage probability is near 0.97. Most of the other methods cover λ values in this range when $y \in [0, 8]$, which gives coverage probabilities closer to 0.95. Figure 5.7 shows the maximum and minimum coverage probabilities of the level γ plausibility interval when $\gamma \in [0.5, 1]$ and $\lambda \in [0, 100]$. Within this range of λ values the minimum coverage probability is close to γ. As the λ range is narrowed, the minimum coverage probability becomes larger for many values of γ due to discreteness. It may be possible to obtain a specific minimum coverage probability for a given λ range by choosing a smaller γ value.

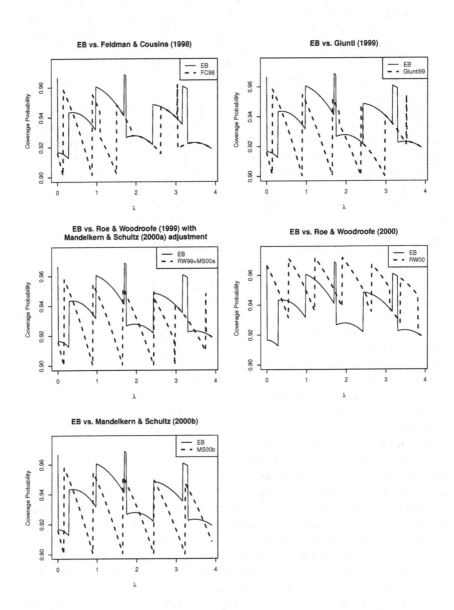

Figure 5.5 *Taken from [82]. Coverage probability of plausibility interval, where EB stands for elastic belief, for Poisson signal rate, λ, compared to the intervals of [85] (top left), [115] (top right), [208] with [172] adjustment (middle left), [209] (middle right), and [173] (bottom left), when $\gamma = 0.9$ and $b = 3$.*

Table 5.1 Taken from [82]. Level 0.9 interval bounds for λ when $b = 3$. Methods: EB — the elastic belief method, Feldman and Cousins [85] (FC98), Giunti [115] (Giunti99), Roe and Woodroofe [208] with Mandelkern and Schultz [172] adjustment (RW99+MS00a), Roe and Woodroofe [209] (RW00), and Mandelkern and Schultz [173] (MS00b).

y	EB Plaus. Int.		FC98 [85]		Giunti99 [115]		RW99+MS00a [172]		RW00 [209]		MS00b [173]	
	Lower	Upper	Lower	Upper	Lower	Upper	Lower	Upper	Lower	Upper	Lower	Upper
0	0	0	0	1.08	0	1.82	0	2.44	0	2.53	0	4.69
1	0	1.74	0	1.88	0	2.42	0	2.95	0	3.09	0	4.69
2	0	3.30	0	3.04	0	3.52	0	3.75	0	3.82	0	4.69
3	0	4.75	0	4.42	0	4.76	0	4.80	0	4.71	0	4.69
4	0	6.15	0	5.60	0	5.69	0	6.01	0	5.74	0	5.60
5	0	7.51	0	6.99	0	7.10	0	7.28	0	6.85	0	7.04
6	0	8.84	0.15	8.47	0.15	8.54	0.16	8.42	0	8.07	0.16	8.64
7	0.29	10.15	0.89	9.53	0.90	9.56	0.9	9.58	0.55	9.29	0.90	9.54
8	0.98	11.43	1.51	10.99	1.66	11.03	1.66	11.02	1.21	10.62	1.66	11.08
9	1.70	12.71	1.88	12.30	2.38	12.30	2.44	12.23	1.90	11.91	2.44	12.30
10	2.43	13.96	2.63	13.50	2.98	13.53	2.98	13.51	2.64	13.24	3.23	13.55
11	3.17	15.21	3.04	14.81	3.52	14.81	3.75	14.77	3.37	14.47	4.03	14.81
12	3.92	16.44	4.01	16.00	4.36	16.03	4.52	16.01	4.14	15.69	4.69	16.05

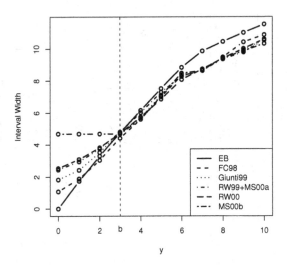

Figure 5.6 *Taken from [82]. Width of level 0.9 plausibility interval (EB — elastic belief)*
for Poisson signal rate, λ, compared to the intervals of Feldman and Cousins [85] (FC98),
Giunti [115] (Giunti99), Roe and Woodroofe [208] with Mandelkern and Schultz [172] ad-
justment (RW99+MS00a), Roe and Woodroofe [209] (RW00), and Mandelkern and Schultz
[173] (MS00b) for y = 0,..., 10 and b = 3.

5.6 Concluding remarks

The theory of IMs allows direct probabilistic inference from data to parameters without introducing priors or relying on asymptotic arguments. The elastic belief method
presented here extends the IM theory to situations where conflict cases can arise
from parameter constraints. As an alternative to the conditioning rule, it achieves
higher efficiency by using conflict cases as evidence for specific parameter values.
This is a reasonable choice when one holds the constraint and model assumptions to
be valid and hence cannot attribute conflict to uncertainty about these assumptions.
The probability represented by bel_x, the belief function obtained from the elastic belief method, is calibrated to a frequency interpretation for any assertion. As functions
of an assertion, likelihood and p-value functions [105, 117] are also available as inferential tools in the constrained Gaussian and Poisson examples, but bel_x has the
advantage of a predictive probability interpretation.

From bel_x it is easy to construct plausibility intervals containing hypothetical
parameter values that are supported by, or at least consistent with, the evidence presented in the data. The two-sided predictive random sets considered here resulted in
two-sided plausibility intervals (Figures 5.2 and 5.4). Although efficient and mathematically convenient, the symmetrical predictive random set (5.16) used in the Poisson example is sometimes larger than necessary. This caused the elastic belief-based

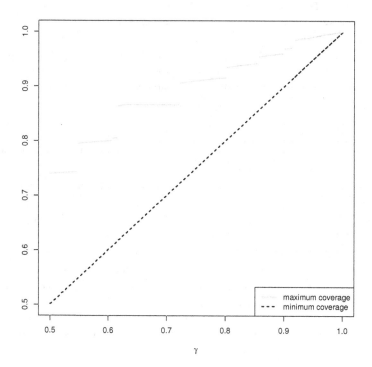

Figure 5.7 *Taken from [82]. Maximum and minimum coverage probabilities for level γ plausibility interval over $\lambda \in [0, 100]$ with $b = 3$.*

plausibility interval to be slightly wider for certain y values than intervals created with other methods. A more efficient IM and narrower plausibility interval may be obtained by considering an assertion-specific predictive random set for each $\{\theta_0\}$ assertion. The authors are currently investigating this approach. One-sided plausibility intervals may be obtained using one-sided predictive random sets. In the Poisson example, a one-sided plausibility interval is expected to have better performance than other methods because the skewness of the Poisson distribution will no longer create the difficulties that arise when using interval length as a criterion.

In the presentation of the elastic belief method, it was assumed that there existed a minimum intersection of the elastic predictive random set and the a-constraint set, $\mathbb{U}_{\mathcal{C},x}$. The elastic predictive random set may be designed so that it is always a closed set (except, possibly, the a-space itself). However, in some situations the constraint set may be problematic. For example, the Gaussian mean could be strictly positive: $\mu \in \mathcal{C} = (0, \infty)$. In this case one could build an IM with $\mathcal{C} = [0, \infty)$ instead. Any mass placed on $\mu = 0$ could be logically interpreted as evidence for 0^+, a point infinitesimally larger than zero.

Finally, the elastic belief method can be used for more general, data-dependent conflict cases. The elastic belief approaches demonstrated here can be extended to situations with nuisance parameters. Before applying the elastic belief method, however, a problem may be simplified by handling nuisance parameters with the marginalization methods. When there are multiple observations, the conditioning methods can reduce the data dimensionality in a manner similar to sufficient statistics. These are the subjects of the next two chapters.

5.7 Exercises

Exercise 5.1. Draw a picture to justify (5.2). Once the intuition is clear, give a proof.

Exercise 5.2. Prove Proposition 5.1.

Exercise 5.3. Show that the condition (5.6) is equivalent to that in (5.8) that defines the natural measure.

Exercise 5.4.　(a) Recall the definition of p-value in Exercise 1.9. Argue that the p-value does not recognize any parameter constraints.

　(b) Is the p-value's independence of any non-trivial parameter constraints problematic? In other words, if p-value is to be interpreted as a measure of the "plausibility" of null hypothesis given data, should it be affected by the particular alternative hypothesis?

(The connection between p-values and IMs is discussed in [181], with the conclusion being that the presence or absence of non-trivial parameter constraints is crucial to the interpretation of p-values.)

Exercise 5.5. For data with model $X \sim N(\theta, 1)$, consider a Bayesian approach where θ is assigned a flat prior. Suppose that θ is known to satisfy $\theta \in [a, b]$ for fixed $a < b$.

　(a) Find the posterior distribution for θ for the given prior and constraints.

　(b) A reasonable validity condition for the posterior is that its distribution function, evaluated at the true $\theta^\star$, be uniformly distributed, as a function of $X \sim N(\theta^\star, 1)$, for any $\theta^\star$. Is the constrained posterior in Part (a) above valid in this sense? (*Hint:* Consider the case $\theta^\star = a$; see, also, [81].)

Exercise 5.6. A data-generating model for a Poisson random variable can be built from the waiting times of a Poisson process. In [67], this model was used to build belief functions for inference about the parameter in the Poisson model, θ. Compared to the association in Section 5.5.2, the Poisson process model poses additional challenges to building an efficient IM. These challenges are present even in situations without parameter constraints.

　(a) Suppose there is an infinite sequence, $T_1, T_2, \ldots$, of independent, exponentially distributed random variables with unit rate. Let $T_0 = 0$ and let

$$S_i = \sum_{j=0}^{i} T_j.$$

Define $Y = \max\{i : S_i \le \theta\}$. Prove that $Y \sim \text{Pois}(\theta)$.

(b) For fixed θ, a realization of Y can be simulated by generating successive T_i from the exponential distribution until

$$S_{i-1} \leq \theta < S_{i-1} + T_i \qquad (5.17)$$

and then taking $y = i - 1$ as the realization. Show that when $Y = y$ is observed, the values of θ, as a set-valued function of S_i is the interval:

$$S_y \leq \theta < S_y + T_{y+1}.$$

That is, this relationship can be used to define an association with the infinite-dimensional auxiliary variable, $T = (T_0, T_1, T_2, \ldots)$.

(c) A valid IM can be created using this a-model by finding a PRS for some function of T. The infinite dimensionality makes it difficult to find an efficient PRS for T directly. However, a conditional IM (Chapter 6) can be created using the finite-dimensional auxiliary variable $(T_1, T_2, \ldots, T_{y+1})$. Show that the dimensionality can be reduced further to the two-dimensional auxiliary variable: (S_y, T_{y+1}).

(d) In order to predict (S_y, T_{y+1}), let:

$$U_1 = G_y(S_y) \quad \text{and} \quad U_2 = 1 - \exp\{-T_{y+1}\},$$

where $U = (U_1, U_2)$ has a uniform mass distribution over $[0, 1]^2$. An efficient predictive random set for U is:

$$S(u) = \{u' : \|u' - h\|_\infty \leq \|u - h\|_\infty\},$$

where $\|t\|_\infty = \max\{|t_1|, |t_2|\}$ and $h = (0.5, 0.5)$. Applying this predictive random set to the association (5.17) gives focal elements of the form:

$$M_y^{(2)}(u, S) = \left[G_y^{-1}(\tfrac{1}{2} - \|u - h\|_\infty), \right.$$
$$\left. G_y^{-1}(\tfrac{1}{2} + \|u - h\|_\infty) + e^{\frac{1}{2} + \|u - h\|_\infty} \right]. \qquad (5.18)$$

Using the association in Section 5.5.2 and predictive random set (5.16) gives the focal elements:

$$M_y^{(1)}(u, S) = \left[G_y^{-1}(\tfrac{1}{2} - |u - \tfrac{1}{2}|), G_{y+1}^{-1}(\tfrac{1}{2} + |u - \tfrac{1}{2}|) \right]. \qquad (5.19)$$

Develop computational method to compute the plausibility interval based upon the Poisson process IM (5.18).

Chapter 6

Conditional Inferential Models

Portions of the material in this chapter are from R. Martin and C. Liu, "Conditional inferential models: combining information for prior-free probabilistic inference," *Journal of the Royal Statistical Society Series B* **77**, 195–217, 2015, reprinted with permission by John Wiley & Sons, Inc.

6.1 Introduction

The basic inferential model (IM) approach presented in Chapter 4 provides a general framework for valid prior-free probabilistic inference. The key idea is, first, to associate observable data X, unknown parameter θ, and an unobservable auxiliary variable U and, second, to predict the unobserved value $u^\star$ of U using a valid predictive random set. The focus in Chapter 4 was on the case where θ and $u^\star$ are of the same dimension. But there are many problems, e.g., iid data from scalar parameter models, where the dimension of the auxiliary variable is greater than that of the parameter. In such cases, efficiency can be gained—recall the efficiency principle in Chapter 3—by first reducing the dimension of the auxiliary variable to be predicted, though it is not obvious how this should be done in general. Here we focus our attention on an auxiliary variable dimension reduction step based on conditioning, first presented in [182]. The key observation is that, typically, certain functions of the auxiliary variables are fully observed. By conditioning on those observed characteristics of the auxiliary variable, we can effectively reduce the dimension of the unobserved characteristics to be predicted. A motivating example, demonstrating the efficiency gain from dimension reduction, along with the detailed developments are presented in Section 6.2. The proposed dimension-reduction approach, based on conditioning, can be viewed as a general tool for combining information about θ across samples, a counterpart to the Bayes theorem and sufficiency. With the lower-dimensional auxiliary variable, we proceed to construct what is called a *conditional IM*. We prove a validity theorem that establishes a desirable calibration property of the conditional IM, and facilitates a common interpretation across users and experiments.

Finding the dimension-reduced representation is sometimes a familiar task. For example, when the minimal sufficient statistic has dimension matching that of the parameter, the conditional IM is exactly that obtained by working directly with said statistic. In other cases, finding the lower-dimensional representation is not so simple, analogous to finding ancillary statistics in the classical context. For this, in Section 6.3, we propose a new differential equation-driven technique for identifying observed characteristics of the auxiliary variable. Two classical conditional inference problems are worked out in Section 6.4, one showing how the proposed differential equation technique leads to an additional dimension reduction beyond what ordinary sufficiency provides. So, besides the development of conditional IMs, the proposed framework also casts new light on the familiar notion of sufficiency, as well as Fisher's attractive but elusive ideas on ancillary statistics and conditional inference.

In some cases, however, it may not be possible to produce a valid conditional IM with these somewhat standard techniques. For this, we propose an extension of the conditional IM framework, in Section 6.5, which allows the lower-dimensional auxiliary variable representation to depend on θ in a certain sense. We refer to these as *local conditional IMs*, and we describe their construction and prove a validity theorem. An important example of such a problem is the bivariate normal model with known means and variances but unknown correlation. For this example, we construct a local conditional IM based on a modification of our differential equations technique, and provide the results of a simulation study that shows that our conditional plausibility intervals outperform the classical r^*-driven asymptotically approximate confidence intervals [9, 100] in both small and large samples.

6.2 Conditional IMs

6.2.1 Motivation

In the case of a scalar auxiliary variable, construction of efficient predictive random sets is relatively easy. However, rarely does the model directly admit a scalar auxiliary variable representation. To see this, suppose $X_1, \ldots, X_n$ are independent $N(\theta, 1)$ with unknown mean θ. In vector notation, an association is $X = \theta 1_n + U$, where 1_n is an n-vector of unity, and $U \sim N_n(0, I)$. At first look, it seems that one must predict an n-dimensional auxiliary variable U. But efficient prediction of U would be challenging, even for moderate n, so reducing its dimension—ideally to one—would be a desirable first step. After reducing the dimension to one, choosing efficient predictive random sets is as easy as in the scalar auxiliary variable case considered in [180].

The basic point is that one pays a price, in terms of efficiency, for predicting higher-dimensional auxiliary variables. To see this better, we shall take a closer look at the normal mean problem above, with $n = 2$. That is, we have a baseline association

$$X_1 = \theta + U_1 \quad \text{and} \quad X_2 = \theta + U_2,$$

where U_1, U_2 are independent $N(0, 1)$. To make things simple, consider the following change of variables: $Y_1 = X_1 + X_2$ and $Y_2 = X_1 - X_2$. In the new variables, we have

$$Y_1 = 2\theta + V_1 \quad \text{and} \quad Y_2 = V_2,$$

where V_1, V_2 are iid $N(0, 2)$. This completes the A-step. Following the basic procedure described in Chapter 4, for the P-step, we should predict the pair (V_1, V_2) with a predictive random set S. A simple L_∞ generalization of the default predictive random set to the case of a two-dimensional auxiliary variable is a random square:

$$S = \{(v_1, v_2) : \max(|v_1|, |v_2|) \leq \max(|V_1|, |V_2|)\}, \quad V_1, V_2 \overset{iid}{\sim} N(0, 2).$$

For a singleton assertion $\{\theta\}$, the C-step gives plausibility function

$$\mathsf{pl}_y(\theta) = \frac{1 - G(2^{-1/2} \max\{|y_1 - 2\theta|, |y_2|\})^2}{1 - G(2^{-1/2}|y_2|)^2}, \tag{6.1}$$

where $G(z) = 1 - 2(1 - \Phi(z))$ is the distribution function for the modulus of a $N(0, 1)$ random variable; see Exercise 6.1. The unusual form here is due to the conditioning to remove conflict cases where $\Theta_y(S) = \varnothing$; see Chapter 5.

As an alternative approach, note that the value of V_2 is known once Y_2 is observed. So rather than trying to predict this component, as in the approach just described, we might condition on this observed value, to sharpen our uncertainty for predicting V_1. Since V_1 and V_2 are actually independent, it suffices to work with the marginal distribution, $V_1 \sim N(0, 2)$. For the A-step, we get $Y_1 = 2\theta + V_1$ and, for the P-step, we use a default predictive random set $S = \{v_1 : |v_1| \leq V_1\}$, where $V_1 \sim N(0, 2)$. For the same singleton assertion, the C-step this time gives plausibility function

$$\mathsf{pl}_y(\theta) = 1 - |2\Phi(2^{-1/2}(y_1 - 2\theta)) - 1|.$$

The claim is that inference based on the latter IM formulation is more efficient than that based on the former. To check this, we consider the sampling distribution of $\mathsf{pl}_Y(0)$ in the case where $Y = (Y_1, Y_2)$ is an independent $N(0, 2)$ random vector. Figure 6.1 shows a quantile plot of the two simulated samples. By the validity theorem, the plausibilities are both stochastically no smaller than $\mathsf{Unif}(0, 1)$. However, we see that plausibilities for the reduced, one-dimensional predictive random set are exactly $\mathsf{Unif}(0, 1)$ distributed, while those based on the two-dimensional predictive random set tend to be considerably larger. The larger plausibility means less efficiency, e.g., wider plausibility intervals, so the IM based on the reduced one-dimensional predictive random set is preferred. This difference in efficiency is explained by the fact that the two-dimensional predictive random set for (V_1, V_2) corresponds to a larger-than-necessary predictive random set for V_1; the conflict cases in the two-dimensional case have little to no effect on efficiency.

In the remainder of this section, we will give a general prescription for increasing efficiency by reducing the dimension. The key is that, in general, some functions of the original auxiliary variable are fully observed, like $V_2 = U_1 - U_2$ in this simple example. Then the strategy is to condition on what is fully observed to sharpen prediction of what is not observed. Since this "conditioning to sharpen inference" strategy is commonly used in statistics, similar considerations are natural in the IM framework. In what follows, we explain IM conditioning in more detail, and give a formal justification (Theorem 6.1).

6.2.2 Dimension reduction via conditioning

Like in Chapter 4, consider a "baseline association" of the form

$$X = a(\theta, U), \quad U \sim \mathsf{P}_U. \tag{6.2}$$

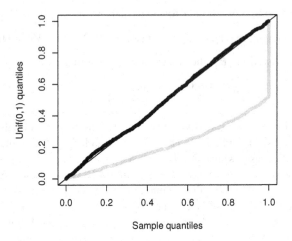

Figure 6.1 *Taken from [182]. Quantile plot of the two plausibility functions* $\mathrm{pl}_Y(0)$ *defined in Section 6.2.1. Gray points correspond to the two-dimensional predictive random set; black points correspond to the one-dimensional predictive random set.*

Here we propose a conditioning strategy, whereby a simultaneous information aggregation and dimension reduction is achieved, that results in an overall gain in efficiency. The intuition is that some functions of the unobserved $u^\star$ are actually observed, so these characteristics do not need to be predicted. Focusing only on the unobserved characteristics of $u^\star$ leads directly to a dimension reduction. However, knowledge about the observed characteristics helps to better predict those unobserved characteristics, so information is accumulated and prediction is sharpened. The general strategy is as follows:

- Identify an observed characteristic, $\eta(U)$, of the auxiliary variable U whose distribution is free (or at least mostly free) of θ;
- Define a conditional association that relates an unobserved characteristic, $\tau(U)$, of the auxiliary variable U to θ and some function $T(X)$ of X.

The second step is familiar, as it often relates to working with, say, a minimal sufficient statistic. The first step, however, is less familiar and generally more difficult; see Section 6.3.

To make this formal, suppose that $x \mapsto (T(x), H(x))$ and $u \mapsto (\psi_T(u), \psi_H(u))$ are one-to-one functions. Suppose that the relationship $x = a(u, \theta)$ in the baseline association (6.2) can be decomposed as

$$H(x) = \eta(u), \tag{6.3a}$$
$$T(x) = a_T(\tau(u), \theta). \tag{6.3b}$$

This decomposition immediately suggests an alternative association. Let $(V_T, V_H) \in \mathbb{V}_T \times \mathbb{V}_H$ be the image of U under (τ, η), and let $\mathsf{P}_{V_T \mid h}$ be the conditional distribution of V_T, given $V_H = h$, where $h \in H(\mathbb{X})$. Since $H(x)$ provides no information about θ, we can take a new association

$$T(X) = a_T(V_T, \theta), \quad V_T \sim P_{V_T|H(x)}. \tag{6.4}$$

We shall refer to this as a *conditional association*. This alternative association can be understood via a certain hierarchical representation of the sampling model; see Remark 6.2. The important point is that τ can often be chosen so that V_T is of lower dimension than U. In fact, V_T will often have dimension the same as that of θ. In addition to providing a summary of the data, as in the classical context, this auxiliary variable dimension reduction has a unique advantage in the IM context: efficient predictive random sets for the lower-dimensional V_T are easier to construct. Furthermore, the conditioning aspect sharpens our predictive ability, improving efficiency even more. We saw these gains in efficiency in the simple example in Section 6.2.1. Some further remarks on this conditional association, and its connections to Fisher's sufficiency and the Bayes theorem, are collected in Section 6.2.3.

Once a decomposition (6.3) is available, construction of the corresponding IM follows exactly as in Chapter 4. To simplify the presentation later on, here we restate the three-step construction of a *conditional IM*.

A-step. Associate $T(x)$ and θ with the new auxiliary variable $v_T = \tau(u)$ to get the collection of sets $\Theta_{T(x)}(v_T) = \{\theta : T(x) = a_T(v_T, \theta)\}$, $v_T \in \mathbb{V}_T$, based on (6.4).

P-step. Fix $h = H(x)$. Predict the unobserved value $v_T^\star$ of V_T with a *conditionally admissible* predictive random set $\mathcal{S} \sim P_{\mathcal{S}|h}$ (see Section 6.2.4).

C-step. Combine results of the A- and P-steps to get

$$\Theta_{T(x)}(\mathcal{S}) = \bigcup_{v_T \in \mathcal{S}} \Theta_{T(x)}(v_T) \subseteq \Theta. \tag{6.5}$$

Then the corresponding conditional belief and plausibility functions are given by

$$\begin{aligned} \mathsf{cbel}_{T(x)|h}(A; \mathcal{S}) &= P_{\mathcal{S}|h}\{\Theta_{T(x)}(\mathcal{S}) \subseteq A \mid \Theta_{T(x)}(\mathcal{S}) \neq \varnothing\} \\ \mathsf{cpl}_{T(x)|h}(A; \mathcal{S}) &= 1 - \mathsf{cbel}_{T(x)|h}(A^c; \mathcal{S}). \end{aligned} \tag{6.6}$$

These functions can be used for inference on θ just like in Section 4.4.4.

A quick remark on notation: in subsequent chapters, we will drop the "c" in cbel but here we want to distinguish the IM output after conditioning from the basic IM output without conditioning. Similarly, in Chapter 7 we introduce a marginal belief function mbel, and that "m" will also be dropped in the later chapters.

When a decomposition (6.3) is available, the conditional association (6.4) and the corresponding conditional IM analysis is intuitively quite reasonable. One could ask, however, if there is any drawback to using IMs built from (6.4) instead of (6.2). The following theorem shows that there is no loss in shifting focus from the baseline association to the dimension-reduced conditional association.

Theorem 6.1. *Suppose the baseline association* (6.2) *admits a decomposition of the form* (6.3). *Let $\mathcal{S}$ be a predictive random set for U in the baseline association with the property that $P_{\mathcal{S}}\{\Theta_x(\mathcal{S}) \neq \varnothing\} > 0$ for all x. Then there exists a predictive random set $\mathcal{S}_{H(x)}$ for $V_T = \tau(U)$, that depends on $H(x)$, in the conditional association such that $\mathsf{bel}_x(A; \mathcal{S}) = \mathsf{cbel}_{T(x)}(A; \mathcal{S}_{H(x)})$ for all x and all assertions $A \subseteq \Theta$.*

Proof. For the given predictive random set $\mathcal{S}$ for U in the baseline association, the corresponding random set $\Theta_x(\mathcal{S})$ in the C-step can be written as

$$
\begin{aligned}
\Theta_x(\mathcal{S}) &= \bigcup_{u \in \mathcal{S}} \{\theta : T(x) = a_T(\tau(u), \theta), H(x) = \eta(u)\} \\
&= \bigcup_{u \in \mathcal{S}} \left[\{\theta : T(x) = a_T(\tau(u), \theta)\} \cap \{\theta : H(x) = \eta(u)\} \right] \\
&= \bigcup_{u \in \mathcal{R}_{H(x)}} \{\theta : T(x) = a_T(\tau(u), \theta)\} \\
&= \Theta_{T(x)}(\tau(\mathcal{R}_{H(x)})),
\end{aligned}
$$

where $\mathcal{R}_{H(x)} := \mathcal{S} \cap \{u : \eta(u) = H(x)\}$. That is, given $\mathcal{S}$, the random sets in the baseline C-step match the random sets in the conditional C-step, but with a modified predictive random set $\mathcal{S}_{H(x)} := \tau(\mathcal{R}_{H(x)})$. Since $\mathsf{P}_{\mathcal{S}}\{\Theta_x(\mathcal{S}) \neq \varnothing\} > 0$ for all x, the two belief functions

$$
\begin{aligned}
\mathrm{bel}_x(A; \mathcal{S}) &= \mathsf{P}_{\mathcal{S}}\{\Theta_x(\mathcal{S}) \subseteq A \mid \Theta_x(\mathcal{S}) \neq \varnothing\}, \\
\mathrm{cbel}_{T(x)}(A; \mathcal{S}_{H(x)}) &= \mathsf{P}_{\mathcal{S}}\{\Theta_{T(x)}(\mathcal{S}_{H(x)}) \subseteq A \mid \Theta_{T(x)}(\mathcal{S}_{H(x)}) \neq \varnothing\},
\end{aligned}
$$

are well-defined conditional probabilities, i.e., no Borel paradox issues. Moreover, by the equality of $\Theta_x(\mathcal{S})$ and $\Theta_{T(x)}(\mathcal{S}_{H(x)})$, it is clear that

$$
\mathrm{bel}_x(A; \mathcal{S}) = \mathrm{cbel}_{T(x)}(A; \mathcal{S}_{H(x)}), \quad \text{for all } x \text{ and all } A \subseteq \Theta.
$$

So, for any predictive random set for U in the baseline association, there is a corresponding predictive random set for $V_T = \tau(U)$ in the conditional association, generally depending on the observed value $H(x)$ of $V_H = \eta(U)$, such that the baseline and conditional belief functions are the same. $\qquad\square$

The result says that, as a starting point, the conditional association is as good as the baseline association in the sense that any belief function that obtains from the latter can be matched by one that obtains from the former. Therefore, the best conditional IM can be no worse than the best baseline IM. However, as we saw in Section 6.2.1, by working with predictive random sets for the lower-dimensional auxiliary variable in the conditional association, we can expect greater efficiency, so the best conditional IM will generally be much better than the best baseline IM.

The theorem's assumption that $\mathcal{S}$ satisfies $\mathsf{P}_{\mathcal{S}}\{\Theta_x(\mathcal{S}) \neq \varnothing\} > 0$, which guarantees that the belief and plausibility functions are well-defined, is relatively mild. Indeed, we can check that the assumption holds for the square predictive random set used in the illustrative example of Section 6.2.1. Indeed, in that context, we had

$$
\Theta_y(\mathcal{S}) = \{\theta : \max(|y_1 - 2\theta|, |y_2|) \leq \max(|V_1|, |V_2|)\}, \quad V_1, V_2 \overset{\mathrm{iid}}{\sim} N(0, 2).
$$

One can check that $\Theta_y(\mathcal{S}) \neq \varnothing$ if and only if $\max(|V_1|, |V_2|) \geq |y_2|$. The latter event has positive $\mathsf{P}_{(V_1, V_2)}$-probability for all y_2, which verifies the theorem's assumption.

6.2.3   Remarks

Remark 6.1. More general decompositions in (6.3) are possible. That is, one may replace "$H(x) = \eta(u)$" in (6.3a) with "$c(x,u) = 0$" for a function c. However, this more general "non-separable" case may not fit into the context of the conditional validity theorem; see Theorem 6.2. See Sections 6.2.4 and 6.5.

Remark 6.2. The decomposition (6.3) boils down to a particular hierarchical representation of the sampling model for X. Indeed, for functions H and T as in (6.3), with $V_H = \eta(U)$, and $V_T = \tau(U)$, data $X \sim \mathsf{P}_{X|\theta}$ can be simulated as follows.

1. Get (V_T, V_H) by sampling $V_H \sim \mathsf{P}_{V_H}$ and $V_T \mid V_H \sim \mathsf{P}_{V_T|V_H}$;
2. Obtain X by solving the system $H(X) = V_H$ and $T(X) = a_T(V_T, \theta)$.

This hierarchical model representation also provides the following insight: when $X = x$ is observed, so too is the value of V_H, and this knowledge can be used to update the auxiliary variable distribution, analogous to the Bayes theorem.

Remark 6.3. There are clearly some close connections between the conditional IM and Fisher's notion of sufficiency. At a very high level, both theories provide a sort of dimension reduction. The key difference between the two is that sufficiency focuses on reducing the dimension of the observable data, while our approach focuses on reducing the dimension of the unobservable auxiliary variable. Although the conditional IM can, in some cases, correspond to a sufficient statistic-type of reduction, this is not necessary; see the remarks at the end of Section 6.4.1. Proper conditioning appears to be more important than the use of sufficient statistics (Exercise 6.2). In fact, in some cases, it is possible, within the IM framework, to reduce the dimension further than that which is provided by sufficiency; see Section 6.5.

Remark 6.4. As we mentioned previously, conditional IMs have some connections to the Bayesian approach, in particular, in how information is combined across samples. In fact, it can be shown that the Bayes solution is a special case of conditional IMs. To see this, consider a simple but generic example. The Bayes model, cast in terms of associations, is of the following form:

$$\theta = U_0, \quad U_0 \sim \mathsf{P}_{U_0} \quad \text{and} \quad X = a(U_0, U_1), \quad U_1 \sim \mathsf{P}_{U_1},$$

where P_U for $U = (U_0, U_1)$ is such that U_1 is conditionally independent given U_0. Here P_{U_0} is like the prior, and the distribution induced by $u_1 \mapsto a(\theta, u_1)$ given $U_0 = \theta$ determines the likelihood. It is clear that the function $a(U_0, U_1)$ is fully observed, so the conditional IM strategy would employ the conditional distribution of U_0 given the observed value x of $a(U_0, U_1)$. It is not hard to see that the belief function based on the "naive" predictive random set $\mathcal{S} = \{U_0\}$ is exactly the Bayesian posterior distribution function obtained via Bayes formula. So in any problem with a known prior distribution, the Bayes solution can be obtained as a special case of the conditional IM. No non-naive predictive random set is needed here because the naive IM itself is valid; this is consistent with the simple corresponding fact for posterior probabilities under a Bayes model with known prior.

Remark 6.5. As a follow-up to Remark 6.4, since a full prior is not required to construct a conditional IM, it is possible to develop an inferential framework based on

conditional IMs and "partial prior information." For example, valid prior information may be available for some but not all components of θ. Incorporating the prior
information where it is available while remaining prior-free where it is not can be
obtained by slight extension of the argument in the previous remark. This important
application of conditional IMs deserves further investigation. See, also, [262].

6.2.4 Validity of conditional IMs

Here we extend the validity results in Chapter 4 to the conditional IM context. The
main obstacle is that the distribution function P_S, determined by the conditional
distribution $P_{V_T|H(x)}$ in (6.3), depends on data through the value $H(x)$. This is handled
in Theorem 6.2 below by conditioning on the observed value of $H(X)$.

Fix $h \in H(\mathbb{X})$, and let $\mathbb{S}_h$ be a collection of closed $P_{V_T|h}$-measurable subsets of $\mathbb{V}_T$
that contains both $\varnothing$ and $\mathbb{V}_T$. As before, we also assume that $\mathbb{S}_h$ is nested in the sense
that either $S \subseteq S'$ or $S' \subseteq S$ for all $S, S' \in \mathbb{S}_h$. Then S is a *conditionally admissible*
predictive random set, given h, if its distribution $P_{S|h}$ satisfies

$$P_{S|h}\{S \subseteq K\} = \sup_{S \in \mathbb{S}_h : S \subseteq K} P_{V_T|h}\{S\}, \quad K \subseteq \mathbb{V}_T. \tag{6.7}$$

In this case, the distribution of S depends on the particular h. We now have the
following extension of the validity theorem to the case of conditional IMs.

Theorem 6.2. *For any h, suppose that S is conditionally admissible, given h, with
distribution $P_{S|h}$ as in (6.7). If $\Theta_{T(x)}(S) \neq \varnothing$ with $P_{S|h}$-probability 1 for all x such
that $H(x) = h$, then the conditional IM is conditionally valid, i.e., for any $A \subseteq \Theta$,*

$$\sup_{\theta \notin A} P_{X|\theta}\{\mathrm{cbel}_{T(X)|h}(A; S) \geq 1 - \alpha \mid H(X) = h\} \leq \alpha, \quad \forall \, \alpha \in (0, 1). \tag{6.8}$$

Proof. Take any $\theta \notin A$ as the true value of the parameter. Next, note that $T(X) =$
$a_T(V_T, \theta)$, with $V_T \sim P_{V_T|h}$, characterizes the conditional distribution of X, given
$H(X) = h$. Since $A \subset \{\theta\}^c$, monotonicity of the belief function gives

$$\mathrm{cbel}_{T(X)|h}(A; S) \leq \mathrm{cbel}_{T(X)|h}(\{\theta\}^c; S) = P_{S|h}\{\Theta_{T(X)}(S) \not\ni \theta\} = Q_{S|h}(V_T).$$

By the result in Exercise 6.3, the right-hand side above is stochastically no larger
than $\mathrm{Unif}(0, 1)$. This, in turn, implies the same of the left-hand side $\mathrm{cbel}_{T(X)|h}(A; S)$,
as a function of $X \sim P_{X|\theta}$, given $H(X) = h$. Therefore,

$$P_{X|\theta}\{\mathrm{cbel}_{T(X)|h}(A; S) \geq 1 - \alpha \mid H(X) = h\} \leq P\{\mathrm{Unif}(0, 1) \geq 1 - \alpha\} = \alpha.$$

Taking supremum over $\theta \notin A$ proves (6.8). $\square$

Now is a good time to recall Remark 6.1. More general decompositions of the
baseline association are allowed in the discussion in Section 6.2.2, but only for the
"separable" version (6.3a) is it possible to prove a conditional validity theorem. The
point is that a condition like $c(X, U) = 0$ does not identify a fixed subset of $\mathbb{X}$ on
which probability calculations can be restricted—the subspace would depend on U.

Since the calibration property in Theorem 6.2 holds for all assertions A, we may translate (6.8) to a statement in terms of the corresponding plausibility function:

$$\sup_{\theta \in A} \mathsf{P}_{X|\theta}\{\mathsf{cpl}_{T(X)|h}(A;\mathcal{S}) \leq \alpha \mid H(X) = h\} \leq \alpha, \quad \forall \, \alpha \in (0,1). \qquad (6.9)$$

So, in addition to providing an objective scale for interpreting the conditional belief and plausibility function values, (6.9) provides desirable properties of conditional IM-based frequentist procedures. For example, if $h = H(x)$ is observed, the conditional $100(1-\alpha)\%$ plausibility region for θ is $\{\theta : \mathsf{cpl}_{T(x)|h}(\theta;\mathcal{S}) > \alpha\}$. Then, by (6.9), the conditional coverage probability is $\mathsf{P}_{X|\theta}\{\mathsf{cpl}_{T(X)|h}(\theta;\mathcal{S}) > \alpha \mid H(X) = h\} \geq 1 - \alpha$. This is a more meaningful coverage probability since it is conditioned on a particular aspect of the observed data, namely, $H(x) = h$. In other words, the probability calculation focuses on a relevant subset $\{x : H(x) = h\}$ of the sample space. In some cases, though, conditional validity is the same as ordinary validity.

Corollary 6.1. *Suppose that the predictive random set $\mathcal{S}$ does not depend on the observed $H(x) = h$, so that $\mathsf{P}_{\mathcal{S}|h} \equiv \mathsf{P}_{\mathcal{S}}$ and $\mathsf{cbel}_{T(x)|h} \equiv \mathsf{cbel}_{T(x)}$. Then under the conditions of Theorem 6.2, the conditional IM is unconditionally valid, i.e., for any $A \subseteq \Theta$,*

$$\sup_{\theta \notin A} \mathsf{P}_{X|\theta}\{\mathsf{cbel}_{T(X)}(A;\mathcal{S}) \geq 1 - \alpha\} \leq \alpha, \quad \forall \, \alpha \in (0,1).$$

Proof. Since the distribution of $\mathcal{S}$ is free of h in this case, the belief function $\mathsf{cbel}_{T(X)|h} \equiv \mathsf{cbel}_{T(X)}$ is also free of h. Therefore, before taking supremum in the last line of the proof of Theorem 6.2, we can take expectation over h to remove the conditioning, so that the validity property holds unconditionally, as in Chapter 4. □

Two possible ways the condition of Corollary 6.1 may hold are as follows. First, in the P-step, the user may specify $\mathcal{S}$ directly without dependence on the observed $H(x) = h$; see Section 6.4.1. Second, it could happen that V_T and V_H are statistically independent, in which case the distribution $\mathsf{P}_{\mathcal{S}}$ for $\mathcal{S}$ is determined by the marginal distribution of V_T, which does not depend on h.

6.3   Finding conditional associations

6.3.1   Familiar things: Likelihood and symmetry

In many problems, finding a decomposition (6.3) and the corresponding conditional association is easy to do. In general, the definition of sufficiency implies that we can define a conditional association via, say, the marginal distribution of the minimal sufficient statistic; see Section 6.4.3. In standard problems, such as full-rank exponential families, minimal sufficient statistics are easily obtained, so this is probably the simplest approach. This, of course, includes both discrete and continuous problems. Similarly, if the problem has a group structure, invariance considerations can be used to find a decomposition; see Section 6.4.1. But one can consider other conditional associations if desirable. For example, when the minimal sufficient statistic has dimension larger than that of the parameter, as in curved exponential families, e.g., $N(\theta, \theta^2)$, then some special conditioning can potentially further reduce the dimension; see Section 6.4.2.

6.3.2 A new differential equations-based technique

Here we describe a novel technique for finding conditional associations, based on differential equations. The method can be used for going directly from the baseline association to something lower-dimensional. In fact, in those nice problems mentioned above, it is easy to check that this differential equation-based technique reproduces the solutions based on minimal sufficiency, group invariance, etc. However, in our experience, this new approach is especially powerful in cases where the familiar things fail to give a fully satisfactory reduction. In such cases, the differential equation-based technique can provide a further dimension reduction, beyond what sufficiency alone can give.

For concreteness, suppose $\Theta \subseteq \mathbb{R}$; the multi-parameter case can be handled similarly, as in Chapter 8.3. The intuition is that τ should map $\mathbb{U} \subseteq \mathbb{R}^n$ to Θ, so that $V_T = \tau(U)$ is one-dimensional, like θ. Moreover, η should map $\mathbb{U}$ into a $(n-1)$-dimensional manifold in $\mathbb{R}^n$, and be insensitive to changes in θ in the following sense. For baseline association $x = a(\theta, u)$, suppose that $u_{x,\theta}$ is the unique solution for u. Then for fixed x, we require that $\eta(u_{x,\theta})$ be constant in θ. In other words, we require that $\partial u_{x,\theta}/\partial\theta$ exists and

$$
\underset{n\times 1}{0} = \frac{\partial\eta(u_{x,\theta})}{\partial\theta} = \underset{n\times n,\, \text{rank } n-1}{\frac{\partial\eta(u)}{\partial u}\bigg|_{u=u_{x,\theta}}} \cdot \underset{n\times 1}{\frac{\partial u_{x,\theta}}{\partial\theta}}. \tag{6.10}
$$

It is clear from the construction that, if a solution η of this partial differential equation exists, then the value of $\eta(U)$ is fully observed, i.e., there is a corresponding function H, not depending on θ, such that $H(X) = \eta(U)$. So, with appropriate choice of τ, the solution η of (6.10) determines the decomposition (6.3). A different but related use of θ-derivatives of the association is presented in [97].

Formal theory on existence of solutions and on solving the differential equation system (6.10) is available. For example, the *method of characteristics* described in [200] is powerful tool for solving such systems. However, such formalities here will take us too far off track. Examples of this method in action are given in Section 6.4.2, 6.5.4, and Chapter 8. In most cases, this differential equations method is applied after an initial step based on sufficiency provides an unsatisfactory dimension reduction.

6.4 Three detailed examples

6.4.1 A Student-t location problem

Suppose $X_1, \ldots, X_n$ is an independent sample from a Student-t distribution $t_\nu(\theta)$, where the degrees of freedom ν is known but the location θ is unknown. This is a somewhat peculiar problem because there is no satisfactory reduction via sufficiency. For the IM approach, start with a baseline association $X = \theta 1_n + U$, with

$U = (U_1, \ldots, U_n)^\top$ and $U_i \sim t_v$, independent, for $i = 1, \ldots, n$. For this location parameter problem, invariance considerations suggest the following decomposition:

$$X - T(X)1_n = U - T(U)1_n \quad \text{and} \quad T(X) = \theta + T(U),$$

where $T(\cdot)$ is the maximum likelihood estimator. Let $V_T = T(U)$ and $V_H = H(U) = U - T(U)1_n$. If h is the observed $H(X)$, then it follows from the result of [8] that the conditional distribution of V_T, given $V_H = h$, has a density

$$f_{v,h}(v_T) = c(v,h) \prod_{i=1}^{n} \{v + (v_T + h_i)^2\}^{-(v+1)/2},$$

where $c(v,h)$ is a normalizing constant that depends only on v and h. If we write $F_{v,h}$ for the distribution function corresponding to the density $f_{v,h}$ above, then a conditional IM for θ can be built based on the following association:

$$T(X) = \theta + F_{v,h}^{-1}(W), \quad W \sim \mathsf{Unif}(0,1).$$

With this conditional association, we are ready for the P- and C-steps. For simplicity, in the P-step we elect to take the predictive random set $\mathcal{S}$ as in (4.8); this also has some theoretical justification since $f_{v,h}$ should be approximately symmetric about $v_T = 0$; see Chapter 4. For the C-step, the random set $\Theta_{T(x)}(\mathcal{S})$ is

$$\left[T(x) - F_{v,h}^{-1}\left(\tfrac{1}{2} + |W - \tfrac{1}{2}|\right), \, T(x) - F_{v,h}^{-1}\left(\tfrac{1}{2} - |W - \tfrac{1}{2}|\right)\right], \quad W \sim \mathsf{Unif}(0,1).$$

From this point, numerical methods can be used to compute the conditional belief and plausibility functions. For example, if $A = \{\theta\}$ is a singleton assertion, then

$$\mathsf{cpl}_{T(x)|h}(\theta; \mathcal{S}) = 1 - \left|1 - 2F_{v,h}(\theta - T(x))\right|,$$

and the corresponding $100(1 - \alpha)\%$ plausibility interval for θ is

$$\{\theta : \mathsf{cpl}_{T(x)|h}(\theta; \mathcal{S}) > \alpha\} = \left(T(x) + F_{v,h}^{-1}(\alpha/2), \, T(x) + F_{v,h}^{-1}(1 - \alpha/2)\right).$$

For illustration, we present the results of a simple simulation study. In particular, for several pairs (n, v), 5000 Monte Carlo samples of size n are obtained from a Student-t distribution with v degrees of freedom and center $\theta = 0$. For each sample, the 95% plausibility interval for θ based on the conditional IM above is obtained. For comparison, we also compute the 95% confidence interval based on the asymptotic normality of the maximum likelihood estimate, and a 95% flat-prior Bayesian credible interval (Exercise 6.4). The results of this simulation are summarized in Table 6.4.1. We find that the results here are almost indistinguishable, so favor must go to the plausibility intervals, since these have guaranteed coverage for all n, while the other two are only asymptotically correct.

We also did the conditional IM calculations with an alternative decomposition, which took $V_T = U_1$ and $V_H = (0, U_2 - U_1, \ldots, U_n - U_1)$. We were surprised to see that the results obtained with this "naive" decomposition were indistinguishable from those shown here based on the arguably more reasonable maximum likelihood-driven decomposition. This suggests that the particular choice of decomposition may not be so important; instead, it is the conditioning part that seems to matter most.

Table 6.1 *Taken from [182]. Coverage probabilities and expected lengths of the 95% intervals for θ in the Student-t example based on, respectively, the conditional IM (CIM), asymptotic normality of the maximum likelihood estimate (MLE), and flat-prior Bayes.*

Method	n	Coverage probability v				Expected length v			
		3	5	10	25	3	5	10	25
CIM	5	0.944	0.949	0.951	0.949	2.28	2.08	1.93	1.83
	10	0.949	0.951	0.952	0.953	1.56	1.45	1.35	1.29
	25	0.953	0.944	0.951	0.949	0.97	0.91	0.85	0.81
	50	0.953	0.951	0.953	0.947	0.68	0.64	0.60	0.58
MLE	5	0.931	0.939	0.940	0.946	2.10	1.99	1.88	1.80
	10	0.953	0.942	0.949	0.941	1.51	1.42	1.334	1.28
	25	0.938	0.948	0.947	0.950	0.96	0.90	0.85	0.81
	50	0.946	0.946	0.954	0.956	0.68	0.64	0.60	0.57
Bayes	5	0.949	0.955	0.946	0.948	2.28	2.08	1.93	1.82
	10	0.960	0.948	0.951	0.942	1.56	1.45	1.35	1.29
	25	0.943	0.949	0.948	0.950	0.97	0.91	0.85	0.81
	50	0.947	0.947	0.955	0.956	0.68	0.64	0.60	0.58

6.4.2 Fisher's problem of the Nile

Suppose two independent exponential samples, namely $X_1 = (X_{11}, \ldots, X_{1n})$ and $X_2 = (X_{21}, \ldots, X_{2n})$, are available, the first with mean θ^{-1} and the second with mean θ. The goal is to make inference on $\theta > 0$. The name comes from an application [95] to fertility of land in the Nile river valley. In this example, the maximum likelihood estimate is not sufficient, so conditioning on an ancillary statistic is recommended.

Sufficiency considerations suggest the following initial dimension reduction step:

$$S(X_1) = \theta^{-1} U_1 \quad \text{and} \quad S(X_2) = \theta U_2, \quad U_1, U_2 \sim \text{Gamma}(n, 1),$$

where $S(X_i) = \sum_{j=1}^{n} X_{ij}$. But efficiency can be gained by considering a further reduction to a scalar auxiliary variable. Here we employ the differential equation technique in Section 6.3.2. Start with $u_{x,\theta} = (\theta S(x_1), \theta^{-1} S(x_2))^{\top}$. Differentiating with respect to θ reveals that our (real valued) conditioning function η must satisfy

$$\frac{\partial \eta(u)}{\partial u}\bigg|_{u=u_{x,\theta}} \begin{pmatrix} S(x_1) \\ -\theta^{-2} S(x_2) \end{pmatrix} = 0.$$

If we take $\eta(u) = \{u_1 u_2\}^{1/2}$, then

$$\frac{\partial \eta(u)}{\partial u}\bigg|_{u=u_{x,\theta}} = \frac{1}{2\{S(x_1)S(x_2)\}^{1/2}} (\theta^{-1} S(x_2), \theta S(x_1))$$

and this satisfies the differential equation above. Therefore, for (6.3), we take

$$H(X) = V_H \quad \text{and} \quad T(X) = \theta V_T, \tag{6.11}$$

where $T(X) = \{S(X_1)/S(X_2)\}^{1/2}$, $H(X) = \{S(X_1)S(X_2)\}^{1/2}$, $V_T = \{U_1/U_2\}^{1/2}$, and $V_H = \{U_1U_2\}^{1/2}$. These quantities are familiar from the classical approach: $T(X)$ is the maximum likelihood estimate of θ, $H(X)$ is an ancillary statistic, and the pair $(T,H)(X)$ is a jointly minimal sufficient statistic [112].

By (6.11) and our general discussion in Section 6.2.2, we can focus on a conditional association based on $T(X) = \theta V_T$. The conditional distribution of V_T given $V_H = h$ is a generalized inverse Gaussian distribution [7] with density function

$$f_h(v_T) = \frac{1}{2v_T K_0(2h)} \exp\{-h(v_T^{-1} + v_T)\}, \tag{6.12}$$

where K_0 is the modified Bessel function of the second kind. As a final simplifying step, write the conditional association as

$$T(X) = \theta F_h^{-1}(W), \quad W \sim \mathsf{Unif}(0,1), \tag{6.13}$$

where F_h is the distribution function corresponding to the density f_h in (6.12). This completes the A-step. If we take $\mathcal{S}$ as in (4.8) for the P-step, then the C-step gives

$$\Theta_{T(x)}(\mathcal{S}) = \left[\frac{T(x)}{F_h^{-1}(\frac{1}{2} + |W - \frac{1}{2}|)}, \frac{T(x)}{F_h^{-1}(\frac{1}{2} - |W - \frac{1}{2}|)}\right], \quad W \sim \mathsf{Unif}(0,1).$$

From this, the conditional belief/plausibility functions are readily evaluated.

For illustration, we display plausibility functions $\mathsf{cpl}_t(\theta; \mathcal{S})$ for two conditional IMs. The first is based on that derived above; the second is based on a similar derivation, but we ignore V_H and simply work with the marginal distribution of V_T in (6.11). Figure 6.2 shows plausibility functions for $T(x) = 0.90$, with $n = 20$ and true $\theta = 1$, sampled from its conditional distribution given h, for two different values of h. In this case, if h is large (i.e., $h > n$), then the bona fide conditional IM has narrower level sets than the naive conditional IM. The opposite is true when h is small (i.e., $h < n$). This is due to the fact that the conditional Fisher information in T is an increasing function in h; see Example 1 in [112]. Therefore, T has more variability when h is small, and this adjustment should be reflected in the plausibility function. The bona fide conditional IM catches this phenomenon while the naive one does not.

6.4.3 A two-parameter gamma problem

Let $X = (X_1, \ldots, X_n)$ be an independent sample from $\mathsf{Gamma}(\theta_1, \theta_2)$, where $\theta_1 > 0$ and $\theta_2 > 0$ are the shape and scale parameters, respectively, both unknown. In this case, we may construct a conditional association based on the marginal distribution of the two-dimensional complete sufficient statistic, which we choose to represent as $T_1 = \sum_{i=1}^n X_i$ and $T_2 = n^{-1}\sum_{i=1}^n \log X_i - \log(T_1/n)$. Then we have a conditional association

$$T_1 = \theta_2 F_{n\theta_1}^{-1}(U_1) \quad \text{and} \quad T_2 = G_{\theta_1}^{-1}(U_2),$$

where U_1, U_2 are independent $\mathsf{Unif}(0,1)$, F_a is the distribution function of $\mathsf{Gamma}(a, 1)$, and G_b is some distribution function without a familiar form. For the

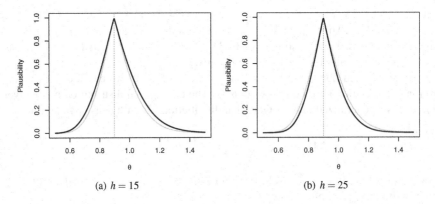

Figure 6.2 *Taken from [182]. Plausibility functions for the conditional IM (black) and the "naive" conditional IM (gray) in the Nile example, with $T = 0.90$, $n = 20$, and the true $\theta = 1$. Gray curves in the two plots are the same since the naive IM does not depend on h.*

P-step, consider an analogue of the default predictive random set (4.8), given by the random square:

$$S = \{(u_1, u_2) : \max(|u_1 - \tfrac{1}{2}|, |u_2 - \tfrac{1}{2}|) \leq \max(|U_1 - \tfrac{1}{2}|, |U_2 - \tfrac{1}{2}|)\},$$

with $U_1, U_2 \overset{iid}{\sim} \mathrm{Unif}(0,1)$. In this case, with observed $t = (t_1, t_2)$, the C-step gives

$$\Theta_t(S) = \{(\theta_1, \theta_2) : \max(|F_{n\theta_1}(t_1/\theta_2) - \tfrac{1}{2}|, |G_{\theta_1}(t_2) - \tfrac{1}{2}|) \leq \max(|U_1 - \tfrac{1}{2}|, |U_2 - \tfrac{1}{2}|)\}.$$

From here, we can write down the plausibility function for a singleton assertion:

$$\mathrm{cpl}_t(\{\theta_1, \theta_2\}; S) = 1 - \max\{|2F_{n\theta_1}(t_1/\theta_2) - 1|, |2G_{\theta_1}(t_2) - 1|\}^2.$$

Evaluating $G_{\theta_1}(\cdot)$ requires Monte Carlo but, since T_2 is θ_2-ancillary, the same Monte Carlo samples can be used for all candidate θ_2 values.

For illustration, we simulated a single sample of size $n = 25$ from a gamma distribution with shape $\theta_1 = 7$ and scale $\theta_2 = 3$. Figure 6.3 displays a sample of size 5000 from a Bayesian posterior distribution for (θ_1, θ_2) based on the Jeffreys prior. Also displayed are the 90% confidence ellipse based on the asymptotic normality of the maximum likelihood estimator, and the 90% conditional IM plausibility region

$$\{(\theta_1, \theta_2) : \mathrm{cpl}_t(\{\theta_1, \theta_2\}; S) > 0.1\}.$$

Besides having guaranteed coverage, the plausibility region captures the non-elliptical shape of the posterior distribution. For larger n, all three regions will have a similar shape.

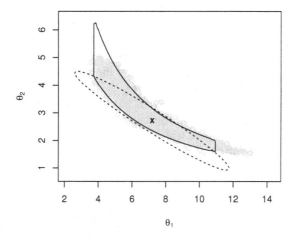

Figure 6.3 *Taken from [182]. A Bayesian posterior sample (gray) based on the Jeffreys prior, the 90% confidence ellipse based on asymptotic normality of the maximum likelihood estimator (dashed), and the 90% conditional IM plausibility region.*

6.5 Local conditional IMs

6.5.1 Motivation: Bivariate normal model

So far we have seen that the conditional IM approach is successful in problems where the baseline association admits a decomposition of the form (6.3). However, as alluded to above, there are interesting and important problems where apparently no such decomposition exists. Next is one such problem, which may be considered as a "benchmark example" for conditional inference; Example 5 in [112].

Suppose $(X_{11}, X_{21}), \ldots, (X_{1n}, X_{2n})$ is an independent sample from a standard bivariate normal distribution with zero means, unit variances, but unknown correlation coefficient $\theta \in (-1, 1)$. A natural first step toward inference on θ is to take advantage of the fact that $X_1 + X_2$ and $X_1 - X_2$ are independent. In particular, by defining

$$X_1 \leftarrow \frac{1}{2}\sum_{i=1}^{n}(X_{1i} + X_{2i})^2 \quad \text{and} \quad X_2 \leftarrow \frac{1}{2}\sum_{i=1}^{n}(X_{1i} - X_{2i})^2,$$

we may rewrite the baseline association as

$$X_1 = (1 + \theta)U_1 \quad \text{and} \quad X_2 = (1 - \theta)U_2, \quad U_1, U_2 \sim \mathsf{ChiSq}(n). \qquad (6.14)$$

Sufficiency justifies this first reduction. Equation (6.14) is equivalent to

$$\frac{X_1}{U_1} + \frac{X_2}{U_2} = 2 \quad \text{and} \quad \frac{X_1}{X_2} = \frac{1+\theta}{1-\theta}\frac{U_1}{U_2}. \qquad (6.15)$$

The first equation depends on data and auxiliary variable—free of θ—while the second depends also on θ. But note that the first expression in (6.15) is not of the form specified in (6.3a). Actually, this first expression is of the more general "non-separable" form $c(X,U) = 0$ described in Remark 6.1. So, although (6.15) provides a suitable decomposition of the baseline association, the requirements of Theorem 6.2 are not met, so the resulting conditional IM may not be valid.

To elaborate on this last point, observe that the distribution for θ obtained via the distribution of (U_1, U_2), given $X_1/U_1 + X_2/U_2 = 2$, is exactly a type of fiducial distribution. As we mentioned in Chapter 2, conditioning on the full data, (X_1, X_2) in this case, for fixed θ, makes the distribution of (U_1, U_2) degenerate. Therefore, the "continue to regard" operation, which treats (U_1, U_2) as independent chi-squares, given data, may be difficult to justify.

6.5.2 Relaxing the separability condition via localization

As described above, the separability in (6.3a) can be too strict, but extending the conditional validity theorem to allow non-separablility appears difficult. The idea here is to relax (6.3a) in a different direction. Specifically, we propose to allow the pair of function (H, η) in (6.3a) to depend, locally, on the parameter. This generalization allows us additional flexibility to reduce the auxiliary variable dimension.

Start by fixing an arbitrary $\theta_0 \in \Theta$. As in Section 6.2.2, consider a pair of functions (T, H_{θ_0}), depending on θ_0, such that $x \mapsto (T(x), H_{\theta_0}(x))$ is one-to-one. Now take the corresponding functions $u \mapsto (\tau(u), \eta_{\theta_0}(u))$, one-to-one, such that the baseline association, at $\theta = \theta_0$, can be decomposed as

$$H_{\theta_0}(X) = \eta_{\theta_0}(U) \quad \text{and} \quad T(X) = a_T(\tau(U), \theta_0). \tag{6.16}$$

That is, (6.16), together with $U \sim \mathsf{P}_U$, describes the sampling distribution $X \sim \mathsf{P}_{X|\theta_0}$. If $H_{\theta_0}(X) = h_0$ is observed, then we can compute the conditional distribution $\mathsf{P}_{V_T|h_0,\theta_0}$ of $V_T = \psi_T(U)$ given $\psi_{H,\theta_0}(U) = h_0$, which, in turn, we can use to construct predictive random sets.

From this point, we may proceed as usual. That is, for the A-step, we get sets $\Theta_{T(x)}(v_T) = \{\theta : T(x) = a_T(v_T, \theta)\}$ just as before. For the P-step, we pick a conditionally admissible predictive random set $\mathcal{S} \sim \mathsf{P}_{S|h_0,\theta_0}$. Finally, the C-step produces the conditional plausibility function

$$\mathsf{cpl}_{T(x)|h_0,\theta_0}(A; \mathcal{S}) = 1 - \mathsf{P}_{S|h_0,\theta_0}\{\Theta_{T(x)}(\mathcal{S}) \subseteq A^c\}, \quad A \subseteq \Theta.$$

We shall refer to the corresponding conditional IM as a *local* conditional IM at $\theta = \theta_0$. The adjective "local" is meant to indicate the dependence of the construction on the particular point θ_0. As we see below, the validity properties of this local conditional IM are, in a certain sense, also local.

6.5.3 Validity of local conditional IMs

The following theorem shows that for each θ_0 value, the local conditional IM at θ_0 is valid for some important assertions depending on the particular θ_0.

Theorem 6.3. *For any θ_0, take $h_0 \in H_{\theta_0}(\mathbb{X})$. Suppose that $S \sim P_{S|h_0,\theta_0}$ is conditionally admissible. If $\Theta_{T(x)}(S) \neq \varnothing$ with $P_{S|h_0,\theta_0}$-probability 1 for all x such that $H_{\theta_0}(x) = h_0$, then the local conditional IM at θ_0 is conditionally valid for $A = \{\theta_0\}$:*

$$P_{X|\theta_0}\{\mathsf{cpl}_{T(X)|h_0,\theta_0}(\theta_0;S) \leq \alpha \mid H_{\theta_0}(X) = h_0\} \leq \alpha, \quad \forall \, \alpha \in (0,1).$$

Proof. See Exercise 6.5.   □

The validity result here is not as strong as in Theorem 6.2, a consequence of the localization. It does, however, imply that the local conditional plausibility region,

$$\{\theta : \mathsf{cpl}_{T(x)|H_\theta(x)}(\theta;S) > \alpha\}, \tag{6.17}$$

has the nominal (conditional) $1 - \alpha$ coverage probability. This theoretical result is confirmed by the simulation experiment in Section 6.5.4 below. Observe that, in the definition of conditional plausibility region (6.17), the plausibility function depends on θ in two places—in the argument (the assertion) and in the local conditional IM itself. The latter structural dependence of the IM on the particular assertion is consistent with the optimality developments described in [180].

6.5.4   Bivariate normal model, revisited

Here we demonstrate that the localization technique can be successfully used to solve the bivariate normal problem described above. Start with the relation in (6.14). Fix θ_0. To construct the functions (H, η_{θ_0}), depending on θ_0, and the corresponding local conditional IM at θ_0, we modify the differential equation approach in Section 6.3.2.

In this case, if we let $u_{x,\theta} = (x_1/(1+\theta), x_2/(1-\theta))^\top$, then we have

$$\frac{\partial u_{x,\theta}}{\partial \theta} = \left(-\frac{x_1}{(1+\theta)^2}, \frac{x_2}{(1-\theta)^2}\right)^\top.$$

For a local conditional IM at θ_0, we propose to choose a real-valued $\eta_{\theta_0}(u)$ such that $\partial \eta_{\theta_0}(u_{x,\theta})$ vanishes at $\theta = \theta_0$. If we take

$$\eta_{\theta_0}(u) = (1 + \theta_0) \log u_1 + (1 - \theta_0) \log u_2, \tag{6.18}$$

then

$$\frac{\partial \eta_{\theta_0}(u)}{\partial u} = \left(\frac{1+\theta_0}{u_1}, \frac{1-\theta_0}{u_2}\right),$$

so the derivative of $\eta_{\theta_0}(u_{x,\theta})$ with respect to θ is

$$\begin{aligned}
\frac{\partial \eta_{\theta_0}(u_{x,\theta})}{\partial \theta} &= \frac{\partial \eta_{\theta_0}(u)}{\partial u}\bigg|_{u=u_{x,\theta}} \cdot \frac{\partial u_{x,\theta}}{\partial \theta} \\
&= -\frac{(1+\theta_0)^2}{x_1} \cdot \frac{x_1}{(1+\theta)^2} + \frac{(1-\theta_0)^2}{x_2} \cdot \frac{x_2}{(1-\theta)^2} \\
&= -\frac{(1+\theta_0)^2}{(1+\theta)^2} + \frac{(1-\theta_0)^2}{(1-\theta)^2}.
\end{aligned}$$

The latter expression clearly evaluates to zero at $\theta = \theta_0$, so η_{θ_0} satisfies the desired differential equation. The corresponding function $H(x) = H_{\theta_0}(x)$ is given by

$$H_{\theta_0}(x) = (1 + \theta_0)\log\{x_1/(1 + \theta_0)\} + (1 - \theta_0)\log\{x_2/(1 - \theta_0)\}.$$

For the local conditional association—the second expression in (6.16)—we take

$$T(X) = z(\theta) + V_T,$$

where $T(x) = \log(x_1/x_2)$, $z(\theta) = \log\{(1 + \theta)/(1 - \theta)\}$, and $V_T = T(U)$. Then $P_{V_T|\theta_0,h_0}$ is the conditional distribution of V_T, given (θ_0, h_0), where h_0 is the observed $H_{\theta_0}(X) = H_{\theta_0}(x)$. This conditional distribution has a density, given by

$$f_{h_0,\theta_0}(v_T) \propto \exp\{-n\theta_0 v_T/2 - \cosh(v_T/2)e^{(h_0 - \theta_0 v_T)/2}\}.$$

If we let F_{h_0,θ_0} denote the corresponding distribution function, then we can describe this conditional association model by

$$T(X) = z(\theta) + F_{h_0,\theta_0}^{-1}(W), \quad W \sim \text{Unif}(0,1).$$

If, for the P-step, we use the predictive random set $\mathcal{S}$ in (4.8), then the local conditional plausibility function is

$$\text{cpl}_{T(x)|h_0,\theta_0}(\theta_0; \mathcal{S}) = 1 - \left| 1 - 2F_{h_0,\theta_0}(T(x) - z(\theta_0)) \right|.$$

A local conditional $100(1 - \alpha)\%$ plausibility interval for θ can be found just as before, by thresholding the plausibility function at α. It follows from Theorem 6.3 that these intervals will have the nominal coverage probabilities.

For illustration, we consider a small simulation example. We compute the local conditional 90% plausibility interval for θ for 5000 Monte Carlo samples where, in each case, the true θ is sampled randomly from $\{0.0, 0.3, 0.6, 0.9\}$. For several values of n, the estimated coverage probabilities and expected lengths are compared, in Table 6.2, to those of the conditional frequentist interval based on the "r^*" approximation [9, 100, 201, 202], and a Bayesian credible interval based on the Jeffreys prior. The general message is that, compared to the other methods, the local conditional IM intervals have exact coverage for all n, though the intervals appear to be slightly longer on average when n is small. But when n is moderate or large, there is no apparent difference in the performance. Since one cannot hope to do much better than the Jeffreys prior Bayes intervals for large n, we see that the local conditional IM results are at least asymptotically efficient, along with being valid for all n.

6.6   Concluding remarks

This chapter extends the basic IM framework laid out in [180] by developing an auxiliary variable dimension reduction strategy. This reduction simultaneously accomplishes two goals. First, it provides a suitable combination of information across samples, and we argue in Remarks 6.3 and 6.4 that Fisher's concept of sufficiency

Table 6.2 *Taken from [182]. Coverage probabilities and expected lengths of 90% interval estimates for θ in the bivariate normal problem based on, respectively, the local conditional IM (LCIM), the r* approach reviewed by [202], and a Jeffreys prior Bayes approach.*

	Coverage probability			Expected length		
n	LCIM	r^*	Bayes	LCIM	r^*	Bayes
10	0.896	0.845	0.880	0.66	0.61	0.62
25	0.895	0.867	0.883	0.42	0.40	0.41
50	0.907	0.897	0.907	0.30	0.30	0.30
100	0.903	0.888	0.896	0.21	0.21	0.21

and Bayes' theorem can both be viewed as special cases of this combination of information via conditioning. Second, this reduction makes construction of efficient predictive random sets considerably simpler. A new differential equation technique is proposed by which an auxiliary variable dimension reduction can be found even in cases where sufficiency fails to give a satisfactory reduction. In addition, as our simulation results in Sections 6.4.1 and 6.5.4 demonstrate, even with a default choice of predictive random set, the conditional IMs are as good or better than those standard likelihood and Bayes methods. This suggests that our proposed method of combining information is efficient. We expect that the conditional IM approach, paired with optimal predictive random sets, will do even better. However, more work is needed on efficient computation of these optimal predictive random sets.

The local conditional IMs considered in Section 6.5 are an important contribution. Indeed, these tools provide a means to reduce the effective dimension even in cases where the minimal sufficient statistic has dimension greater than that of the parameter. For example, in the bivariate normal correlation problem, we identified a one-dimensional auxiliary variable to predict, even though there is no dimension reduction that can be achieved via sufficiency. The idea of focusing on validity locally at a single $\theta = \theta_0$ itself seems to provide an improvement, this is, in fact, a special case of a more general idea. One could measure locality by a general assertion A, not necessarily a singleton $A = \{\theta_0\}$. In this way, one can develop a conditional IM that focuses on validity at a particular assertion A, thus extending the range of application of local conditional IMs. Even this latter extension is a special case of a more general idea, where associations are based on generic functions of (X, θ, U), not necessarily exact formulations of the sampling model. This new idea will be explored elsewhere.

The examples in this paper have focused on continuous distributions. Efficient inference in discrete problems is challenging in any framework, and IMs are no different. For nice discrete problems, e.g., regular exponential families, the IM analysis described herein can be carried out without difficulty. However, when sufficiency considerations alone provide inadequate auxiliary variable dimension reduction, new tools are needed.

6.7   Exercises

Exercise 6.1. Derive the formula in (6.1).

Exercise 6.2. In Remark 6.3 we made a claim that conditioning is more fundamental than sufficiency. This exercise provides some justification for this claim.

(a) Let X_1, X_2 be iid $N(\theta, 1)$. Derive the conditional distribution of X_1 given $X_2 - X_1$.

(b) Show that an equi-tailed confidence interval for θ based on the conditional distribution in Part (a) is the same as that based on the well-known sampling distribution of $\bar{X}$.

(c) Repeat Parts (a) and (b) for the case of X_1, X_2 iid exponential with mean θ. That is, look at the conditional distribution of X_1, given X_2/X_1.

(d) How far can you generalize the idea in the previous parts?

Exercise 6.3. Prove the following stochastic ordering result:

Fix $h \in H(\mathbb{X})$ and take conditionally admissible $S \sim P_{S|h}$ as in Section 6.2.4. Write $Q_{S|h}(v_T) = P_{S|h}\{S \not\ni v_T\}$. Then $Q_{S|h}(V_T)$ is stochastically no larger than $\mathrm{Unif}(0, 1)$ for $V_T \sim P_{V_T|h}$.

(*Hint:* Similar to the proof of Theorem 4.1 in Chapter 4.)

Exercise 6.4. Consider the Student-t location model in Section 6.4.1.

(a) How can we justify the claim that a flat prior for θ is non-informative?

(b) Write down the Bayes posterior distribution for θ, under a flat prior.

(c) Discuss computation of 95% credible intervals.

Exercise 6.5. Prove Theorem 6.3. (*Hint:* See proof of Theorem 6.2.)

Chapter 7

Marginal Inferential Models

Portions of the material in this chapter are from R. Martin and C. Liu, "Marginal inferential models: prior-free probabilistic inference on interest parameters," *Journal of the American Statistical Association*, to appear, 2015, reprinted with permission by the American Statistical Association, www.amstat.org.

7.1 Introduction

In statistical inference problems, it is often the case that only some component or, more generally, some feature of the parameter θ is of interest. For example, in linear regression, with $\theta = (\beta, \sigma^2)$, often only the vector β of slope coefficients is of interest, even though the error variance σ^2 is also unknown. Here we partition θ as $\theta = (\psi, \xi)$, where ψ is the interest parameter and ξ is the nuisance parameter. The goal is to make valid and efficient inference on ψ in the presence of unknown ξ.

In these nuisance parameter problems, a modification of the classical likelihood framework is called for. Frequentists often opt for profile likelihood methods [50], where the unknown ξ is replaced by its conditional maximum likelihood estimate $\hat{\xi}_\psi$. The effect is that the likelihood function involves only ψ, so, under some conditions, point estimates and hypothesis tests with desirable properties can be constructed as usual. The downside, however, is that no uncertainty in ξ is accounted for when it is fixed at its maximum likelihood estimate. A Bayesian style alternative is the marginal likelihood approach, which assumes an *a priori* probability distribution for ξ. The marginal likelihood for ψ is obtained by integrating out ξ. This marginal likelihood inference effectively accounts for uncertainty in ξ, but difficulties arise from the requirement of a prior distribution for ξ. Indeed, suitable reference priors may not be available or there may be marginalization problems [57].

For these difficult problems, something beyond standard frequentist and Bayesian approaches may be needed. The IM approach described in previous chapters is appealing for a variety of reasons, one being that it provides valid prior-free probabilistic inference. As we encountered in Chapter 6, construction of efficient predictive random sets for relatively high-dimensional auxiliary variables can be challenging. So, keeping in mind the efficiency principle of Chapter 3, the goal is to make inference as efficient as possible. In Chapter 6, we took advantage of the fact that certain features of the auxiliary variable were observed, so prediction could be sharpened by conditioning. In the marginal inference problem discussed here, involving nuisance parameters, it turns out that a different sort of dimension-reduction is both possible and necessary. The basic idea is that certain dimensions of the auxiliary variable, corresponding to the nuisance parameters, can be safely ignored. In particular, if the model is regular in the sense of Definition 7.1, then the specific dimension on which to collapse is clear, and a marginal IM obtains. Sections 7.2.4 and 7.2.5 discuss validity and efficiency of marginal IMs in the regular case. Several examples of challenging marginal inference problems are given in Section 7.3.

When the model is not regular, and the interest and nuisance parameters cannot easily be separated, a different strategy is needed. The idea is that non-separability of the interest and nuisance parameters introduces some additional uncertainty, and we handle this by taking larger predictive random sets. In Section 7.4, we describe a marginalization strategy for non-regular problems based on uniformly valid predictive random sets. Details are given for two benchmark examples: the Behrens–Fisher and gamma mean problems. We conclude with a brief discussion in Section 7.5.

7.2 Marginal inferential models

7.2.1 Preview: Normal mean problem

Suppose $X_1, \ldots, X_n$ are independent $N(\mu, \sigma^2)$ observations, with $\theta = (\mu, \sigma^2)$ unknown, but only the mean μ is of interest. Following the conditional IM argument in Chapter 6, the baseline (conditional) association for θ may be taken as

$$\bar{X} = \mu + \sigma n^{-1/2} U_1 \quad \text{and} \quad S = \sigma U_2, \tag{7.1}$$

where $U_1 \sim N(0,1)$ and $(n-1)U_2^2 \sim \text{ChiSq}(n-1)$. This is equivalent to defining an association for (μ, σ^2) based on the minimal sufficient statistic. This association involves two auxiliary variables; but since there is effectively only one parameter, we hope, for the sake of efficiency, to reduce the dimension of the auxiliary variable. We may equivalently write this association as

$$\bar{X} = \mu + S n^{-1/2} U_1 / U_2 \quad \text{and} \quad S = \sigma U_2. \tag{7.2}$$

The second expression in the above display has the following property: for any s, μ, and u_2, there exists a σ such that $s = \sigma u_2$. This implies that, since σ is free to vary, there is no direct information that can be obtained about μ by knowing U_2. Therefore, there is no benefit to retain the second expression in (7.2)—and eventually predict the corresponding auxiliary variable U_2—when μ is the only parameter of interest.

An alternative way to look at this point is as follows. When $(\bar{x}, s)$ is fixed at the observed value, since σ can take any value, we know that the unobserved (U_1, U_2) must lie on exactly one of the u-space curves

$$\frac{u_1}{u_2} = \frac{n^{1/2}(\bar{x} - \mu)}{s},$$

indexed by μ. The "best possible inference" on (μ, σ^2) corresponds to observing the pair (U_1, U_2). In this case, however, the "best possible marginal inference" on

μ obtains if only we observed which of these curves (U_1, U_2) lies on. This curve is only a one-dimensional quantity, compared to the two-dimensional (U_1, U_2), so an auxiliary variable dimension reduction is possible as a result of the fact that only μ is of interest. In this particular case, we can ignore the U_2 component and work with the auxiliary variable $V = U_1/U_2$, whose marginal distribution is a Student-t with $n-1$ degrees of freedom. As this involves a one-dimensional auxiliary variable only, we have simplified the P-step without sacrificing efficiency; see Section 7.2.5.

7.2.2 Regular models and marginalization

The goal of this section is to formalize the arguments given in Section 7.2.1 for the normal mean problem. For $\theta = (\psi, \xi)$, with ψ the interest parameter, the basic idea is to set up a new association between the data X, an auxiliary variable W, and the parameter of interest ψ only. With this, we can achieve an overall efficiency gain since the dimension of W is generally less than that of the original auxiliary variable.

To emphasize that $\theta = (\psi, \xi)$, rewrite the baseline association as

$$p(X; \psi, \xi) = a(U; \psi, \xi), \quad U \sim \mathsf{P}_U. \tag{7.3}$$

Now suppose that there are functions $\bar{p}$, $\bar{a}$, and c, and new auxiliary variables $V = (V_1, V_2)$, with distribution P_V, such that (7.3) can equivalently be written as

$$\bar{p}(X, \psi) = \bar{a}(V_1, \psi) \tag{7.4a}$$
$$c(X, V_2, \psi, \xi) = 0. \tag{7.4b}$$

The equivalence we have in mind here is that a sample X from the sampling model $\mathsf{P}_{X|\psi,\xi}$, for given (ψ, ξ), can be obtained by sampling $V = (V_1, V_2)$ from P_V and solving for X. See Remark 7.1 below for more on the representation (7.4).

The normal mean example in Section 7.2.1, with association (7.1) and auxiliary variables (U_1, U_2), is clearly of the form (7.4), with $V_1 = U_1/U_2$ and $V_2 = U_2$. In the normal example, recall that the key point leading to efficient marginal inference was that observing U_2 does not provide any direct information about the interest parameter μ. For the general case, we need to assume that this condition, which we call "regularity," holds. This assumption holds in many examples, but there are non-regular models and, in such cases, special considerations are required; see Section 7.4.

Definition 7.1. The association (7.3) is *regular* if it can be written in the form (7.4), and the function c satisfies, for any (x, v_2, ψ), there exists a ξ such that $c(x, v_2, \psi, \xi) = 0$.

In the regular case, it is clear that, like in the normal mean example in Section 7.2.1, knowing the exact value of V_2 does not provide any information about the interest parameter ψ, so there is no benefit to retaining the component (7.4b) and eventually trying to predict V_2. Therefore, in the regular case, we propose to construct a marginal IM for ψ with an association based only on (7.4a). That is,

$$\bar{p}(X, \psi) = \bar{a}(V_1, \psi), \quad V_1 \sim \mathsf{P}_{V_1}. \tag{7.5}$$

In regard to efficiency, as in the normal mean example of Section 7.2.1, the key point here is that V_1 is generally of lower dimension than U, so, in the regular case, an auxiliary variable dimension reduction is achieved, thereby increasing efficiency.

From this point, we can follow the usual three steps to construct a marginal IM for ψ. For the A-step, start with the marginal association (7.5) and write

$$\Psi_x(v_1) = \{\psi : \bar{p}(x, \psi) = \bar{a}(v_1, \psi)\}.$$

For the P-step, introduce a valid predictive random set $\mathcal{S}$ for V_1. Combine these results in the C-step to get

$$\Psi_x(\mathcal{S}) = \bigcup_{v_1 \in \mathcal{S}} \Psi_x(v_1).$$

If $\Psi_x(\mathcal{S}) \neq \varnothing$ with $\mathsf{P}_\mathcal{S}$-probability 1, then, for any assertion $A \subseteq \Psi$, the marginal belief and plausibility functions can be computed as follows:

$$\mathsf{mbel}_x(A; \mathcal{S}) = \mathsf{P}_\mathcal{S}\{\Psi_x(\mathcal{S}) \subseteq A\}$$
$$\mathsf{mpl}_x(A; \mathcal{S}) = 1 - \mathsf{mbel}_x(A^c; \mathcal{S}).$$

These functions can be used for inference as discussed in Chapter 4. In particular, we may construct marginal plausibility intervals for ψ using mpl_x. As usual, if $\Psi_x(\mathcal{S}) = \varnothing$ with positive $\mathsf{P}_\mathcal{S}$-probability, then some adjustment to the belief function formula is needed, and this can be done by conditioning or by stretching. The latter method, due to [82] is preferred; see Section 7.3.3.

7.2.3   Remarks

Remark 7.1. When there exists a one-to-one mapping $x \mapsto (T(x), H(x))$ such that the conditional distribution of $T(X)$, given $H(X)$, is free of ξ and the marginal distribution of $H(X)$ is free of ψ, then an association of the form (7.4) is available. These considerations are similar to those presented in [220] and the references therein. Specifically, suppose the distribution of the minimal sufficient statistic factors as $p(t \mid h, \psi)p(h \mid \xi)$, for statistics $T = T(X)$ and $H = H(X)$. In this case, the observed value h of H provides no information about ψ, so we can take (7.4a) to characterize the conditional distribution $p(t \mid h, \psi)$ of T, given $H = h$, and (7.4b) to characterize the marginal distribution $p(h \mid \xi)$ of H. Also, if $\mathsf{P}_{X \mid \psi, \xi}$ is a composite transformation model [11], then T can be taken as a (maximal) ξ-invariant, whose distribution depends on ψ only.

Remark 7.2. Consider a Bayes model with a genuine (not default) prior distribution Π for (ψ, ξ). In this case, we can write an association, in terms of an auxiliary variable $U = (U_0, U_\psi, U_\xi) \sim \mathsf{P}_{U_0} \times \Pi$, as

$$(\psi, \xi) = (U_\psi, U_\xi), \quad p(X; \psi, \xi) = a(U_0; U_\psi, U_\xi).$$

According to the argument in Remark 6.4 in Chapter 6, an initial dimension reduction obtains, so that the baseline association can be re-expressed as

$$(\psi, \xi) = (U_\psi, U_\xi), \quad (U_\psi, U_\xi) \sim \Pi_X,$$

where Π_X is just the usual posterior distribution of (ψ, ξ), given X, obtained from the Bayes formula. Now, by splitting the posterior into the appropriate marginal and conditional distributions, we get a decomposition

$$\psi = U_\psi \quad \text{and} \quad \xi - U_\xi = 0, \quad (U_\psi, U_\xi) \sim \Pi_X.$$

This association is regular, so a marginal association for ψ obtains from its marginal posterior distribution, completing the A-step. The P-step is a bit different in this case and deserves some explanation. When no meaningful prior distribution is available, there is no probability space on which to carry out probability calculations. The use of a random set on the auxiliary variable space provides such a space, and validity ensures that the corresponding belief and plausibility calculations are meaningful. However, if a genuine prior distribution is available, then there is no need to introduce an auxiliary variable, random sets, etc. Therefore, in the genuine-prior Bayes context, we can take a singleton predictive random set in the P-step and the IM belief function obtained in the C-step is exactly the usual Bayesian posterior distribution.

Remark 7.3. The previous discussion, and the properties to be discussed below, suggest that the baseline association, or sampling model, being regular in the sense of Definition 7.1 is a sufficient condition for valid marginalization. In such cases, as in Sections 7.3.1–7.3.2, fiducial and objective Bayes methods are valid for certain assertions, and the marginal IM will often give the same answers; examples with (implicit or explicit) parameter constraints, like that in Section 7.3.3, reveal some differences between IMs and fiducial and Bayes. However, in non-regular problems, Bayes and fiducial marginalization may not be valid; see Section 7.4. In this light, regularity appears to also be a necessary condition for valid marginal inference.

Remark 7.4. Condition (7.4) helps to formalize the discussion in Example 5.1 in [127] for marginalization in the fiducial context by characterizing the set of problems for which his manipulations can be carried out. Perhaps more importantly, in light of the observations in Remark 7.3, it helps to explain why, for valid prior-free probabilistic marginal inference, the marginalization step must be considered *before* producing evidence measures on the parameter space. Indeed, producing fiducial or objective Bayes posterior distributions for (ψ, ξ) and then marginalizing to ψ is not guaranteed to work. The choice of data-generating equation or reference prior must be considered before actually carrying out the marginalization. The same is true for IMs, though identifying the relevant directions in the auxiliary variable space, as discussed in the previous sections, is arguably more natural than, say, constructing reference priors.

7.2.4   Validity of regular marginal IMs

An important question is if, for suitable $\mathcal{S}$, the marginal IM is valid. We give the affirmative answer in the following theorem.

Theorem 7.1. *Suppose that the baseline association* (7.3) *is regular in the sense of Definition 7.1, and that $\mathcal{S}$ is a valid predictive random set for V_1 with the property that $\Psi_x(\mathcal{S}) \neq \varnothing$ with $\mathsf{P}_\mathcal{S}$-probability 1 for all x. Then the marginal IM is valid in the* *sense that, for any $A \subset \Psi$ and any $\alpha \in (0,1)$, the marginal belief function satisfies*

$$\sup_{(\psi,\xi) \in A^c \times \Xi} \mathsf{P}_{X|(\psi,\xi)}\big\{\mathsf{mbel}_X(A;\mathcal{S}) \geq 1 - \alpha\big\} \leq \alpha.$$

Since this holds for all A, we also have

$$\sup_{(\psi,\xi) \in A \times \Xi} \mathsf{P}_{X|(\psi,\xi)}\big\{\mathsf{mpl}_X(A;\mathcal{S}) \leq \alpha\big\} \leq \alpha.$$

Proof. Similar to the validity theorem proofs in Chapters 4 and 6. This result is also covered by the proof of Theorem 7.3 below.   □

Therefore, if the baseline association is regular and the predictive random set is valid, then the marginal IM constructed has the desirable frequency calibration property. In particular, this means that marginal plausibility intervals based on mpl_X will achieve the nominal frequentist coverage probability; we see this exactness property in the examples in Section 7.3. More importantly, this validity property ensures that the output of the marginal IM is meaningful both within and across experiments.

7.2.5 Efficiency of regular marginal IMs

Start with a regular association (7.4), and let $\mathcal{S} \sim \mathsf{P}_{\mathcal{S}}$ be a valid predictive random set for (V_1, V_2) in $\mathbb{V}_1 \times \mathbb{V}_2$. Assume that $\Theta_x(\mathcal{S})$, a random subset of $\Psi \times \Xi$, is non-empty with $\mathsf{P}_{\mathcal{S}}$-probability 1 for all x. This assumption holds, for example, if the dimension of (V_1, V_2) matches that of (ψ, ξ), which can often be achieved by applying the techniques in [182] to the baseline association prior to marginalization.

In the regular case, we have proposed to marginalize over ξ by "ignoring" the second component of the association (7.4b). We claim that an alternative way to view the marginalization step is via an appropriate stretching of the predictive random set. To see this, for a given $\mathcal{S}$, consider an enlarged predictive random set $\bar{\mathcal{S}}$ obtained by stretching $\mathcal{S}$ to fill up the entire V_2-dimension. Equivalently, take $\mathcal{S}_1$ to be the projection of $\mathcal{S}$ to the V_1-dimension, and then set $\bar{\mathcal{S}} = \mathcal{S}_1 \times \mathbb{V}_2$. It is clear that, if $\mathcal{S}$ is valid, then so is $\bar{\mathcal{S}}$, but, as explained in the next paragraph, the larger $\bar{\mathcal{S}}$ cannot be more efficient than $\mathcal{S}$. In the case of marginal inference, however, the bigger predictive random set $\bar{\mathcal{S}}$ is equivalent to $\mathcal{S}$, so stretching/ignoring yields valid marginalization without loss of efficiency; see, also, the discussion below following Theorem 7.2.

We pause here to give a quick high-level review of the notion of IM efficiency. In the present context, we have two valid predictive random sets, $\mathcal{S}$ and $\bar{\mathcal{S}}$. If B is some assertion about (ψ, ξ), then we say that $\mathcal{S}$ is as efficient as $\bar{\mathcal{S}}$ (relative to B) if $\mathsf{pl}_X(B;\mathcal{S})$ is stochastically no larger than $\mathsf{pl}_X(B;\bar{\mathcal{S}})$ as a function of X when $(\psi, \xi) \notin B$. To see the intuition behind this definition, note that a plausibility region for (ψ, ξ) is obtained by keeping those singleton assertions whose plausibility exceeds a threshold. Making the plausibility function as small as possible, subject to the validity constraint, will make the plausibility region smaller and, hence, the inference more efficient. One quick application, for the present case where $\mathcal{S} \subset \bar{\mathcal{S}}$, note that $\Theta_x(\mathcal{S}) \subseteq \Theta_x(\bar{\mathcal{S}})$. Since the bigger set will have a higher probability of intersecting with an assertion, $\mathsf{pl}_x(\cdot;\bar{\mathcal{S}}) \geq \mathsf{pl}_x(\cdot;\mathcal{S})$ and $\bar{\mathcal{S}}$ is no more efficient than $\mathcal{S}$.

For marginal inference, the assertions of interest are of the form $B = A \times \Xi$, where $A \subseteq \Psi$ is the marginal assertion on ψ. In this case, it is easy to check that

$$\mathsf{pl}_x(A \times \Xi; \bar{\mathcal{S}}) = \mathsf{pl}_x(A \times \Xi; \mathcal{S}), \quad \text{for all } x,$$

so $\mathcal{S}$ and $\bar{\mathcal{S}}$ are equivalent for inference on $A \times \Xi$. That is, no efficiency is lost by stretching the predictive random set $\mathcal{S}$ to $\bar{\mathcal{S}}$. A natural question is: why stretch $\mathcal{S}$ to $\bar{\mathcal{S}}$? To answer this, note that $\mathsf{pl}_x(A \times \Xi; \bar{\mathcal{S}})$ is exactly equal to $\mathsf{mpl}_x(A; \mathcal{S}_1)$, the marginal plausibility function for ψ, at A, based on the projection $\mathcal{S}_1$ of the random set $\mathcal{S}$ to the V_1-dimension. Therefore, we see that "ignoring" the auxiliary variable V_2 in (7.4b) is equivalent to stretching the predictive random set for (V_1, V_2) till it fills the V_2-dimension. We summarize this discussion in the following theorem.

Theorem 7.2. *Consider a regular association* (7.4) *and let $\mathcal{S}$ be a valid predictive random set for (V_1, V_2) such that $\Theta_x(\mathcal{S}) \neq \varnothing$ with $\mathsf{P}_\mathcal{S}$-probability 1 for each x. Then the projection $\mathcal{S}_1$ is a valid predictive random set for V_1 and, furthermore, $\mathsf{pl}_x(A \times \Xi; \mathcal{S}) = \mathsf{mpl}_x(A; \mathcal{S}_1)$ for any marginal assertion $A \subseteq \Psi$.*

The theorem says that marginalization by ignoring (7.4b) or stretching a joint predictive random set to fill the V_2-dimension results in no loss of efficiency. The point is that any valid predictive random set for (V_1, V_2) will result in a valid marginal predictive random set for V_1. However, there are advantages to skipping directly to specification of the marginal predictive random set for V_1. First, the lower-dimensional auxiliary variable is easier to work with. Second, a good predictive random set for V_1 will tend to be smaller than the corresponding projection down from a good predictive random set for (V_1, V_2). Figure 7.1 gives an example of this for (V_1, V_2) iid $\mathsf{N}(0, 1)$. Realizations of two predictive random sets (both with coverage probability 0.5) are shown. Clearly, constructing an interval directly for V_1 and then stretching over V_2 (rectangle) is more efficient than the projection of the joint predictive random set for (V_1, V_2). So, by Theorem 7.2, a valid joint IM will lead to a valid marginal IM, but marginalization before constructing a predictive random set, as advocated in Section 7.2.2, is generally more efficient.

7.3 Examples

7.3.1 Bivariate normal correlation

Let $X_1, \ldots, X_n$, with $X_i = (X_{i1}, X_{i2})$, $i = 1, \ldots, n$, be an independent sample from a bivariate normal distribution with marginal means and variances $\xi = (\mu_1, \mu_2, \sigma_1^2, \sigma_2^2)$ and correlation ψ. It is well known that $(\hat{\mu}_1, \hat{\mu}_2, \hat{\sigma}_1^2, \hat{\sigma}_2^2, \hat{\psi})$, with $\hat{\mu}_j$ and $\hat{\sigma}_j^2$, $j = 1, 2$, the marginal sample means and variances, respectively, and

$$\hat{\psi} = \frac{\sum_{i=1}^n (X_{i1} - \bar{X}_1)(X_{i2} - \bar{X}_2)}{\left\{ \sum_{i=1}^n (X_{i1} - \bar{X}_1)^2 \sum_{i=1}^n (X_{i2} - \bar{X}_2)^2 \right\}^{1/2}},$$

the sample correlation, together form a joint minimal sufficient statistic. The argument in Chapter 6 implies that the conditional IM for $\theta = (\psi, \xi)$ can be expressed

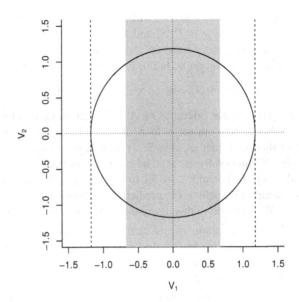

Figure 7.1 *Taken from [183]. Realizations of two predictive random sets for* (V_1, V_2) *iid* $N(0, 1)$*; both have coverage probability 0.5. Circle is for the pair, and the projection down to* V_1*-dimension is shown with dashed lines; gray rectangle is the cylinder obtained by constructing a predictive random set directly for* V_1*.*

in terms of this minimal sufficient statistic. That is, following the initial conditioning step, our baseline association for (ψ, ξ) looks like

$$\hat{\mu}_j = \mu_j + \sigma_j M_j, \quad \hat{\sigma}_j^2 = \sigma_j^2 V_j, \quad \hat{\psi} = a(C, \psi), \quad j = 1, 2,$$

for an appropriate collection of auxiliary variables $U = (M_1, M_2, V_1, V_2, C)$ and a function $a(\cdot, \psi)$ to be specified below in (7.6). This is clearly a regular association, so we get a marginal association for ψ, which is most easily expressed as

$$\hat{\psi} = G_\psi^{-1}(W), \quad W \sim \text{Unif}(0, 1), \tag{7.6}$$

where G_ψ is the distribution function of the sample correlation. Fisher developed fiducial intervals for ψ based on the fiducial density $p(\psi \mid \hat{\psi}) = |\partial G_\psi(\hat{\psi})/\partial \psi|$. In particular, the middle $1 - \alpha$ region of this distribution is a $100(1 - \alpha)\%$ interval estimate for ψ. It is known that this fiducial interval is exact, and also corresponds to the marginal posterior for ψ under the standard objective Bayes prior for $\theta = (\mu_1, \mu_2, \sigma_1^2, \sigma_2^2, \psi)$; see [13]. Interestingly, there is no proper Bayesian prior with the fiducial distribution $p(\psi \mid \hat{\psi})$ as the posterior. However, it is easy to check that, with the default predictive random set (4.8) for U in (7.6), the corresponding marginal plausibility interval for ψ corresponds exactly to the classical fiducial interval.

7.3.2 Ratio of normal means

Let $X_1 \sim N(\psi \xi, 1)$ and $X_2 \sim N(\xi, 1)$ be two independent normal samples, with unknown $\theta = (\psi, \xi)$, and suppose the goal is inference on ψ, the ratio of means. This is the simplest version of the Fieller–Creasy problem [51, 90]. Problems involving ratios of parameters, such as a gamma mean (Section 7.4.2.2), are generally quite challenging and require special considerations.

To start, write the baseline association as

$$X_1 = \psi \xi + U_1 \quad \text{and} \quad X_2 = \xi + U_2, \quad U_1, U_2 \overset{\text{iid}}{\sim} N(0,1).$$

After a bit of algebra, this is clearly equivalent to

$$\frac{X_1 - \psi X_2}{(1 + \psi^2)^{1/2}} = \frac{U_1 - \psi U_2}{(1 + \psi^2)^{1/2}} \quad \text{and} \quad X_2 - \xi - U_2 = 0.$$

We, therefore, have a regular decomposition (7.4), so "ignoring" the part involving ξ gives the marginal association

$$\frac{X_1 - \psi X_2}{(1 + \psi^2)^{1/2}} = V,$$

where $V = (U_1 - \psi U_2)/(1 + \psi^2)^{1/2}$. Since V is a pivot, i.e., $V \sim N(0,1)$ for all ψ, the marginal association can be expressed as

$$\frac{X_1 - \psi X_2}{(1 + \psi^2)^{1/2}} = \Phi^{-1}(W), \quad W \sim \text{Unif}(0,1).$$

For the corresponding marginal IM, the A-step gives

$$\Psi_x(w) = \{\psi : x_1 - \psi x_2 = (1 + \psi^2)^{1/2} \Phi^{-1}(w)\}.$$

Note that this problem has a non-trivial constraint, i.e., $\Psi_x(w)$ is empty for some (x, w) pairs. A similar issue arises in the many-normal-means problem in Section 7.3.3. However, if $\mathcal{S}$ is symmetric about $w = 0.5$, like the default (4.8), then $\Psi_x(\mathcal{S})$ is non-empty with $\mathsf{P}_\mathcal{S}$-probability 1 for all x. Therefore, the marginal IM is valid if $\mathcal{S}$ is valid and symmetric about $w = 0.5$. In fact, for $\mathcal{S}$ in (4.8), the marginal plausibility intervals for ψ are the same as the confidence interval proposed by Fieller. Then validity of our marginal IM provides an alternative proof of the coverage properties of Fieller's interval.

For illustration, using the default predictive random set (4.8) for W, the marginal plausibility function for point assertions about ψ is given by

$$\text{mpl}_x(\psi) = \mathsf{P}_\mathcal{S}\{\Psi_x(\mathcal{S}) \ni \psi\} = 1 - \left| 2\Phi\left(\frac{x_1 - \psi x_2}{(1 + \psi^2)^{1/2}}\right) - 1 \right|.$$

A plot of this marginal plausibility function for two different data sets is provided in Figure 7.2. In Panel (a), we see that the plausibility function has the familiar shape, and the corresponding marginal plausibility region is an interval that contains the

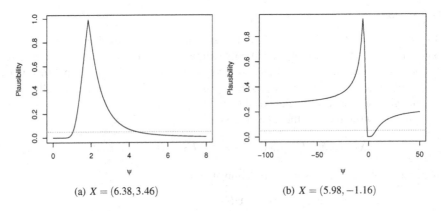

Figure 7.2 Plots of the plausibility function for ψ, the ratio of two normal means, based on the marginal IM with the default predictive random set.

sample ratio $X_1/X_2 = 1.84$. In Panel (b), however, we see that the plausibility function has a very unusual shape, and the corresponding marginal plausibility region is a union of two *unbounded* intervals. This ratio-of-parameters problem is a challenging one [16] and, according to the theorem in [118] and the fact that our marginal plausibility region is exact, we expect to see an unbounded plausibility region. However, while an unbounded confidence region is not very useful, one can still extract some information from the marginal plausibility function, which is a clear advantage of the IM approach; one can also evaluate the marginal belief and plausibility functions at other assertions besides the set of point assertions considered here. Finally, this example also shows that there are problems/data sets for which one cannot construct an exact confidence distribution as in [262]; in this case, it is clear that stacking up exact (one-sided) confidence intervals for ψ in will not yield a proper distribution on the real line—it puts positive mass on $\pm\infty$.

7.3.3 Many-normal means

Suppose $X \sim \mathsf{N}_n(\theta, I_n)$, where X and θ are the n-vectors of observations and means, respectively. Assume $\theta \neq 0$ and write θ as (ψ, ξ), where $\psi = \|\theta\|$ is the length of θ and $\xi = \theta/\|\theta\|$ is the unit vector in the direction of θ.

A baseline association for this problem is

$$X = \psi\xi + U, \quad U \sim \mathsf{N}_n(0, I_n). \tag{7.7}$$

Since ψ, the parameter of interest in our case, is a scalar, it is inefficient to predict the n-vector U. Fortunately, the marginalization strategy discussed above will yield a lower dimensional auxiliary variable.

Take M to be an orthonormal matrix with $\xi^\top$ as its first row, and define $V = MU$. This transformation does not alter the distribution, i.e., both U and V are $\mathsf{N}_n(0, I_n)$, and the baseline association (7.7) can be re-expressed as

$$\|X\|^2 = (\psi + V_1)^2 + \|V_{2:n}\|^2 \quad \text{and} \quad \frac{X - M^{-1}V}{\|X - M^{-1}V\|} = \xi.$$

This is of the regular form (7.4), so the left-most equation above gives a marginal IM for ψ. We make one more change of auxiliary variable, $W = F_{n,\psi}((\psi + V_1)^2 + \|V_{2:n}\|^2)$, where $F_{n,\psi}$ is the distribution function of a non-central chi-square with n degrees of freedom and non-centrality parameter ψ^2. The new marginal association is $\|X\|^2 = F_{n,\psi}^{-1}(W)$, with $W \sim \text{Unif}(0,1)$, and the A-, P-, and C-steps can proceed as usual. In particular, for the P-step, we can use the predictive random set $\mathcal{S}$ in (4.8). However, the set $\Psi_x(w) = \{\psi : \|x\|^2 = F_{n,\psi}(w)\}$ is empty for w in a set of positive measure so, if we take $\mathcal{S}$ to as in (4.8), then the conditions of Theorem 7.1 are violated. To remedy this, and to construct a valid marginal IM, the preferred technique is to use an elastic predictive random set [82]. These details, discussed next, are interesting and allow us to highlight a particular difference between IMs and fiducial.

The default predictive random set $\mathcal{S}$ is determined by a sample $W \sim \text{Unif}(0,1)$, so write $\mathcal{S} = \mathcal{S}_W$. Then $\Psi_x(\mathcal{S}_W)$ is empty if and only if

$$F_{n,0}(\|x\|^2) < \tfrac{1}{2} \quad \text{and} \quad |W - \tfrac{1}{2}| < \tfrac{1}{2} - F_{n,0}(\|x\|^2).$$

To avoid this, [82] propose to stretch $\mathcal{S}_W$ just enough that $\Psi_x(\mathcal{S}_W)$ is non-empty. In the case where $F_{n,0}(\|x\|^2) < \tfrac{1}{2}$, we have

$$\Psi_x(\mathcal{S}_W) = \begin{cases} \{0\} & \text{if } |W - \tfrac{1}{2}| \leq \tfrac{1}{2} - F_{n,0}(\|x\|^2), \\ \{\psi : |F_{n,\psi}(\|x\|^2) - \tfrac{1}{2}| \leq |W - \tfrac{1}{2}|\} & \text{otherwise.} \end{cases}$$

So, for the case $F_{n,0}(\|x\|^2) < \tfrac{1}{2}$, the marginal plausibility function based on the elastic version of the default predictive random set is $\text{mpl}_x(0) = 1$ and, for $\psi > 0$,

$$\text{mpl}_x(\psi) = 2F_{n,0}(\|x\|^2) - \max\{|2F_{n,\psi}(\|x\|^2) - 1| + 2F_{n,0}(\|x\|^2) - 1, 0\};$$

the corresponding $100(1 - \alpha)\%$ marginal plausibility interval is $[0, \hat{\psi}(x, \tfrac{\alpha}{2})]$, where $\hat{\psi}(x, q)$ solves the equation $F_{n,\psi}(\|x\|^2) = q$. When $F_{n,0}(\|x\|^2) > \tfrac{1}{2}$, $\Psi_x(\mathcal{S}_W)$ is non-empty with probability 1, so no adjustments to the default predictive random set are needed. In that case, the marginal plausibility function is

$$\text{mpl}_x(\psi) = 1 - |2F_{n,\psi}(\|x\|^2) - 1|,$$

and the corresponding $100(1 - \alpha)\%$ plausibility interval for ψ is

$$\{\psi : \tfrac{\alpha}{2} \leq F_{n,\psi}(\|x\|^2) \leq 1 - \tfrac{\alpha}{2}\}.$$

Chapter 5 shows that an IM obtained using elastic predictive random sets are valid, so the coverage probability of the plausibility interval is $1 - \alpha$ for all ψ.

For comparison, the generalized fiducial approach outlined in Example 5.1 of

[127] uses the a same association $\|X\|^2 = F_{n,\psi}^{-1}(W)$ with $W \sim \text{Unif}(0,1)$. As in the discussion above, something must be done to avoid the conflict cases. Following the standard fiducial "continue to regard" logic, it seems that one should condition on the event $\{W \le F_{n,0}(\|x\|^2)\}$. In this case, the corresponding central $100(1 - \alpha)\%$ fiducial interval for ψ is

$$\left\{ \psi : \frac{\alpha}{2} \le \frac{F_{n,\psi}(\|x\|^2)}{F_{n,0}(\|x\|^2)} \le 1 - \frac{\alpha}{2} \right\}. \tag{7.8}$$

The recommended objective Bayes approach, using the prior $\pi(\theta) \propto \|\theta\|^{-(n-1)}$ [235], is more difficult computationally but its credible intervals are similar to those in the previous display; see Exercise 7.4. However, this fiducial approach is less efficient compared to the marginal IM solution presented above. Alternative fiducial solutions, such as that in Example 5 of [128], which employ a sort of stretching, similar to our use of an elastic predictive random set (see Chapter 5), give results comparable to our marginal IM.

7.4 Marginal IMs for non-regular models

7.4.1 Motivation and a general strategy

The previous sections focused on the case where the sampling model admits a regular baseline association (7.4). However, there are important problems which are not regular and new techniques are needed for such cases.

Our strategy here is based on the idea of marginalization via predictive random sets. That is, marginalization can be accomplished by using a predictive random set for U in the baseline association that spans the entire "non-interesting" dimension in the transformed auxiliary variable space. But since we are now focusing on the non-regular model case, some further adjustments are needed. Start with an association of the form

$$\bar{p}(X; \psi) = Z_1(\xi), \tag{7.9a}$$
$$c(X, Z_2, \psi, \xi) = 0. \tag{7.9b}$$

See the examples in Section 7.4.2. This is similar to (7.4) except that the distribution of $Z_1(\xi)$ above depends on the nuisance parameter ξ. Therefore, we cannot hope to eliminate ξ by simply "ignoring" (7.9b) as we did in the regular case.

If ξ were known, then a valid predictive random set could be introduced for $Z_1(\xi)$. But this predictive random set for $Z_1(\xi)$ would generally be valid only for the particular ξ in question. Since ξ is unknown in our context, this predictive random set would need to be enlarged in order to retain validity for all possible ξ values. This suggests a notion of *uniformly valid* predictive random sets.

Definition 7.2. A predictive random set $\mathcal{S}$ for $Z_1(\xi)$ is *uniformly valid* if it is valid for all ξ, i.e., $\mathsf{P}_{\mathcal{S}}\{\mathcal{S} \not\ni Z_1(\xi)\}$ is stochastically no larger than $\text{Unif}(0,1)$, as a function of $Z_1(\xi) \sim \mathsf{P}_{Z_1(\xi)}$, for all ξ.

For this more general non-regular model case, we have an analogue of the validity result in Theorem 7.1 for the case of a regular association. We shall refer to the resulting IM for ψ as a *generalized marginal IM*.

Theorem 7.3. *Let S be a uniformly valid predictive random set for $Z_1(\xi)$ in (7.9a), such that $\Psi_x(S) \neq \varnothing$ with P_S-probability 1 for all x. Then the corresponding generalized marginal IM for ψ is valid in the sense of Theorem 7.1 for all ξ.*

Proof. The assumed representation of the sampling model $X \sim P_{X|(\psi,\xi)}$ implies that there exists a corresponding $Z_1(\psi) \sim P_{Z_1(\xi)}$. That is, probability calculations with respect to the distribution of X and are equivalent to probability calculations with respect to the distribution of $Z_1(\xi)$. Take $\psi \notin A$, so that $A \subseteq \{\psi\}^c$. Then we have

$$\mathsf{mbel}_X(A;S) \leq \mathsf{mbel}_X(\{\psi\}^c;S) = P_S\{\Psi_X(S) \not\ni \psi\} = P_S\{S \not\ni Z_1(\xi)\}.$$

The assumed uniform validity implies that the right-hand side is stochastically no larger than $\mathsf{Unif}(0,1)$ as a function of $Z_1(\xi)$. This implies the same of the left-hand side as a function of X. Therefore,

$$\sup_{\psi \notin A} P_{X|(\psi,\xi)}\{\mathsf{mbel}_X(A;S) \geq 1-\alpha\} \leq \alpha, \quad \forall\, \alpha \in (0,1).$$

This holds for all ξ, completing the proof. □

There are a variety of ways to construct uniformly valid predictive random sets, but here we present just one idea, which is appropriate for singleton assertions and construction of marginal plausibility intervals. Efficient inference for other kinds of assertions may require different considerations. Our strategy here is based on the idea of replacing a nuisance parameter-dependent auxiliary variable $Z_1(\xi)$ by a nuisance parameter-independent type of stochastic bound.

Definition 7.3. Let Z be a random variable with distribution function F_Z and median zero. Another random variable $Z^\star$, with distribution function $F_{Z^\star}$ and median zero is said to be *stochastically fatter* than Z if the distribution function satisfy the constraint

$$F_Z(z) < F_{Z^\star}(z), \quad z < 0 \quad \text{and} \quad F_Z(z) > F_{Z^\star}(z), \quad z > 0.$$

That is, $Z^\star$ has heavier tails than Z in both directions.

For example, the Student-t distribution with ν degrees of freedom is stochastically fatter than $N(0,1)$ for all finite ν. Two more examples are given in Section 7.4.2.

Our proposal for constructing a generalized marginal IM for ψ is based on the following idea: get a uniformly valid predictive random set for $Z_1(\xi)$ by first finding a new auxiliary variable $Z_1^\star$ that is stochastically fatter than $Z_1(\xi)$ for all ξ, and then introducing an ordinarily valid predictive random set for $Z_1^\star$. The resulting predictive random set for $Z_1(\xi)$ is necessarily uniformly valid. In this way, marginalization in non-regular models can be achieved through the choice predictive random set.

Since the dimension-reduction techniques presented in the regular model case may be easier to understand and implement, it would be insightful to formulate the non-regular problem in this way also. For this, consider replacing (7.9a) with

$$p(X, \psi) = Z_1^\star. \tag{7.10}$$

After making this substitution, the two parameters ψ and ξ have been separated in (7.10) and (7.9b), so the decomposition is regular. Then, by the theory above, the marginal IM should now depend only on (7.10). The key idea driving this strategy is that a valid predictive random set for Z_1^* is necessarily uniformly valid for $Z_1(\xi)$.

7.4.2 Examples

7.4.2.1 Behrens–Fisher problem

The Behrens–Fisher problem concerns inference on the difference between two normal means, based on two independent samples, when the standard deviations are completely unknown. It turns out that there are no exact tests/confidence intervals that do not depend on the order in which the data is processed. Standard solutions are given by [137, 214], and various approximations are available, e.g., [256, 257]. For a review of these and other procedures, see [106, 114, 149].

Suppose independent samples $X_{11}, \ldots, X_{1n_1}$ and $X_{21}, \ldots, X_{2n_2}$ are available from the populations $N(\mu_1, \sigma_1^2)$ and $N(\mu_2, \sigma_2^2)$, respectively. Summarize the data sets with $\bar{X}_k$ and S_k, $k = 1, 2$, the respective sample means and standard deviations. The parameter of interest is $\psi = \mu_2 - \mu_1$. The problem is simple when σ_1 and σ_2 are known, or unknown but proportional. For the general case, however, there is no simple solution. Here we derive a generalized marginal IM solution for this problem.

The basic sampling model is of the location-scale variety, so the general results in Chapter 6 suggest that we may immediately reduce to a lower-dimensional model based on the sufficient statistics. That is, we take as our baseline association

$$\bar{X}_k = \mu_k + \sigma_k n_k^{-1/2} U_{1k}, \quad \text{and} \quad S_k = \sigma_k U_{2k}, \quad k = 1, 2, \tag{7.11}$$

where the auxiliary variables are independent with $U_{1k} \sim N(0,1)$ and $(n_k - 1)U_{2k}^2 \sim \text{ChiSq}(n_k - 1)$ for $k = 1, 2$. To incorporate $\psi = \mu_2 - \mu_1$, combine the set of constraints in (7.11) for μ_1 and μ_2 to get $\bar{Y} = \psi + \sigma_2 n_2^{-1/2} U_{12} - \sigma_1 n_1^{-1/2} U_{11}$, where $\bar{Y} = \bar{X}_2 - \bar{X}_1$. Define $f(\sigma_1, \sigma_2) = (\sigma_1^2/n_1 + \sigma_2^2/n_2)^{1/2}$ and note that

$$\sigma_2 n_2^{-1/2} U_{12} - \sigma_1 n_1^{-1/2} U_{11} \quad \text{and} \quad f(\sigma_1, \sigma_2) U_1$$

are equal in distribution, where $U_1 \sim N(0,1)$. Making a change of auxiliary variables leads to a new and simpler baseline association for the Behrens–Fisher problem:

$$\bar{Y} = \psi + f(\sigma_1, \sigma_2) U_1 \quad \text{and} \quad S_k = \sigma_k U_{2k}, \quad k = 1, 2. \tag{7.12}$$

Since $S_k = \sigma_k U_{2k}$ for $k = 1, 2$, we may next rewrite (7.12) as

$$\frac{\bar{Y} - \psi}{f(S_1, S_2)} = \frac{f(\sigma_1, \sigma_2)}{f(\sigma_1 U_{21}, \sigma_2 U_{22})} U_1, \quad \text{and} \quad S_k = \sigma_k U_{2k}, \quad k = 1, 2. \tag{7.13}$$

If the right-hand side were free of (σ_1, σ_2), the association would be regular and we could apply the techniques presented in Sections 7.2 and 7.3. Instead, we have a decomposition like (7.9), so we shall follow the ideas presented in Section 7.4.1.

Toward this, define $\xi = \xi(\sigma_1, \sigma_2) = (1 + n_1\sigma_1^2/n_2\sigma_2^2)^{-1}$, which takes values in $(0,1)$. Make another change of auxiliary variable

$$Z_1(\xi) = \frac{U_1}{\{\xi U_{21}^2 + (1-\xi)U_{22}^2\}^{1/2}} \quad \text{and} \quad Z_2 = \frac{U_{22}^2}{U_{21}^2}.$$

Then (7.13) can be rewritten as

$$\frac{\bar{Y} - \psi}{f(S_1, S_2)} = Z_1(\xi) \quad \text{and} \quad \frac{n_1 S_2^2}{n_2 S_1^2} = Z_2 \frac{1-\xi}{\xi}. \tag{7.14}$$

[137] shows that $Z_1^* \sim t_{n_1 \wedge n_2 - 1}$ is stochastically fatter than $Z_1(\xi)$ for all ξ. Let G denote the distribution function of Z_1^*, and let $W \sim \mathrm{Unif}(0,1)$. Then the corresponding version of (7.10) is

$$\frac{\bar{Y} - \psi}{f(S_1, S_2)} = G^{-1}(W).$$

We can get a uniformly valid predictive random set for $Z_1(\xi)$ by choosing an ordinarily valid predictive random set for W. If we use the default predictive random set in (4.8) for W, then our generalized marginal IM plausibility intervals for ψ match up with the confidence intervals in [137, 214]. Also, the validity result from Theorem 7.3 gives an alternative proof of the conservative coverage properties of the Hsu–Scheffé interval. Finally, when both n_1 and n_2 are large, the bound Z_1^* on $Z_1(\xi)$ is tight. So, at least for large samples, the generalized marginal IM is efficient.

7.4.2.2   Gamma mean

Let $X_1, \ldots, X_n$ be observations from a gamma distribution with unknown shape parameter $\alpha > 0$ and unknown mean ψ. Here the goal is inference on the mean. This problem has received attention, for it involves an exponential family model where a ratio of canonical parameters is of interest and there is no simple way to do marginalization. Likelihood-based solutions are presented in [104] and [153] take a different approach. Here we present a generalized marginal IM solution.

The gamma model admits a two-dimensional minimal sufficient statistic for $\theta = (\psi, \alpha)$, which we will take as

$$T_1 = \log\left(\frac{1}{n}\sum_{i=1}^n X_i\right) \quad \text{and} \quad T_2 = \frac{1}{n}\sum_{i=1}^n \log X_i.$$

The most natural choice of association between data, θ, and auxiliary variables is

$$T_1 = \log\left(\frac{1}{n}\sum_{i=1}^n U_i'\right) + \log\frac{\psi}{\alpha} \quad \text{and} \quad T_2 = \frac{1}{n}\sum_{i=1}^n \log U_i' + \log\frac{\psi}{\alpha},$$

where $U_1', \ldots, U_n'$ are iid gamma random variables with both shape and mean equal to α. This association can be simplified by writing

$$T_1 - \log\psi = U_1(\alpha) \quad \text{and} \quad T_2 - \log\psi = U_2(\alpha),$$

where $U_1(\alpha)$ and $U_2(\alpha)$ are distributed as, respectively,

$$\log\left(\frac{1}{n}\sum_{i=1}^{n}\frac{U_i'}{\alpha}\right) \quad \text{and} \quad \frac{1}{n}\sum_{i=1}^{n}\log\frac{U_i'}{\alpha}.$$

That the distributions of $U_1(\alpha)$ and $U_2(\alpha)$ depend on the nuisance parameter α makes this a non-regular problem.

Next, define $V_1 = U_1(\alpha)$ and $V_2 = U_1(\alpha) - U_2(\alpha)$. For notational simplicity, we have omitted the dependence of (V_1, V_2) on α. It is easy to check that $n\alpha e^{V_1}$ has a gamma distribution with shape parameter $n\alpha$; write $F_{n\alpha}$ for the corresponding gamma distribution function. Let $\kappa(V_2)$ be an estimator of α based on V_2 alone. This estimator could be a moment estimator or perhaps a maximum likelihood estimator based on the the marginal distribution of V_2; we give explicit estimators later. Suppose that this estimator is consistent, i.e., $\kappa(V_2) \to \alpha$ in probability, with respect to the marginal distribution of V_2, as $n \to \infty$. Then, as $n \to \infty$,

$$\Phi^{-1}\big(F_{n\kappa(V_2)}(n\kappa(V_2)e^{V_1})\big) \to N(0,1), \quad \text{in distribution,}$$

under the joint distribution of V_1 and V_2, for any $\alpha > 0$.

The limit distribution result above is the beginning, rather than the end, of our analysis. Indeed, consider the new association

$$\Phi^{-1}\big(F_{n\kappa(T_1-T_2)}(n\kappa(T_1-T_2)e^{T_1-\log\psi})\big) = \Phi^{-1}\big(F_{n\kappa(V_2)}(n\kappa(V_2)e^{V_1})\big).$$

Let $Z_1(\alpha)$ be a random variable with the same distribution as the quantity on the right-hand side. We know that $Z_1(\alpha) \to N(0,1)$, in distribution, as $n \to \infty$, for all α.

Take $\kappa(V_2)$ to be a moment-based estimator defined as follows. The expectation of V_2 is equal to $g(n\alpha) - g(\alpha) - \log(n)$, where g is the digamma function. Then $\kappa(v_2)$ is defined as the solution for α in the equation $v_2 = g(n\alpha) - g(\alpha) - \log(n)$. Similar equations appear in [104, 140] in a likelihood context. Consistency of this moment estimator, as $n \to \infty$, is straightforward. Moreover, for this $\kappa(V_2)$, we can derive limiting distributions for $Z_1(\alpha)$ as $\alpha \to \{0, \infty\}$. Indeed, [140] shows that $2n\alpha V_2$ converges in distribution to $\mathsf{ChiSq}(2n-2)$ and $\mathsf{ChiSq}(n-1)$ when $\alpha \to 0$ and $\alpha \to \infty$, respectively. So, from this and the asymptotic approximation $g(x) = \log x - 1/(2x)$ for large x, it follows that $Z_1(\alpha) \to t_{n-1}$, in distribution, as $\alpha \to \infty$. A corresponding limit distribution as $\alpha \to 0$ is available, but we will not need this.

We have a version of (7.9a) given by

$$\Phi^{-1}\big(F_{n\kappa(T_1-T_2)}(n\kappa(T_1-T_2)e^{T_1-\log\psi})\big) = Z_1(\alpha),$$

and the goal is to find a uniformly valid predictive random set for $Z_1(\alpha)$. We will proceed by finding a random variable $Z_1^\star$ that is stochastically fatter than $Z_1(\alpha)$ for all α. Unfortunately, the limit distribution t_{n-1} is not a suitable bound. But it turns out that a relatively simple adjustment will do the trick. First, take $\hat{v}$ as the projection of $2n\kappa(V_2)V_2$ onto $[n-1, 2(n-1)]$. Then, we define $\kappa^\star(V_2) = m(\hat{v})/2nV_2$, where $m(\hat{v})$ is the median of the $\mathsf{ChiSq}(\hat{v})$ distribution. With this adjusted estimator, we claim that the $Z_1^\star$ is stochastically fatter than $Z_1(\alpha)$ for all α, where

$$Z_1^\star \sim 0.5\,t_{n-1}^+ + 0.5\,(c_n\,t_{n-1})^-. \tag{7.15}$$

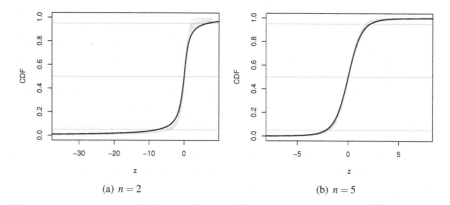

Figure 7.3 *Taken from [183]. Distribution functions of $Z_1(\alpha)$, based on the adjusted $\kappa^*(V_2)$, over a range of α. Heavy line is the bound* (7.15).

Here t_{n-1}^+ is the positive half t distribution, and $(c_n t_{n-1})^-$ is the negative half scaled-t distribution. The scaling factor c_n on the negative side is given by

$$c_n = \{2(n-1)/m(2n-2)\}^{1/2}.$$

Theoretical results justifying the claimed bound are available but, for brevity, we will only show a picture. Figure 7.3 shows the distribution of $Z_1(\alpha)$, based on the adjusted $\kappa^*(V_2)$, for $n = 2$ and $n = 5$, for a range of α, along with the distribution function corresponding to the mixture in (7.15). The claim that Z_1^* is stochastically fatter than $Z_1(\alpha)$ is clear from the picture; in fact, the bound is quite tight for n as small as 5.

If G is the distribution function for the mixture in (7.15), and $W \sim \text{Unif}(0,1)$, then we can get a uniformly valid predictive random set for $Z_1(\alpha)$ by choosing a ordinarily valid predictive random set for W, such as the default (4.8). From here, constructing the generalized marginal IM for ψ is straightforward.

For illustration, we consider data on survival times of rats exposed to radiation given in [104], modeled as an independent gamma sample with mean ψ. The 95% generalized marginal IM plausibility interval for ψ is $(96.9, 134.4)$. Second- and third-order likelihood-based 95% confidence interval for ψ are $(97.0, 134.7)$ and $(97.2, 134.2)$, respectively. The third-order likelihood interval is the shortest, but the plausibility interval has guaranteed coverage for all n, so a direct comparison is difficult. In simulations (not shown), for $n = 2$, the generalized marginal IM is valid, but conservative; see Figure 7.3(a). For larger n, the likelihood and IM methods are comparable.

7.5   Concluding remarks

This chapter focused on the problem of inference in the presence of nuisance parameters, and proposed a new strategy for auxiliary variable dimension reduction within the IM framework. This reduction in dimension generally improves inference. The regular versus non-regular classification introduced here shows which problems admit exact and efficient marginalization. In the regular case, this marginalization can be accomplished efficiently using the techniques describe herein. For non-regular problems, we propose a strategy based on uniformly valid predictive random sets, and one technique to construct these random sets using stochastic bounds. While this simple strategy maintains validity of the marginal IM, these can be conservative when n is small. Therefore, work is needed to develop more efficient marginalization strategies in non-regular problems.

The dimension reduction considerations here are of critical importance for all statisticians working on high-dimensional problems. In our present context, we have information that only a component of the parameter vector is of interest, and so we should use this information to reduce the dimension of the problem. More generally, in particular in high-dimensional applications, there is information available about θ, such as sparsity, and the goal is to incorporate this information and improve efficiency. There are a variety of ways one can accomplish this, but all amount to a kind of dimension reduction. So it is possible that the dimension reduction considerations here can help shed light on this important issue in modern statistical problems.

7.6 Exercises

Exercise 7.1. Consider the many-normal-means problem from Section 7.3.3, i.e., $X \sim N_n(\theta, I_n)$, in the case of $n = 2$. Find a regular marginal IM for inference about the direction of θ, i.e., $\theta/\|\theta\|$.

Exercise 7.2. Suppose that $X_1, \ldots, X_n$ are iid $N(\mu, \sigma^2)$, where both μ and σ^2 are unknown. Construct a marginal IM for inference on $\psi = \mu/\sigma$, the standardized mean or signal-to-noise ratio. (Hint: see [165].)

Exercise 7.3. Suppose that the observed data consists of two independent binomial counts, i.e., $X_1 \sim \text{Bin}(n_1, \theta_1)$ and $X_2 \sim \text{Bin}(n_2, \theta_2)$, where (n_1, n_2) is known but (θ_1, θ_2) is unknown. The goal is inference on the *odds ratio* $\psi = \frac{\theta_1}{1-\theta_1} / \frac{\theta_2}{1-\theta_2}$. Of particular interest is the assertion that $\psi = 1$. Develop a marginal IM approach for inference on ψ and compare your method, via simulations, to Fisher's exact test, which is a conditional frequentist approach.

Exercise 7.4. Consider the normal mean problem in Section 7.3.3.

(a) Perform a simulation study to assess the coverage probability of the fiducial intervals for $\psi = \|\theta\|$ given in (7.8). Use $n = 2, 5, 10$ and $\psi = 0.1, 0.5, 1.0$. Does it make a difference how the length ψ is allocated among the individual entries in the mean vector?

(b) Compare the coverage probability results in Part (a) to those obtained from the marginal IM plausibility intervals described here and/or the more sophisticated fiducial intervals described in Example 5.1 of [127].

(c) How do the objective Bayes intervals for ψ, based on the prior $\pi(\theta) \propto \|\theta\|^{-(n-1)}$ [235], compare?

Exercise 7.5. A "naive" marginalization strategy is to construct a (joint) plausibility function for the full parameter θ, using a nested predictive random set, and then profile out the nuisance parameter by maximizing over η for each fixed value of the interest parameter ψ.

(a) Provide some theoretical justification for this idea. (*Hint:* Use the fact that the predictive random set is nested.)

(b) Re-do any one of the examples in this chapter using this alternative marginalization approach. Do your results agree with those presented here?

Exercise 7.6. An alternative construction of a marginal IM for the Behrens–Fisher problem is discussed in [165]. Apply this alternative marginalization strategy to the gamma mean model discussed in Section 7.4.2.2.

Chapter 8

Normal Linear Models

8.1 Introduction

It is difficult to argue with the claim that normal linear models—which include linear regression, analysis of variance (ANOVA), mixed-effect models, etc—are the most widely used models in applied statistics. For this reason, we think it is important to discuss these models in some detail. Of particular interest here is, first, some general auxiliary variable dimension reduction strategies. In the examples considered here, this corresponds to specifying a lower-dimensional association for the sufficient statistics. Once all the general dimension reduction has been carried out, the question becomes whether a further dimension reduction can be carried out due to the fact that only some feature of the unknown parameter is of interest. For this we will apply various marginalization techniques as discussed in Chapter 7 to further reduce the auxiliary variable dimension.

In Section 8.2, we consider the basic linear regression model, which contains the ANOVA models as a special case. For this model there is not much to say here since the conditioning and marginalization steps are standard, basically matching those existing approaches in the literature. We will, however, return to the basic regression problem in Chapter 10 in the context of model/variable selection, a challenging modern problem in statistics. Section 8.3 considers a normal linear mixed model with two variance components. After marginalizing out the fixed effects, it is known that the dimension of the minimal sufficient statistic is, in general, greater than the dimension of the parameter, which is two. In this case, a local conditional IM can be constructed that will reduce the auxiliary variable dimension down to that of the parameter, without loss of efficiency. We specifically focus on the problem of inference on the heritability coefficient, which is a scalar function of the pair of variance components. For this, we construct a special local conditional IM and demonstrate, with both theory and numerical examples, that the results are both valid and efficient. Some remarks about extensions and open problems in the mixed model example are discussed in Section 8.4.

8.2 Linear regression

8.2.1 The model and baseline association

To set notation, consider the following specification of the linear regression model:

$$Y = X\beta + \sigma Z, \tag{8.1}$$

where $Y = (Y_1, \ldots, Y_n)^\top$ is a n-vector of response variables, X is an $n \times p$ matrix of predictor variables, $\beta = (\beta_1, \ldots, \beta_p)^\top$ is a p-vector of regression coefficients, σ^2 is the residual variance, and $Z = (Z_1, \ldots, Z_n)^\top$ is a n-vector of standard Gaussian errors, i.e., $Z \sim N_n(0, I_n)$, a n-dimensional Gaussian distribution with mean zero and identity covariance matrix. We assume that $p < n$ and that X is fixed with non-singular $X^\top X$. Without loss of generality, we will assume that Y and the columns of X have been centered, so that we can ignore the intercept term.

In our IM language, (8.1) is the baseline association, connecting observable data Y (and X), unknown parameters $\lambda = (\beta, \sigma)$, and unobservable auxiliary variables Z. The goal is, as usual, to accurately predict the unobservable auxiliary variable with a random set. But first we want to reduce the dimension of Z as much as possible.

8.2.2 General dimension-reduction steps

For the regression problem at hand, the auxiliary variable Z is n-dimensional while $\lambda = (\beta, \sigma^2)$ is $(p+1)$-dimensional, and $p+1 \leq n$. This suggests the possibility of reducing the dimension of the auxiliary variable, and the techniques in Chapter 6 are appropriate. Since the regression problem admits a $(p+1)$-dimensional minimal sufficient statistic, the IM dimension reduction is straightforward. Let $\hat{\beta}$ be the least-squares estimator of β, and $\hat{\sigma}$ the corresponding estimator of σ based on the residual sum of squares. Then $(\hat{\beta}, \hat{\sigma})$ is a joint minimal sufficient statistic for (β, σ), and we can identify a lower-dimensional (conditional) association:

$$\hat{\beta} = \beta + \sigma V_1 \quad \text{and} \quad \hat{\sigma} = \sigma V_2 \tag{8.2}$$

where $V = (V_1, V_2)$ are independent and satisfy

$$V_1 \sim N_p(0, M) \quad \text{and} \quad (n-p-1)V_2^2 \sim \text{ChiSq}(n-p-1), \tag{8.3}$$

with $M = (X^\top X)^{-1}$. The key point is that we have replaced the n-dimensional auxiliary variable Z with a $(p+1)$-dimensional auxiliary variable V. This reduction in dimension will help to simplify the task of accurate prediction of the auxiliary variable. Note that the above conditional association can also be obtained by applying the differential equation technique discussed in Chapter 6.3.2; see Exercise 8.1.

8.2.3 Marginal IM for all regression coefficients

It is often the case in applications that interest is in the regression coefficients, β, and not in the error variance, σ^2. In such cases, it is possible to further reduce the dimension of the auxiliary variable via marginalization. Note that the (conditional) association (8.2) can be rewritten as

$$\hat{\beta} = \beta + \hat{\sigma}(V_1/V_2) \quad \text{and} \quad \hat{\sigma} = \sigma V_2.$$

The general theory described in Chapter 7, and also in [183], says that the second equation above—the one that is free of β—can be ignored. This leaves a marginal association involving a p-dimensional auxiliary variable $W = V_1/V_2$. That is,

$$\hat{\beta} = \beta + \hat{\sigma}W, \quad W \sim t_p(0, M; n - p - 1), \tag{8.4}$$

where the distribution of W is a p-dimensional Student-t, with $n - p - 1$ degrees of freedom, centered at the origin, and scale matrix $M = (X^\top X)^{-1}$. Again, the key point is that the $(p+1)$-dimensional auxiliary variable in (8.2) has been replaced by a p-dimensional auxiliary variable W. No further dimension reduction is possible, and we take (8.4) as our starting point for constructing an IM for inference on β.

For the P-step, based on the distribution of the auxiliary variable W, a reasonable generalization of the default predictive random set discussed in previous chapters is one with contours determined by thresholding the p-variate scaled Student-t density. That is, a default predictive random set S is given by

$$S = \{w \in \mathbb{R}^p : w^\top Mw \le W^\top MW\}, \quad W \sim t_p(0, M; n - p - 1). \tag{8.5}$$

Exercise 8.2 invites the reader to show that, for the IM obtained from this choice of predictive random set, the corresponding plausibility for a singleton assertion $A = \{b\}$ for fixed $b \in \mathbb{R}^p$ is exactly the p-value of the most-powerful F-test [159]. Moreover, if b is allowed to vary, we obtain a plausibility function for singleton assertions, and thresholding this function at level $\alpha \in (0,1)$ gives a $100(1 - \alpha)\%$ plausibility ellipsoid, and this matches the standard F-confidence ellipsoid. Recall, however, that even though the particular form of the plausibility region is familiar, the interpretation of a region of "sufficiently plausible" parameter values, given the observed data, is unique to the IM framework.

8.2.4 *Marginal IM for the error variance*

Suppose that the parameter of interest is the error variance, σ^2. To construct a marginal IM for σ^2, go back the baseline association (8.2). Just like in the previous subsection, we can eliminate σ^2 from the first expression, so by the general theory in Chapter 7, the first expression can be ignored, leaving a very simple marginal association:

$$\hat{\sigma} = \sigma V_2,$$

where V_2 has the distribution identified in (8.3). If we let F be the distribution function of V_2, then we may rewrite this A-step as

$$\hat{\sigma} = \sigma F^{-1}(W), \quad W \sim \text{Unif}(0, 1),$$

and, for the P-step, we can consider a default predictive random set for the uniform W; see Exercise 8.4 for an alternative strategy. For a singleton assertion $A = \{s^2\}$ for σ^2, the plausibility function is

$$\mathsf{pl}_Y(s^2) = 1 - \left|2F(\hat{\sigma}/s) - 1\right|,$$

and plausibility intervals based on thresholding this plausibility function agree with the standard confidence intervals for σ^2 in the linear model literature.

8.3 Linear mixed effect models

8.3.1 The model and a baseline association

Normal linear mixed effect models are useful in a variety of biological, physical, and social scientific applications with variability coming from multiple sources; see [147] and [219]. In this section, based on the work in [41], we focus on the case of two variance components, and the general model can be written as

$$Y = X\beta + Z\alpha + \varepsilon, \tag{8.6}$$

where Y is a n-vector of response variables, X and Z are design matrices for the fixed and random effects of dimension $n \times p$ and $n \times a$, respectively, β is a p-vector of unknown parameters, α is a normal random a-vector with mean 0 and covariance matrix $\sigma_\alpha^2 A$, and ε is a normal random n-vector with mean 0 and covariance matrix $\sigma_\varepsilon^2 I_n$. Here, $\sigma^2 = (\sigma_\alpha^2, \sigma_\varepsilon^2)$ is the pair of variance components. Assume A is known and X is of (full) rank $p < n$. The unknown parameters in this model are the p fixed-effect coefficients β and the two variance components σ^2, so the full parameter space is $(p+2)$-dimensional.

The goal of the IM framework, as discussed in previous chapters, is valid prior-free probabilistic inference. This is facilitated by first associating the observable data and unknown parameters to a set of unobservable auxiliary variables. For example, the marginal distribution of Y from (8.6) is

$$Y \sim N_n(X\beta, \sigma_\varepsilon^2 I_n + \sigma_\alpha^2 ZAZ^\top), \tag{8.7}$$

which can be written in association form:

$$Y = X\beta + (\sigma_\varepsilon^2 I_n + \sigma_\alpha^2 ZAZ^\top)^{1/2} U, \quad U \sim N_n(0, I_n). \tag{8.8}$$

That is, observable data Y and unknown parameters (β, σ^2) are associated with unobservable auxiliary variables U, in this case, a n-vector of independent standard normal random variables. Given this association, we introduce a predictive random set for U and then combine with the observed value of Y to get the IM and, in particular, the corresponding plausibility function. As before, the plausibility function assigns, to assertions about the parameter of interest, a measure of the plausibility that the assertion is true. It can be used to design exact frequentist tests or confidence regions; recall Section 4.4.4.

Notice, in the association (8.8), that the auxiliary variable U is, in general, of a higher dimension than that of the parameter (β, σ^2). Motivated by the efficiency principle of Chapter 3, we want to try to reduce the dimension of U as much as possible. This dimension reduction is obtained via a conditioning operation, as described in Chapter 6. Here we apply these techniques, as well as the marginalization techniques in Chapter 7, to get a dimension-reduced association for the certain features of the full parameter (β, σ^2).

8.3.2 General dimension-reduction steps

Start with the linear mixed model (8.6). Following the setup in [71], let K be a $n \times (n-p)$ matrix such that $KK^\top = I_n - X(X^\top X)^{-1}X^\top$ and $K^\top K = I_{n-p}$. Next, let $B = (X^\top X)^{-1}X^\top$. Then $y \mapsto (K^\top y, By)$ is a one-to-one mapping. Moreover, the distribution of $K^\top Y$ depends on $(\sigma_\alpha^2, \sigma_\varepsilon^2)$ only, and the distribution of BY depends on (β, σ^2), with β a location parameter. In particular, from (8.7), we get

$$K^\top Y \sim N_{n-p}(0, \sigma_\varepsilon^2 I_{n-p} + \sigma_\alpha^2 G) \quad \text{and} \quad BY \sim N_p(\beta, C_\sigma),$$

where $G = K^\top ZAZ^\top K$ and C_σ is a $p \times p$ covariance matrix of a known form that depends on $\sigma^2 = (\sigma_\alpha^2, \sigma_\varepsilon^2)$; its precise form is not important. From this point, the general theory in Chapter 7 allows us to marginalize over β by simply deleting the BY component. Therefore, a marginal association for $(\sigma_\alpha^2, \sigma_\varepsilon^2)$ is

$$K^\top Y = (\sigma_\varepsilon^2 I_{n-p} + \sigma_\alpha^2 G)^{1/2} U_2, \quad U_2 \sim N_{n-p}(0, I_{n-p}).$$

This marginalization reduces the auxiliary variable dimension from n to $n-p$.

In the marginal association for $(\sigma_\alpha^2, \sigma_\varepsilon^2)$ above, there are $n-p$ auxiliary variables but only two parameters. Classical results on sufficient statistics in mixed effects model that will facilitate further dimension reduction. For the matrix G defined above, let $\lambda_1 > \cdots > \lambda_L \geq 0$ denote the (distinct) ordered eigenvalues with multiplicities $r_1, \ldots, r_L$, respectively. Let $P = [P_1, \ldots, P_L]$ be a $(n-p) \times (n-p)$ orthogonal matrix such that $P^\top GP$ is diagonal with eigenvalues $\{\lambda_\ell : \ell = 1, \ldots, L\}$, in their multiplicities, on the diagonal. For P_ℓ, a $(n-p) \times r_\ell$ matrix, define

$$S_\ell = Y^\top KP_\ell P_\ell^\top K^\top Y, \quad \ell = 1, \ldots, L.$$

Olsen et al. [195] showed that $(S_1, \ldots, S_L)$ is a minimal sufficient statistic for $(\sigma_\alpha^2, \sigma_\varepsilon^2)$. Moreover, the distribution of $(S_1, \ldots, S_L)$ is determined by

$$S_\ell = (\lambda_\ell \sigma_\alpha^2 + \sigma_\varepsilon^2)V_\ell, \quad V_\ell \sim \text{ChiSq}(r_\ell), \quad \text{independent}, \quad \ell = 1, \ldots, L. \quad (8.9)$$

This reduces the auxiliary variable dimension from $n-p$ to L. We take (8.9) as our "baseline association."

Even in this reduced baseline association, there are L auxiliary variables but only two parameters, which means there is room for even further dimension reduction. The next section shows how to reduce to a scalar auxiliary variable when the parameter of interest is the pair of variance components $\sigma^2 = (\sigma_\alpha^2, \sigma_\varepsilon^2)$ or the scalar heritability coefficient discussed in Section 8.3.4.

8.3.3   Marginal IM for the variance components

For simplicity, here we will assume that $\lambda_L = 0$; this assumption often holds, as in the three examples discussed in Section 8.3.4.4 below, but there are models, such as the full animal model [132], where it may fail. To start, for a given $S = s$ and σ^2, we can solve for the auxiliary variable v in the above baseline association:

$$v_{s,\sigma^2,\ell} = \frac{s_\ell}{\lambda_\ell \sigma_\alpha^2 + \sigma_\varepsilon^2}, \quad \ell = 1,\ldots,L-1, \quad v_{s,\sigma^2,L} = \frac{s_L}{\sigma_\varepsilon^2}.$$

Differentiating this expression with respect to both components of σ^2 gives an $L \times 2$ matrix $\partial v_{s,\sigma^2}/\partial \sigma^2 = \mathrm{diag}\{v_{s,\sigma^2}\}W(\sigma^2)$, where the rows of $W(\sigma^2)$ are given by

$$w_\ell(\sigma^2) = \left(-\frac{\lambda_\ell}{\lambda_\ell \sigma_\alpha^2 + \sigma_\varepsilon^2}, -\frac{1}{\lambda_\ell \sigma_\alpha^2 + \sigma_\varepsilon^2}\right), \quad \ell = 1,\ldots,L-1,$$

$$w_L(\theta) = (0, -1/\sigma_\varepsilon^2).$$

Choose an arbitrary localization point $\sigma_0^2 = (\sigma_{0\alpha}^2, \sigma_{0\varepsilon}^2)$. The goal is to find a function $\eta_{\sigma_0^2}(v)$ that satisfies the differential equation

$$\underbrace{\left.\frac{\partial \eta_{\sigma_0^2}(u)}{\partial v}\right|_{v=v_{s,\sigma^2}}}_{(L-2)\times L} \cdot \underbrace{\mathrm{diag}\{v_{s,\sigma^2}\}}_{L\times L} \cdot \underbrace{W(\sigma^2)}_{L\times 2} = \underbrace{0}_{(L-2)\times 2} \quad \text{at } \sigma^2 = \sigma_0^2. \qquad (8.10)$$

The method of characteristics [200] suggests the logarithmic function

$$\eta_{\sigma_0^2}(v)^\top = (\log v_1, \cdots, \log v_L)\Pi(\sigma_0^2)^\top,$$

where $\Pi(\sigma_0^2)$ is a $(L-2) \times L$ matrix with rows orthogonal to the columns of $W(\sigma_0^2)$. Since $\Pi(\sigma_0^2)W(\sigma_0^2)$ vanishes, it is easy to check that (8.10) holds for this $\eta_{\sigma_0^2}$. Then the corresponding $H_{\sigma_0^2}(s)$ satisfies

$$H_{\sigma_0^2}(s)^\top = \left(\log \frac{s_1}{\lambda_1 \sigma_{0\alpha}^2 + \sigma_{0\varepsilon}^2}, \cdots, \log \frac{s_{L-1}}{\lambda_{L-1} \sigma_{0\alpha}^2 + \sigma_{0\varepsilon}^2}, \log \frac{s_L}{\sigma_{0\varepsilon}^2}\right)\Pi(\sigma_0^2)^\top.$$

Take two orthogonal L-vectors which are not orthogonal to the columns of $W(\sigma_0^2)$. One of these vectors should be $(0,\ldots,0,1)$, so that one component of the conditional association will be free of σ_α^2. The other vector can be, say, $(1,\ldots,1,0)$. Then define the mapping $\tau(v)$, taking values in $\mathbb{R}^2$, via the equation

$$\begin{pmatrix} \eta_{\sigma_0^2}(v) \\ \tau(v) \end{pmatrix} = \begin{pmatrix} \Pi(\sigma_0^2) \\ 1 \cdots 1\, 0 \\ 0 \cdots 0\, 1 \end{pmatrix} \begin{pmatrix} \log v_1 \\ \vdots \\ \log v_L \end{pmatrix}.$$

This, in turn, defines the two-dimensional $(T_1, T_2)(s)$ to be used in the conditional association. In particular, the conditional association is given by

$$\sum_{\ell=1}^{L-1} \log S_\ell = \sum_{\ell=1}^{L-1} \log(\lambda_\ell \sigma_\alpha^2 + \sigma_\varepsilon^2) + \sum_{\ell=1}^{L-1} \log V_\ell, \qquad \log S_L = \log \sigma_\varepsilon^2 + \log V_L,$$

and we set $T_1 = \sum_{\ell=1}^{L-1} \log S_\ell$, $T_2 = \log S_L$, $\tau(V)_1 = \sum_{\ell=1}^{L-1} \log V_\ell$, and $\tau(V)_2 = \log V_L$. Furthermore, since this representation is linear on the log scale, and $V_1, \ldots, V_L$ are independent chi-square, the conditional distribution of $\tau(V)$, given $\eta_{\sigma_0^2}(V) = H_{\sigma_0^2}(s)$, can be readily found numerically.

A small numerical illustration of this IM approach for a one-way random effects model is presented in [182]. From the "joint" IM for $(\sigma_\alpha^2, \sigma_\varepsilon^2)$, a natural question is if "marginal" IMs for σ_α^2 and σ_ε^2 can be obtained and, indeed, there are naive approaches. For example, a marginal plausibility function for σ_α^2 can be obtained by maximizing a plausibility function for $(\sigma_\alpha^2, \sigma_\varepsilon^2)$ over the σ_ε^2-margin, but we have found that this is not a particularly efficient approach. Therefore, some more careful marginalization techniques are needed. Unfortunately, the model is not "regular" in the sense of Definition 7.1, so marginalization is not straightforward and further work is needed. There are, however, certain functions of $(\sigma_\alpha^2, \sigma_\varepsilon^2)$ for which direct marginalization is possible, as we discuss next.

8.3.4   Marginal IM for the heritability coefficient

8.3.4.1   Definition and motivation

In biological applications, the quantities α and ε in (8.6) denote the genetic and environmental effects, respectively. Given that "a central question in biology is whether observed variation in a particular trait is due to environmental or biological factors" [242], the *heritability coefficient*, $\rho = \sigma_\alpha^2/(\sigma_\alpha^2 + \sigma_\varepsilon^2)$, which represents the proportion of phenotypic variance attributed to variation in genotypic values, is a fundamentally important quantity. Indeed, mixed-effect models and inference on the heritability coefficient has been applied recently in genome-wide association studies [119, 265].

Given the importance of the heritability coefficient, there are a number of methods available to construct confidence intervals for ρ or, equivalently, for $\psi = \sigma_\alpha^2/\sigma_\varepsilon^2$. When the design is balanced, [120, 219] give a confidence intervals for ψ and other quantities. When the design is possibly unbalanced, as we assume here, the problem is more challenging; in particular, exact ANOVA-based confidence intervals generally are not available. [243] gave intervals for ψ in the unbalanced case, and subsequent contributions include [35, 87, 132, 157]. Bayesian [109, 110, 259] and fiducial methods [43, 71, 93] are also available. Here we present an IM approach for inference on ρ, which is provably exact and empirically more efficient.

8.3.4.2   Further dimension reduction

Note, here we do not need to assume that $\lambda_L = 0$ as we did previously. For the moment, it will be convenient to work with the variance ratio, $\psi = \sigma_\alpha^2/\sigma_\varepsilon^2$. Since $\psi = \rho/(1 - \rho)$ is a one-to-one function of ρ, the two parametrizations are equivalent. Rewrite the baseline association (8.9) as

$$S_\ell = \sigma_\varepsilon^2(\lambda_\ell \psi + 1)V_\ell, \quad V_\ell \sim \mathsf{ChiSq}(r_\ell), \quad \text{independent}, \quad \ell = 1, \ldots, L. \quad (8.11)$$

If we make the following transformations,

$$X_\ell = (S_\ell/r_\ell)/(S_L/r_L), \quad \ell = 1,\ldots,L-1, \qquad X_L = S_L,$$
$$U_\ell = (V_\ell/r_\ell)/(V_L/r_L), \quad \ell = 1,\ldots,L-1, \qquad U_L = V_L.$$

then the association (8.11) becomes

$$X_\ell = \frac{\lambda_\ell \psi + 1}{\lambda_L \psi + 1} U_\ell, \quad \ell = 1,\ldots,L-1, \qquad X_L = \sigma_\varepsilon^2 (\lambda_L \psi + 1) U_L.$$

Since for every (X,U,ψ), there exists a σ_ε^2 that solves the right-most equation, it follows from the general theory in Chapter 7 that a marginal association for ψ is obtained by deleting the component above involving σ_ε^2. In particular, a marginal association for ψ is

$$X_\ell = \frac{\lambda_\ell \psi + 1}{\lambda_L \psi + 1} U_\ell, \quad \ell = 1,\ldots,L-1.$$

If we write

$$f_\ell(\rho) = \frac{1 + \rho(\lambda_\ell - 1)}{1 + \rho(\lambda_L - 1)}, \quad \ell = 1,\ldots,L-1,$$

then we get a marginal association for ρ of the form

$$X_\ell = f_\ell(\rho) U_\ell, \quad \ell = 1,\ldots,L-1. \tag{8.12}$$

Marginalization reduces the auxiliary variable dimension by 1. Further dimension reduction will be considered next. Note that the new auxiliary variable U is a multivariate F-distributed random vector [2].

The goal is to construct a local conditional IM for ρ as described in Chapter 6. Select a fixed value ρ_0; more on this choice below. To reduce the dimension of the auxiliary variable U in (8.12) from $L-1$ to 1, we construct two pairs of functions, (T,H_0) and (τ,η_0). We insist that $x \mapsto (T(x),H_0(x))$ and $u \mapsto (\tau(u),\eta_0(u))$ are both one-to-one, and $H_0 = H_{\rho_0}$ and $\eta_0 = \eta_{\rho_0}$ are allowed to depend on the selected ρ_0. We want η_0 to be insensitive to changes in the auxiliary variable, and the strategy is to use the partial differential equation techniques like those in the previous section.

Write the association (8.12) so that u is a function of x and ρ, i.e.,

$$u_\ell(x,\rho) = x_\ell / f_\ell(\rho), \quad \ell = 1,\ldots,L-1.$$

We want to choose the function η_0 such that the partial derivative of $\eta_0(u(x,\rho))$ with respect to ρ vanishes at $\rho = \rho_0$. By the chain rule, we have

$$\frac{\partial \eta_0(u(x,\rho))}{\partial \rho} = \frac{\partial \eta_0(u)}{\partial u}\bigg|_{u=u(x,\rho)} \frac{\partial u(x,\rho)}{\partial \rho},$$

so our goal is to find η_0 to solve the following partial differential equation:

$$\frac{\partial \eta_0(u)}{\partial u}\bigg|_{u=u(x,\rho)} \frac{\partial u(x,\rho)}{\partial \rho} = 0, \quad \text{at } \rho = \rho_0;$$

here $\partial u/\partial \rho$ is a $(L-1) \times 1$ vector and $\partial \eta_0/\partial u$ is a $(L-2) \times (L-1)$ matrix of rank $L-2$. Toward solving this partial differential equation, first we get that the partial derivative of $u_\ell(x,\rho)$ with respect to ρ satisfies

$$\frac{\partial u_\ell(x,\rho)}{\partial \rho} = -\frac{f_\ell'(\rho)}{f_\ell(\rho)^2} x_\ell = -g(\rho) u_\ell(x,\rho),$$

where $g(\rho) = \{\frac{\partial}{\partial \rho} \log f_1(\rho), \ldots, \frac{\partial}{\partial \rho} \log f_{L-1}(\rho)\}^\top$. This simplifies the relevant partial differential equation to the following:

$$\left.\frac{\partial \eta_0(u)}{\partial u}\right|_{u=u(x,\rho)} \operatorname{diag}\{u(x,\rho)\} g(\rho) = 0, \quad \text{at } \rho = \rho_0, \tag{8.13}$$

where $\operatorname{diag}(a)$ is a diagonal matrix constructed from a vector a. The method of characteristics [200] identifies a logarithmic function of the form

$$\eta_0(u)^\top = (\log u_1, \ldots, \log u_{L-1}) M_0^\top, \tag{8.14}$$

where $M_0 = M_{\rho_0}$ is a $(L-2) \times (L-1)$ matrix with rows orthogonal to $g(\rho)$ at $\rho = \rho_0$. For example, since the matrix that projects to the orthogonal complement of the column space of $g(\rho_0)$ has rank $L-2$, we can take M_0 to be a matrix whose $L-2$ rows form a basis for that space. For M_0 defined in this way, it is easy to check that η_0 in (8.14) is indeed a solution to the partial differential equation (8.13).

Now that we have the η_0 function, it remains to specify τ and (T, H_0). The easiest to specify next is $H_0(x)$, the value of $\eta(u(x,\rho_0))$, as a function of x:

$$H_0(x)^\top = \left(\log \frac{x_1}{f_1(\rho_0)}, \ldots, \log \frac{x_{L-1}}{f_{L-1}(\rho_0)}\right) M_0^\top.$$

As we describe below, the goal is to condition on the observed value of $H_0(X)$.

Next, we define T and τ to supplement H_0 and η_0, respectively. In particular, take a $(L-1)$-vector $w(\rho)$ which is not orthogonal to $g(\rho)$ at $\rho = \rho_0$. It is easy to check that the entries in $g(\rho)$ are strictly positive for all ρ. Therefore, we can take $w(\rho)$ independent of ρ; for example, $w(\rho) \equiv 1_{L-1}$ is not orthogonal to $g(\rho)$. Now set

$$\begin{pmatrix} \tau(u) \\ \eta_0(u) \end{pmatrix} = \begin{pmatrix} 1_{L-1}^\top \\ M_0 \end{pmatrix} \begin{pmatrix} \log u_1 \\ \vdots \\ \log u_{L-1} \end{pmatrix}. \tag{8.15}$$

This is a log-linear transformation, and the linear part is non-singular, so this is a one-to-one mapping. Finally, we take T as

$$T(x) = \sum_{\ell=1}^{L-1} \log x_\ell.$$

Since $(T(x), H_0(x))$ is log-linear, just like $(\tau(u), \eta_0(u))$, it is also one-to-one.

We can now write the conditional association for ρ. For the given ρ_0, the mapping $x \mapsto (T(x), H_0(x))$ describes a split of our previous association (8.12) into two pieces:

$$\sum_{\ell=1}^{L-1} \log X_\ell = \sum_{\ell=1}^{L-1} \log f_\ell(\rho) + \sum_{\ell=1}^{L-1} \log U_\ell \quad \text{and} \quad H_0(X) = \eta_0(U).$$

The first piece carries direct information about ρ. The second piece plays a conditioning role, correcting for the fact that some information was lost in reducing the $(L-1)$-dimensional X to a one-dimensional $T(X)$. To complete the specification, write $\phi(\rho) = \sum_{\ell=1}^{L-1} \log f_\ell(\rho)$ and $V = \sum_{\ell=1}^{L-1} \log U_\ell$. Then we have

$$T(X) = \phi(\rho) + V, \quad V \sim P_{V|h_0, \rho_0}, \tag{8.16}$$

where $P_{V|h_0, \rho_0}$ is the conditional distribution of $\tau(U)$, given that $\eta_0(U)$ equals the observed value h_0 of $H_0(X)$.

To summarize, (8.16) completes the association step that describes the connection between observable data, unknown parameter of interest, and unobservable auxiliary variables. Of particular interest is that this association involves only a one-dimensional auxiliary variable compared to the association (8.11) obtained from the minimal sufficient statistics that involves an L-dimensional auxiliary variable. This dimension reduction will come in handy for the choice of a predictive random set in the following section. The price we paid for this dimension reduction was the choice of a particular localization point ρ_0. In the next section, we employ a trick to side-step this issue when the goal is to construct plausibility intervals for ρ.

8.3.4.3 Exact validity

Having reduced the auxiliary variable to a scalar in (8.16), the choice of an efficient predictive random set is now relatively simple. Consider the following version of the "default" predictive random set:

$$S = \{v : |v - \mu_0| \le |V - \mu_0|\}, \quad V \sim P_{V|h_0, \rho_0}. \tag{8.17}$$

This S, with distribution $P_{S|h_0, \rho_0}$, is a random interval, centered at the mean μ_0 of the conditional distribution $P_{V|h_0, \rho_0}$. For the contour function of S, defined as

$$\gamma_S(v) = P_{S|h_0, \rho_0}(S \ni v),$$

it is easy to check that

$$\gamma_S(v) = P_{V|h_0, \rho_0}\{|V - \mu_0| \ge |v - \mu_0|\} = 1 - F_{h_0, \rho_0}(|v - \mu_0|),$$

where F_{h_0, ρ_0} is the distribution function of $|V - \mu_0|$ for $V \sim P_{V|h_0, \rho_0}$. From the construction above, it is clear that it is a continuous distribution. Then, $|V - \mu_0|$ is a continuous random variable, so $\gamma_S(V) \sim \text{Unif}(0, 1)$. Therefore, S in (8.17) is valid, so our IM output for ρ has a meaningful scale for interpretation. In particular, the marginal plausibility intervals for ρ, defined below, will be exact.

Let $X = x$ be the observations in (8.12). The association step yields the collection of sets $R_x(v) = \{\rho : T(x) = \phi(\rho) + v\}$ indexed by v. Combine these with the predictive random set above to get an enlarged x-dependent random set:

$$R_x(\mathcal{S}) = \bigcup_{v \in \mathcal{S}} R_x(v). \tag{8.18}$$

Now, for a given assertion $\{\rho = r\}$, we compute the plausibility function,

$$\mathsf{pl}_{x|h_0,\rho_0}(r) = \mathsf{P}_{\mathcal{S}|h_0,\rho_0}\{R_x(\mathcal{S}) \ni r\},$$

the probability that the random set $R_x(\mathcal{S})$ contains the asserted value r of ρ. A simple calculation shows that

$$\mathsf{pl}_{x|h_0,\rho_0}(r) = \gamma_{\mathcal{S}}\left(T(x) - \phi(r)\right) = 1 - F_{h_0,\rho_0}(|T(x) - \phi(r) - \mu_0|),$$

where F_{h_0,ρ_0} is the distribution function defined above. The above display shows that the plausibility function can be expressed directly in terms of the distribution of $\mathcal{S}$, without needing to go through the construction of $R_x(\mathcal{S})$ as in (8.18).

Recall the question of how to choose the localization point ρ_0. Following [182], we propose to choose ρ_0 to match the value of ρ specified by the singleton assertion. That is, we propose to let the localization point depend on the assertion. All the elements in the plausibility function above with a 0 subscript, denoting dependence on ρ_0, are changed in an obvious way to get a new plausibility function

$$\mathsf{pl}_{x|h_\rho,\rho}(\rho) = 1 - F_{h_\rho,\rho}(|T(x) - \phi(\rho) - \mu_\rho|). \tag{8.19}$$

We treat this as a function of ρ to be used for inference. In particular, we can construct a $100(1 - \alpha)\%$ plausibility interval for ρ as follows:

$$\Pi_\alpha(x) = \{\rho : \mathsf{pl}_{x|h_\rho,\rho}(\rho) > \alpha\}. \tag{8.20}$$

The plausibility function, and the corresponding plausibility region, are easy to compute. Moreover, the calibration of the predictive random set leads to exact plausibility function-based confidence intervals, as we now show.

We need some notation for the sampling distribution of X, given all the relevant parameters. Recall that the distribution of X actually depends on $(\sigma_\alpha^2, \sigma_\varepsilon^2)$ or, equivalently, $(\rho, \sigma_\varepsilon^2)$. The error variance σ_ε^2 is a nuisance parameter, but σ_ε^2 still appears in the sampling model for X. We write this sampling distribution as $\mathsf{P}_{X|\rho,\sigma_\varepsilon^2}$.

Proposition 8.1. *Take the association (8.16) and the default predictive random set $\mathcal{S}$. Then for any ρ, any value h_ρ of H_ρ, and any σ_ε^2, the plausibility function satisfies*

$$\mathsf{P}_{X|\rho,\sigma_\varepsilon^2}\{\mathsf{pl}_{X|h_\rho,\rho}(\rho) \leq \alpha \mid H_\rho(X) = h_\rho\} = \alpha, \quad \forall\,\alpha \in (0,1). \tag{8.21}$$

Proof. See Exercise 8.5.   □

Averaging the left-hand side of (8.21) over h_ρ, with respect to the distribution of $H_\rho(X)$, and using an iterated expectation gives the following unconditional version of Proposition 8.1.

Corollary 8.1. *Under the conditions of Proposition 8.1, for any* $(\rho, \sigma_\varepsilon^2)$,

$$\mathsf{P}_{X|\rho,\sigma_\varepsilon^2}\{\mathsf{pl}_{X|H_\rho(X),\rho}(\rho) \le \alpha\} = \alpha, \quad \forall\, \alpha \in (0,1).$$

Since we have proper calibration of the plausibility function, both conditionally and unconditionally, coverage probability results for the plausibility interval (8.20) are also available. This justifies our choice to call $\Pi_\alpha(x)$ a $100(1-\alpha)\%$ plausibility interval, i.e., the frequentist coverage probability of Π_α is exactly $1 - \alpha$.

Corollary 8.2. *The coverage probability of* $\Pi_\alpha(X)$ *in* (8.20) *is exactly* $1 - \alpha$.

8.3.4.4 Numerical results

Evaluation of the plausibility function in (8.19) requires the distribution function $F_{h_\rho,\rho}$ of $|V - \mu_\rho|$ corresponding to the conditional distribution $\mathsf{P}_{V|h_\rho,\rho}$ of $V = \tau(U)$, given $\eta_\rho(U) = H_\rho(X)$. This conditional distribution is not of a convenient form, so numerical methods are needed. For ρ fixed, since the transformation (8.15) from U to $(\tau(U), \eta_\rho(U))$ is of a log-linear form, and the density function of U can be written in closed-form, we can evaluate the joint density for $(\tau(U), \eta_\rho(U))$ and, hence, the conditional density of $V = \tau(U)$. Numerical integration is used to evaluate the normalizing constant, the mean μ_ρ, and the distribution function $F_{h_\rho,\rho}$.

For our simulation study, we consider a standard one-way random effects model:

$$y_{ij} = \mu + \alpha_i + \varepsilon_{ij}, \quad i = 1, \ldots, a, \quad j = 1, \ldots, n_i,$$

where $\alpha_1, \ldots, \alpha_a$ are independent with common distribution $N(0, \sigma_\alpha^2)$, and the ε_{ij}s are independent with common distribution $N(0, \sigma_\varepsilon^2)$; the α_is and ε_{ij}s are also mutually independent. Our goal is to compare the proposed IM-based plausibility intervals for ρ with the confidence intervals based on several competing methods. Of course, the properties of the various intervals depend on the design, in this case, the within-group sample sizes $n_1, \ldots, n_a$, and the values of $(\sigma_\alpha^2, \sigma_\varepsilon^2)$. Our focus here is on cases with small sample sizes, namely, where the total sample size $n = n_1 + \cdots + n_a$ is fixed at 15. The three design patterns $(n_1, \ldots, n_a)$ considered are: $(1,1,1,1,1,10)$, $(2,4,4,5)$, and $(2,3,10)$. The nine $(\sigma_\alpha^2, \sigma_\varepsilon^2)$ pairs considered are: $(0.05, 10)$, $(0.1, 10)$, $(0.5, 10)$, $(1, 10)$, $(0.5, 2)$, $(1, 1)$, $(2, 0.5)$, $(5, 0.2)$, and $(10, 0.1)$. Without loss of generality, we set $\mu = 0$.

For each design pattern and pair of $(\sigma_\alpha^2, \sigma_\varepsilon^2)$, 1000 independent data sets were generated and 95% two-sided interval estimates for ρ were computed based on the exact method of [35], the fiducial method of [71], and the proposed IM method. Empirical coverage probabilities and average length of the confidence interval under each setting were compared to investigate the performance of each method. Besides these three methods, we also implemented Bayesian and profile likelihood approaches. The three aforementioned methods all gave better intervals than the Bayesian method, and the profile likelihood method was very unstable with small sample sizes, often having very high coverage with very wide intervals or very low coverage with very narrow intervals. So, these results are not reported.

A summary of the simulation results is displayed in Figure 8.1. Panel (a) displays

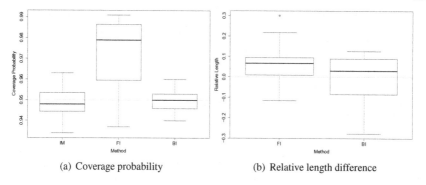

(a) Coverage probability   (b) Relative length difference

Figure 8.1 *Taken from [41]. Simulation results from Section 8.3.4.4. BI corresponds to the exact method of [35], and FI corresponds to the fiducial method of [71].*

the coverage probabilities, and Panel (b) displays the relative length difference, which is defined as the length of the particular interval minus the length of the IM interval, scaled by the length of the IM interval. As we expect from Corollary 8.2, the IM plausibility intervals have coverage at the 95% level. We also see that the IM intervals tend to be shorter than the fiducial and Burch–Iyer confidence intervals. The fiducial intervals have coverage probability exceeding the nominal 95% level, but this comes at the expense of longer intervals on average. Overall, the IM method performs well compared to these existing methods. We also replicated the simulation study in [71], which involves larger sample sizes and a broader range of imbalance, and the relative comparisons between these three methods are the same as here.

Example 8.1. Equal numbers of subjects are tested under each standard and test preparations and a blank dose under a $(2K + 1)$-point symmetrical slope-ratio assay. The response, on logarithmic scale, is assumed to depend linearly on the dose level. A modified balanced incomplete block design with $2K' + 1$ $(K' < K)$ block size is introduced by [52]. The ith dose levels for standard and test preparations are represented by s_i and t_i, $i = 1, \ldots, K$. Under this design, the dose will be equally spaced and listed in ascending order. A balanced incomplete block design with K doses of the standard preparation inside K' blocks is constructed and used as the basic design. Then a modified design is constructed by adding a blank dose and K' doses of the test preparation into every block, under the rule that dose t_i should accompany s_i in every blocks. The model developed by Das and Kulkarni can be written as

$$y_{ijm} = \mu + \beta_j x_{ij} + \alpha_m + \varepsilon_{ijm}, \quad i \in \{s, t, c\}, \quad j = 1, \ldots, k, \quad m = 1, \ldots, b$$

where $y_{sjm}, y_{tjm}, y_{cjm}$ represent observation response in the mth block for the jth dose of standard preparation, test preparation, and blank dose; x_{sj} and x_{tj} represent the jth dose level for standard and test preparation; x_{cj} is zero by default, α_m denotes the mth block effect; and ε_{ijm} denotes independent random errors with common distribution $N(0, \sigma_\varepsilon^2)$. We consider the random block effects and assume that α_m are independent with common distribution $N(0, \sigma_\alpha^2)$. Independence of α_m and ε_{ijm} is also assumed.

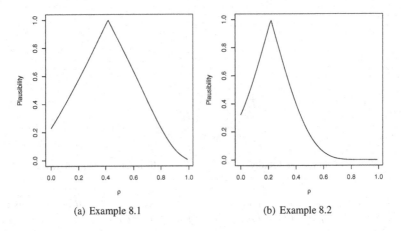

(a) Example 8.1   (b) Example 8.2

Figure 8.2 *Taken from [41]. Plausibility functions for* ρ *in the two examples.*

Table 8.1 *Taken from [41]. Upper bounds on the interval estimates for* ρ *(lower bounds are all zero) based on the three methods in the two real-data examples.*

	Example 8.1		Example 8.2	
Method	90%	95%	90%	95%
Burch–Iyer	0.913	0.956	0.567	0.615
Fiducial	0.916	0.957	0.466	0.530
IM	0.881	0.924	0.554	0.597

We analyze data coming from a nine-point slope-ratio assay on riboflavin content of yeast, with two replications in each dose; see Table 2 in [71] for the design and data. For this design, we have $L = 3$ distinct eigenvalues, namely, 4.55, 1, and 0, with multiplicities 1, 1, and 10, respectively. A plot of the plausibility function for ρ is shown in Figure 8.2(a). The function exceeds 0.2 at $\rho = 0$, which implies that 90% and 95% plausibility intervals include zero. The left panel of Table 8.1 shows the 90% and 95% interval estimates for ρ based on the Burch–Iyer, fiducial, and IM methods. In this case, the IM intervals are provably exact and shorter.

Example 8.2. Harville and Fenech [132] analyzed data on birth weights of lambs. These data consist of the weights information at the birth of 62 single-birth male lambs, and were collected from three selection lines and two control lines. Each lamb was the offspring of one of the 23 rams and each lamb had a distinct dam. Age of the dam was also recorded and separated into three categories, numbered 1 (1–2 years), 2 (2–3 years), and 3 (over 3 years). A linear mixed model for these data is

$$y_{ijkl} = \mu + \beta_i + \pi_j + \alpha_{jk} + \varepsilon_{ijkl},$$

where y_{ijkl} represents the weight of the lth offspring of the kth sire in the jth population lines and of a dam in the ith age category; β_i represents the ith level age effect; π_j represents the jth line effects; α_{jk} denotes random sire effects and are assumed to be independently distributed as $N(0, \sigma_\alpha^2)$; and random errors denoted by ε_{ijkl} are supposed to be independently distributed as $N(0, \sigma_\varepsilon^2)$. Furthermore, the α_{jk}s and ε_{ijkl}s are assumed to be independent. In this case, $L = 18$, $\lambda_1 = 5.09$, $\lambda_L = 0$ and $r_L = 37$; all non-zero eigenvalues have multiplicity 1 except $\lambda_8 = 2$ with multiplicity 2. A plot of the plausibility function for ρ is shown in Figure 8.2(b). As in the previous example, the plausibility function is positive at $\rho = 0$, which means that plausibility intervals with any reasonable level will contain $\rho = 0$. We also used each of the three methods considered above to compute 90% and 95% interval estimates for ρ. The results are shown in the right panel of Table 8.1. In this case, IM gives a shorter interval compared to Burch–Iyer. The fiducial interval, however, is shorter than both exact intervals. We expect the IM interval to be most efficient, so we explore the relative performance a bit further by simulating 1000 independent data sets from the fitted model in this case, i.e., with $\hat{\sigma}_\alpha^2 = 0.767$ and $\hat{\sigma}_\varepsilon^2 = 2.763$ as the true values. In these simulations, the fiducial and IM coverage probabilities were 0.944 and 0.954, respectively, both within an acceptable range of the nominal level, but the average lengths of the intervals are 0.488 and 0.456. That is, the IM intervals tend to be shorter than the fiducial intervals in problems similar to this example, as we would expect.

8.4   Concluding remarks

This chapter considered a few special cases of an important class of normal linear models. The treatment here is certainly far from exhaustive, and Chapter 10 explores the problem of variable selection in the normal linear regression model. Additional examples that deserve further attention include ANOVA, and we expect that the details provided here and in Chapter 10 can provide some new insights on this problem, in particular, multiple comparisons. The reader can likely fill in at least the basic details about the ANOVA problem based on the presentation in this chapter, but some specific questions may deserve special attention in the future.

The mixed-effects model discussed in Section 8.3 is both interesting and challenging, and our analysis here is only a first step. We mentioned in passing that finding marginal IMs for the individual variance components is challenging; the point is that the marginal association for σ_α^2 or σ_ε^2 is not "regular" in the sense of Chapter 7. The generalized marginalization techniques described in Section 7.4 are currently being considered, but we expect that even more sophisticated marginalization techniques will be needed; see Chapter 12. A goal is to develop IM-based machinery that can handle the generality of the fiducial-based approach in [43].

Another important question is if the mixed effect model methodology presented herein can be applied in more complex and high-dimensional mixed-effect models. In genome-wide association studies, for example, mixed models are useful and the heritability coefficient is the target, but the dimensions of the problem are extremely large. We expect that, conceptually, the techniques described here will carry over to the more complex scenario. However, there will be computational challenges to overcome, as with all approaches [144, 271]. This, along with the incorporation of optimal predictive random sets, as in Chapter 4, is a topic for further research.

8.5   Exercises

Exercise 8.1.   Apply the differential equation-driven techniques described in Chapter 6.3.2 to obtain an alternative construction of the conditional association (8.2).

Exercise 8.2.   Consider the problem of inference on all regression coefficients β. A marginal IM for β is described in Section 8.2.3. Take the "default" predictive random set in (8.5). Find the plausibility of the singleton assertion $A = \{b\}$ for fixed $b \in \mathbb{R}^p$. (*Hint:* It involves an F-distribution calculation.)

Exercise 8.3.   Consider the linear regression model of Section 8.2. The goal is to construct an IM for marginal inference on a single regression coefficient, say, β_1.

(a) Derive a marginal IM for β_1.

(b) Simulate data from a regression model with dimension $p > 1$, and plot the plausibility function for singleton assertions about β_1.

(c) On the same plot, overlay the plausibility function obtained by "projecting" the joint IM for β in Section 8.2.3 down to the β_1-margin, via optimization; see Exercise 7.5. Comment on the relative efficiency of the two marginal IMs in terms of the spread of their respective plausibility functions.

Exercise 8.4.   Consider the ordinary linear regression model of Section 8.2, and suppose that σ^2 is the parameter of interest. Following the ideas in Chapter 4, derive an optimal score-balanced predictive random set for inference on σ^2. Compare this to the IM derived in Section 8.2.4 based on a default predictive random set.

Exercise 8.5.   Prove Proposition 8.1.

Chapter 9

Prediction of Future Observations

Portions of the material in this chapter are from R. Martin and R. Lingham, "Prior-free probabilistic prediction of future observations," *Technometrics*, to appear, 2015, reprinted with permission by the American Statistical Association, www.amstat. org.

9.1 Introduction

The prediction of future observations based on the information available in a given sample is a fundamental problem in statistics. For example, in engineering applications, such as computer networking, one might want to predict, to some degree of certainty, the time at which the current system might fail, in order to have resources available to fix it. Despite this being a fundamental problem, the available literature does not seem to give any clear guidelines about how to approach a prediction problem in general. From a frequentist point of view, there are a host of techniques available for constructing prediction intervals in specific examples; see, for example, the book [123] and the papers [21, 89, 152, 250]. Some general approaches to the frequentist prediction problem are presented in [12, 155]. From a Bayesian point of view, if a prior distribution is available for the unknown parameter, then the prediction problem is conceptually straightforward. The Bayesian model admits a joint distribution for the observed data and future data, so a conditional distribution for the latter given the former—the Bayesian predictive distribution—is the natural tool. Applications of Bayesian prediction are presented in [125]. The catch is that often there is no clear choice of prior. Default, or non-informative priors can be used but, in that case, it is not clear that the resulting inference will be meaningful in either a personal probabilistic or frequentist sense. A (generalized) fiducial approach for prediction is presented in [249] which, at a high-level, can be viewed as a sort of compromise to the frequentist and Bayesian approaches. They propose a very natural predictive distribution that obtains from the usual fiducial distribution for the parameter [126, 127]. They show that the prediction intervals obtained from the fiducial predictive distribution are asymptotically correct and perform well, in examples, compared to existing prediction intervals.

The fiducial approach is attractive because no prior distributions are required. However, like objective Bayesian posterior or predictive distributions, fiducial distributions may not be calibrated for meaningful probabilistic inference, except possibly in the limit [165]. The IM framework, discussed in Chapter 4, in general, provides prior-free probabilistic inference on unknown parameters, through the use of auxiliary variables and predictive random sets. A consequence of this special handling of the auxiliary variables is that the inferential output is calibrated both for meaningful interpretation in a given problem and for constructing procedures with desirable frequentist properties.

In this chapter, we provide an IM-based solution to the problem of predicting future observations. The critical observation that drives the approach here is that predicting future observations is a marginal inference problem, one in which the full parameter itself is a nuisance parameter to be marginalized out. With this view, we apply the general marginalization principles from Chapter 7 to eliminate the nuisance parameter, directly providing a marginal IM for the future observations. In Section 9.2.4 we give general conditions under which the resulting IM for prediction is valid, and discuss under what circumstances these conditions hold, in what cases they can be weakened, and other consequences. The key point is that the plausibility function obtained from a valid IM provides a probabilistic summary of the information in the observed data concerning the future data to be predicted; this function can be plotted to provide some visual summary. Moreover, the validity theorem demonstrates that the predictive plausibility interval, defined in (9.8), has the nominal frequentist coverage for all finite samples, not just in the limit. Our focus here is on the case of predicting a univariate future observable, but the multivariate case, discussed briefly in Section 9.4.3, requires some additional considerations. Several practical examples of prediction in the IM context are worked out in Section 9.3. These examples involve a variety of common models, and prediction problems in quality control, environmental, system breakdown, and disease count applications are considered. To compare our IM-based solution to other existing methods, we focus on the frequentist performance of our prediction intervals. In all the examples we consider, the IM intervals are competitive with the existing methods. The take-away message is that the IM approach provides an easily implementable and general method for constructing meaningful prior-free probabilistic summaries of the information in observed data for inference or prediction; that these summaries can be converted to frequentist procedures with fixed-n performance guarantees and comparable efficiencies compared to existing methods is an added bonus.

9.2 Inferential models for prediction

9.2.1 Preview for prediction

Before getting into the general details about the prediction problem, we present a relatively simple example as a preview of our proposed IM approach. Consider a homogeneous Poisson process $\{N(t) : t \geq 0\}$ with rate $\theta > 0$. The arrival times $T_0, T_1, T_2, \ldots$ are such that $T_0 \equiv 0$ and the inter-arrival times $T_i - T_{i-1}$, $i \geq 1$, are independent exponential random variables with rate θ.

If the sampling scheme is to wait for the n-th arrival, then the sufficient statistic for θ in this model is $Y = T_n$, the last arrival time. Based on the arguments in

Chapter 6 and [182], the baseline association for θ is

$$Y = (1/\theta)G_n^{-1}(U), \quad U \sim \mathsf{Unif}(0,1),$$

where G_n is the $\mathsf{Gamma}(n,1)$ distribution function. If inference on θ was the goal, then this would complete the A-step. However, suppose the goal is to predict $\tilde{Y} = T_{n+k}$, the time of the $(n+k)$-th arrival, for some fixed integer $k \geq 1$. Then θ itself is a nuisance parameter, and the quantity of interest is $\tilde{Y}$. From the baseline association above, we can easily solve for θ in terms of (Y,U), i.e.,

$$\theta(Y,U) = G_n^{-1}(U)/Y.$$

Since $\tilde{Y} = T_{n+k}$, for given $Y = T_n$, equals Y plus an independent gamma random variable with shape k and rate θ, following the approach described in Chapter 7 and [183], we have a marginal association for $\tilde{Y}$ given by

$$\tilde{Y} = Y + \frac{1}{\theta(Y,U)}G_k^{-1}(\tilde{U}) = Y\left(1 + \frac{G_k^{-1}(\tilde{U})}{G_n^{-1}(U)}\right).$$

This completes the A-step for prediction. If R denotes the ratio in the far right-hand side above, then R has a generalized gamma ratio distribution [48] with density function $f(r) \propto (1+r)^{-(n+k)}$, $r > 0$. If F is the corresponding distribution function, then we may rewrite the marginal association as

$$\tilde{Y} = Y\{1 + F^{-1}(W)\}, \quad W \sim \mathsf{Unif}(0,1).$$

Thus, we have successfully marginalized out the unknown parameter, directly associating the quantity to be predicted, $\tilde{Y}$, to the observed data, Y, and an auxiliary variable, W. Then, the general IM principles [180, 182, 183] can be applied directly. In particular, we apply the P- and C-steps to the association for $\tilde{Y}$, resulting in prior-free probabilistic prediction of the future arrival time. The next two subsections will describe the proposed approach in more detail, and our examples in Section 9.3 will demonstrate its generality, its quality performance, and its simplicity.

9.2.2 General setup and the A-step

In the prediction problem, there is observed data Y and future data $\tilde{Y}$ to be predicted; the two are linked together through a common parameter θ. Here we assume that $\tilde{Y}$ is a scalar, though it could be a function of several future observations; see Section 9.4.3 for discussion on the multivariate prediction problem. Write the sampling model $\mathsf{P}_{Y|\theta}$ for Y in association form:

$$Y = a(\theta,U), \quad U \sim \mathsf{P}_U, \tag{9.1}$$

where P_U is known and free of θ. We call this the "baseline" association, and it connects observable data Y and unknown parameter θ to an unobservable auxiliary variable U. Despite its simple form, the baseline association is quite general, i.e., it covers cases outside the structural models in [99]; see Sections 9.3.3–9.3.4. For example, for any iid model with a smooth distribution function F_θ, take the i-th component of $a(\theta, U)$ to be $F_\theta^{-1}(U_i)$ for $U_i \sim \mathsf{Unif}(0, 1)$. Intuitively, any model that can be simulated has a form (9.1).

As our first step, assume that this baseline association can be re-expressed as

$$T(Y) = b(\theta, \tau(U)) \quad \text{and} \quad H(Y) = \eta(U),$$

for functions (T, H) and (τ, η) such that $y \mapsto (T(y), H(y))$ and $u \mapsto (\tau(u), \eta(u))$ are one-to-one. A key feature of this decomposition is that a solution $\theta = \theta(y, v)$ of the equation $T(y) = b(\theta, v)$ is available for all (y, v). By conditioning on the observed value, $H(Y)$, of $\eta(U)$, this association can then be reduced as follows:

$$T(Y) = b(\theta, V), \quad V \equiv \tau(U) \sim \mathsf{P}_{\tau(U)|\eta(U)=H(Y)}.$$

Chapter 6 showed that such a decomposition exists in broad generality. For simplicity, we assume here that $\tau(U)$ and $\eta(U)$ are independent, so the conditioning can be dropped, i.e., $\mathsf{P}_{\tau(U)|\eta(U)=H(Y)} \equiv \mathsf{P}_{\tau(U)}$. This assumption holds for many problems, including those in Section 9.3. Dependence in this context is only a technical complication, not conceptual, so we focus here on the simpler case of independent $\tau(U)$ and $\eta(U)$; the dependent case is discussed further in Section 9.4.2.

For the observed data Y and the future data $\tilde{Y}$, write a joint association:

$$T(Y) = b(\theta, V) \quad \text{and} \quad \tilde{Y} = \tilde{a}(\theta, \tilde{U}),$$

where $(V, \tilde{U}) \sim \mathsf{P}_{(V, \tilde{U})}$. When Y and $\tilde{Y}$ are independent, V and $\tilde{U}$ are likewise independent, but in time series problems, for example, the auxiliary variables will be correlated. The use of "$\tilde{a}$" for the mapping instead of simply "a" is to cover the case where Y and $\tilde{Y}$ are related through a common parameter θ, but possibly have different distributions. For example, Y might be an iid normal sample, while $\tilde{Y}$ is the maximum of ten future normal samples; similarly, in a regression context, Y and $\tilde{Y}$ might have different values of the predictor variables.

Solving for θ in the first equation and plugging in to the second gives

$$T(Y) = b(\theta, V) \quad \text{and} \quad \tilde{Y} = \tilde{a}\big(\theta(Y, V), \tilde{U}\big),$$

Since prediction is a marginal inference problem, where θ itself is the nuisance parameter, it follows from the general theory in Chapter 7 that the first equation in the above display can be ignored. This leaves a marginal association for $\tilde{Y}$:

$$\tilde{Y} = \tilde{a}\big(\theta(Y, V), \tilde{U}\big). \tag{9.2}$$

This marginalization has some similarities to the Bayesian and fiducial predictive distributions. That is, the model for $\tilde{Y}$ in (9.2) is that of a mixture of the distribution of $\theta(Y, V)$, for fixed Y, with the distribution of $\tilde{a}(\theta, \tilde{U})$ for fixed θ. This, of course, is not the "true" distribution of $\tilde{Y}$ given Y; the idea is that the future observable $\tilde{Y}$ is being modeled as a Y-dependent function of $(V, \tilde{U})$. We claim that equation (9.2) describes a sort of predictive distribution of $\tilde{Y}$ for a given Y, similar to the frequentist predictive distributions in, e.g., [155]. To see this better, let G_Y be the distribution of the right-hand side of (9.2) as a function of $(V, \tilde{U})$ for fixed Y. Then, in the case where this is an absolutely continuous distribution, (9.2) can be rewritten as

$$\tilde{Y} = G_Y^{-1}(W), \quad W \sim P_W = \mathsf{Unif}(0,1), \tag{9.3}$$

so G_Y plays the role of a predictive distribution for $\tilde{Y}$. This completes the A-step in the construction of the IM for prediction of future data $\tilde{Y}$.

Though (9.2) has some connection to Bayesian and fiducial prediction, it differs from parametric bootstrap prediction. The difference is that the quantity $\theta(Y,V)$ plugged in is not fixed. That is, we consider the distribution of $\tilde{a}(\theta(Y,V),\tilde{U})$ as a function of $(V,\tilde{U})$, not the distribution of $\tilde{a}(\hat{\theta}_Y,\tilde{U})$, as a function of $\tilde{U}$, for fixed $\hat{\theta}_Y$.

9.2.3 P- and C-steps

After the A-step in (9.3), the P-step requires specification of a suitable predictive random set $S \sim P_S$ for W. Recall that the choice of predictive random set can depend on the assertion A of interest [180]. There are three kinds of assertions about $\tilde{Y}$ that will be of interest here in the prediction problem: two one-sided assertions, and a singleton assertion. Given a predictive random set and an assertion of interest, the C-step proceeds by combining the A- and P-step results. Chapter 4 gives a general explanation, but here this amounts to computing the *plausibility of A*, i.e.,

$$\mathsf{pl}_Y(A) = P_S\{G_Y^{-1}(S) \cap A \neq \varnothing\}.$$

Next we discuss, in turn, the P- and C-steps for each of these kinds of assertions.

- *Right-sided.* A right-sided assertion is of the form $A = \{\tilde{Y} > \tilde{y}\}$ for a fixed $\tilde{y}$. For this assertion, by Theorem 4.4, the optimal predictive random set is one-sided: $S = [0, W]$ for $W \sim \mathsf{Unif}(0,1)$. In this case, the C-step gives the plausibility function

$$\mathsf{pl}_Y(A) = P_S\{G_Y^{-1}(S) \cap A \neq \varnothing\} = 1 - G_Y(\tilde{y}) \tag{9.4}$$

  The plausibility function is a non-increasing function of $\tilde{y}$; see Figure 9.1(a) described in Section 9.3.2. Hence the prediction region (9.8) based on the plausibility function in (9.4) will be an upper prediction bound for $\tilde{Y}$.

- *Left-sided.* A left-sided assertion is of the form $A = \{\tilde{Y} \leq \tilde{y}\}$ for a fixed $\tilde{y}$. Similar to the right-sided case, the optimal predictive random set is $S = [W, 1]$ for $W \sim \mathsf{Unif}(0,1)$. Then the C-step gives the plausibility function

$$\mathsf{pl}_Y(A) = P_S\{G_Y^{-1}(S) \cap A \neq \varnothing\} = G_Y(\tilde{y}) \tag{9.5}$$

  The plausibility function is a non-decreasing function of $\tilde{y}$; see Figure 9.2(a). Hence the prediction region (9.8) based on the plausibility function in (9.5) will be a lower prediction bound for $\tilde{Y}$.

- *Singleton.* A singleton assertion is of the form $A = \{\tilde{Y} = \tilde{y}\}$ for a fixed $\tilde{y}$. The optimal predictive random set worked out in [180] for this assertion is complicated, but a natural choice that is suitable in most cases (and optimal in some cases) is the "default" predictive random set $S = \{w : |w - 0.5| \leq |W - 0.5|\}$, for $W \sim \text{Unif}(0, 1)$. Then the C-step gives the plausibility function

$$\text{pl}_Y(A) = \text{P}_S\{G_Y^{-1}(S) \cap A \neq \varnothing\} = 1 - |2G_Y(\tilde{y}) - 1| \qquad (9.6)$$

The prediction region (9.8) based on the plausibility function in (9.6) will be a two-sided prediction bound for $\tilde{Y}$.

It is important to note that, although the general P- and C-steps may appear rather technical, implementation of the IM approach for prediction requires only that one be able to evaluate, either analytically or numerically, the distribution function G_Y. Section 9.3 gives several examples and applications to demonstrate that our IM-based plausibility intervals are good general tools for the prediction problem, and that such intervals are often better than what other methods provide.

9.2.4 Prediction validity

Here we give the main distributional property of the plausibility function for prediction. The key requirement is a mild condition on the predictive random set S. Recall the contour function $\gamma_S(w) = \text{P}_S(S \ni w)$, and we say that the predictive random set S is valid if

$$\gamma_S(W) \geq_{\text{st}} \text{Unif}(0, 1) \quad \text{when } W \sim \text{P}_W, \qquad (9.7)$$

where $\geq_{\text{st}}$ means "stochastically no smaller than." The three assertions A described in Section 9.2.3 depend on a generic $\tilde{y}$. Here we write $\text{pl}_Y(\tilde{y})$ for the plausibility function for such an assertion; the specific kind of assertion will be clear from the context.

Theorem 9.1. *For the marginal association (9.3) for $\tilde{Y}$, let $S \sim \text{P}_S$ be a valid predictive random set for $W \sim \text{Unif}(0, 1)$, i.e., (9.7) holds, which is non-empty with P_S-probability 1. If $G_Y(\tilde{Y}) \sim \text{Unif}(0, 1)$ for $(Y, \tilde{Y}) \sim \text{P}_{(Y,\tilde{Y})|\theta}$ for all θ, then*

$$\sup_\theta \text{P}_{(Y,\tilde{Y})|\theta}\{\text{pl}_Y(\tilde{Y}) \leq \alpha\} \leq \alpha, \quad \forall \, \alpha \in (0, 1).$$

This holds whether $\text{pl}_Y(\tilde{Y})$ is based on right-sided, left-sided, or singleton assertions.

Proof. See Exercise 9.1. □

The following sequence of remarks discusses the assumptions, interpretations, and various extensions of Theorem 9.1. See, also, Section 9.4.

Remark 9.1. Chapter 4 argues that validity gives the plausibility function a scale on which the numerical values can be interpreted. For example, like in the familiar case of p-values, if the plausibility function is small, e.g., $\text{pl}_y(\tilde{y}) < 0.05$, then, for the given $Y = y$, the value $\tilde{y}$ is not a plausible prediction; see, also, Remark 9.2.

Remark 9.2. A consequence of Theorem 9.1 is that the set

$$\{\tilde{y} : \mathsf{pl}_y(\tilde{y}) > \alpha\} \tag{9.8}$$

is a $100(1-\alpha)\%$ prediction plausibility interval/region, i.e., the probability that $\tilde{Y}$ falls inside the region (9.8) is at least $1-\alpha$ under the joint distribution of $(Y,\tilde{Y})$ for any parameter value θ. Then, for the three kinds of assertions, namely, right, left, and singleton, discussed in Section 9.2.3, one gets $100(1-\alpha)\%$ upper, lower, and two-sided prediction intervals, respectively. Moreover, the region (9.8) has the following desirable interpretation: each point $\tilde{y}$ it contains is individually sufficiently plausible. No frequentist, Bayes, or fiducial prediction interval assigns such a meaning to the individual elements it contains.

Remark 9.3. Suppose that S is such that $f_S(V) \sim \mathsf{Unif}(0,1)$ for $V \sim \mathsf{Unif}(0,1)$. Then $\mathsf{pl}_Y(\tilde{Y}) \sim \mathsf{Unif}(0,1)$ as a function of $(Y,\tilde{Y}) \sim \mathsf{P}_{(Y,\tilde{Y})|\theta}$ for all θ. These conditions hold in many examples (see Section 9.3) and they imply that the plausibility region in (9.8) has exact prediction coverage, $1-\alpha$, not just conservative.

Remark 9.4. A natural question is: under what conditions does $G_Y(\tilde{Y}) \sim \mathsf{Unif}(0,1)$ hold? An important example is the case we shall call "separable," where the effect of Y on the right-hand side of (9.2) can be separated from the auxiliary variables, i.e., (9.2) can be rewritten as $p(Y,\tilde{Y}) = \phi(V,\tilde{U})$ for some functions p and ϕ. In the language of [155], the quantity $p(Y,\tilde{Y})$ is an exact pivot. Many problems with a group transformation structure [72] are separable, and are covered by Theorem 9.1. Some of the examples in Section 9.3 are of this type, but the numerical results even for the non-separable models (see Sections 9.3.3–9.3.4) suggest that the validity result holds broadly. Section 9.4.1 has more discussion on the non-separable case.

Remark 9.5. An advantage of the IM's handling of the auxiliary variables, revealed in the previous remarks, is that one has finite-sample control on the prediction coverage. The fiducial approach to prediction, on the other hand, can only guarantee asymptotic control of frequentist prediction coverage, e.g., Theorem 1 in [249].

Remark 9.6. The uniformity condition in Theorem 9.1 can be relaxed to a stochastic ordering condition, but then the conclusion holds only for certain predictive random sets and certain assertions. See Exercise 9.2.

9.3 Examples and applications

9.3.1 *Normal models*

Let $Y = (Y_1, \dots, Y_n)$ be an iid sample from a $\mathsf{N}(\mu, \sigma^2)$ population, where $\theta = (\mu, \sigma)$ is unknown. Our first goal is to predict the next independent observation $\tilde{Y} = Y_{n+1}$. To start, consider the baseline association involving the original data

$$Y_i = \mu + \sigma Z_i, \quad i = 1, \dots, n,$$

where $Z_1, \dots, Z_n$ are iid $\mathsf{N}(0,1)$. Based on the arguments in Chapter 6, a conditional IM for $\theta = (\mu, \sigma)$ has association

$$\tilde{Y} = \mu + \sigma n^{-1/2} U_1 \quad \text{and} \quad S = \sigma U_2,$$

where $\bar{Y} = n^{-1} \sum_{i=1}^{n} Y_i$ is the sample mean, $S^2 = (n-1)^{-1} \sum_{i=1}^{n} (Y_i - \bar{Y})^2$ is the sample variance, $U_1 \sim \mathsf{N}(0,1)$, and $(n-1)U_2^2 \sim \mathsf{ChiSq}(n-1)$, with U_1 and U_2 independent. Then it is easy to see that

$$\theta(Y,U) = \left(\mu(Y,U), \sigma(Y,U)\right) = \left(\bar{Y} - \frac{S}{n^{1/2}} \frac{U_1}{U_2}, \frac{S}{U_2}\right).$$

For the next observation $\tilde{Y} = Y_{n+1}$, the association is just like the baseline association above, i.e., $\tilde{Y} = \mu + \sigma\tilde{U}$, where $\tilde{U}$ is independent of (U_1, U_2). As discussed above, we can insert $\theta(Y,U)$ in place of θ in this association to get a marginal association for $\tilde{Y}$:

$$\tilde{Y} = \bar{Y} - \frac{S}{n^{1/2}} \frac{U_1}{U_2} + \frac{S}{U_2}\tilde{U} = \bar{Y} + S\left(\frac{1}{n^{1/2}} \frac{U_1}{U_2} - \frac{\tilde{U}}{U_2}\right). \qquad (9.9)$$

This is clearly one of those separable cases as described in Remark 9.4. Also,

$$V = \frac{1}{n^{1/2}} \frac{U_1}{U_2} - \frac{\tilde{U}}{U_2}$$

is distributed as $(n^{-1}+1)^{1/2}\mathsf{t}(n-1)$, with distribution function F_n. Then the marginal association (9.9) can be written as $\tilde{Y} = \bar{Y} + SF_n^{-1}(W)$, with $W \sim \mathsf{Unif}(0,1)$. If we are interested in a two-sided prediction interval, then, as in Section 9.2.3, we take a singleton assertion $A = \{\tilde{Y} = \tilde{y}\}$ and get the following plausibility function:

$$\mathsf{pl}_Y(\tilde{y}) = 1 - \left|2F_n\left(\frac{\tilde{y} - \bar{Y}}{S}\right) - 1\right|.$$

Then the corresponding two-sided $100(1-\alpha)\%$ plausibility interval (9.8) for $\tilde{Y}$ is

$$\bar{Y} \pm t_{n-1,1-\alpha/2}^{\star} S(1+n^{-1})^{1/2},$$

where $t_{\nu,p}^{\star}$ is the $100p$th percentile of the t-distribution with ν degrees of freedom. This is exactly the classical Student-t prediction interval discussed in, e.g., [108].

The ideas just discussed extend quite naturally to the case of normal linear regression. The details of the IM calculations would be similar to those presented in [249] for the fiducial case; see Exercise 9.3.

As a more detailed example, Odeh [194] gives a quality control application involving sprinkler systems for fire prevention. In this application, based on a sample of $n = 20$ sprinklers, whose activation temperatures are normally distributed, the goal is to give a two-sided prediction interval for the temperature at which at least $k = 36$ of $m = 40$ new sprinklers will activate. In other words, the goal is to predict the temperature at which at least k of the m new sprinklers will activate. The IM methodology can be used for this problem. Let $\tilde{Y}$ be the k-th largest of m future independent normal observations $Y_{n+1}, \ldots, Y_{n+m}$. The corresponding association for $\tilde{Y}$ is

$$\tilde{Y} = \mu + \sigma\tilde{U}, \quad \text{where} \quad \tilde{U} = k\text{-th largest of } U_{n+1}, \ldots, U_{n+m},$$

and $U_{n+1}, \ldots, U_{n+m}$ are iid $\mathsf{N}(0,1)$. Then the marginal association for $\tilde{Y}$ can be written exactly as in (9.9) and the problem is still separable. The only difference here is that $V = (n^{-1/2}U_1 - \tilde{U})/U_2$ has a non-standard distribution. As before, write $\tilde{Y} = \bar{Y} + SF_{n,m,k}^{-1}(W)$, where $F_{n,m,k}$ is the distribution function of V, and $W \sim \text{Unif}(0,1)$. The distribution $F_{n,m,k}$ can be simulated and, therefore, one can easily get a Monte Carlo approximation of the plausibility function (9.6) for $\tilde{Y}$ and, in turn, a two-sided prediction interval. The IM prediction interval for $\tilde{Y}$ in this normal prediction problem is the same as the fiducial interval in [249] and the interval in [89].

9.3.2 Log-normal models

Let $Y = (Y_1, \ldots, Y_n)$ be an iid sample from a log-normal population, with unkown parameter $\theta = (\mu, \sigma)$. Log-normal models are frequently used in environmental statistics [196]. In this case, $X = (X_1, \ldots, X_n)$, with $X_i = \log(Y_i)$, will be an iid $\text{N}(\mu, \sigma^2)$ sample, and the prediction problem can proceed as in Section 9.3.1 above. In particular, predicting the next observation $\tilde{Y} = Y_{n+1}$ is straightforward, so we focus here on something more challenging. Consider, as in [21], the problem of finding the upper prediction limit for the arithmetic mean of m future log-normal observations, i.e., $\tilde{Y} = m^{-1}\sum_{j=1}^{m} Y_{n+j}$. Working on the log-scale, with the X_i's, we can first reduce dimension according to sufficiency and then solve for θ as follows:

$$\theta(Y, U) = \big(\mu(Y, U), \sigma(Y, U)\big) = \Big(\bar{X} - \frac{S}{n^{1/2}} \frac{U_1}{U_2}, \frac{S}{U_2}\Big),$$

where $\bar{X} = n^{-1}\sum_{i=1}^{n} X_i$, $S^2 = (n-1)^{-1}\sum_{i=1}^{n}(X_i - \bar{X})^2$, $U_1 \sim \text{N}(0,1)$, and $(n-1)U_2^2 \sim \text{ChiSq}(n-1)$, with U_1, U_2 being independent. The marginal association for $\tilde{Y}$, the arithmetic mean of m future log-normal observations, is

$$\tilde{Y} = \frac{1}{m}\sum_{j=1}^{m} e^{\log Y_{n+j}} = \frac{1}{m}\sum_{j=1}^{m} \exp\Big\{\Big(\bar{X} - \frac{S}{n^{1/2}}\frac{U_1}{U_2}\Big) + \Big(\frac{S}{U_2}\tilde{U}_{n+j}\Big)\Big\},$$

where $\tilde{U} = (\tilde{U}_{n+1}, \ldots, \tilde{U}_{n+m})$ are iid $\text{N}(0,1)$, independent of U_1 and U_2. Here we use this association for $\tilde{Y}$ to construct an upper plausibility prediction limit.

The above association is not of the separable form in Remark 9.4. However, for a given Y, if G_Y is the distribution of the right-hand side in the previous display, then the marginal association for $\tilde{Y}$ can be written in the form $\tilde{Y} = G_Y^{-1}(W)$, for $W \sim \text{Unif}(0,1)$, just as in (9.3), completing the A-step. Since we seek to determine an upper prediction limit, the plausibility function for $\tilde{Y}$ is given by (9.4).

For illustration, we consider an environmental study presented in [21] concerning lead concentration in soil. It is a "brownfield" investigation in which a now-closed plating facility was being investigated for future industrial use. In April 1996, $m = 5$ soil borings were installed to delineate the extent of lead-impacted soil at the portion of the facility that may have been used for plating. An important environmental question, which Bhaumik and Gibbons [21] addressed using frequentist prediction methods, is to determine whether the on-site mean lead concentration at this area of the facility exceeded background. To facilitate this determination, $n = 15$ off-site soil samples were collected in areas that were uninfluenced by the activities at the facility. The data are reproduced in Table 9.3.2. Using the Shapiro–Wilk normality test,

Table 9.1 *Taken from [179]. Lead measurements (mg/kg) for soil boring samples in two locations, one off-site and one on-site.*

Off-site	26	63	3	70	16	5	1	57	5	3	24	2	1	48	3
On-site	50	82	95	103	88										

Bhaumik and Gibbons ascertained that a log-normal model provides adequate fit to this data. Our main goal in this application is therefore to demonstrate that the on-site concentrations, on average, do not significantly exceed the background. To this end, we will use the IM framework discussed above to produce an upper prediction limit for the arithmetic mean of lead contents, $\tilde{Y}$, of $m = 5$ on-site soil samples based on the $n = 15$ off-site soil samples, Y, and then we will compare it to the arithmetic mean of the data collected on the on-site lead concentration.

The plausibility for $\tilde{Y}$, for right-sided assertions $A = \{\tilde{Y} > \tilde{y}\}$, as a function of $\tilde{y}$, is shown in Figure 9.1(a). Those $\tilde{y}$ values with plausibility function exceeding 0.05 provide an upper prediction bound for $\tilde{Y}$ which, in this case, is 136.16 mg/kg. For comparison, [21] provide the bound 152.26 mg/kg based on their Gram–Charlier approximation, and [148] provides the bound 139.30 mg/kg based on a Bayesian approach. All three prediction bounds contain the realized arithmetic mean of the on-site data in Table 9.3.2, which was 83.6 mg/kg. We therefore conclude that the on-site concentrations do not significantly exceed the background. However, since smaller upper prediction limits are more precise, our IM-based bound is preferred. An additional advantage of our IM-bound is that it, per Remark 9.2, also has a clearer interpretation than the above Bayesian and frequentist bounds.

To check the prediction performance for settings similar to the soil example, we take 5000 samples of size $n = 15$ from a log-normal distribution with $\mu = 2.173$ and $\sigma^2 = 2.3808$, the maximum likelihood estimates based on the off-site data in Table 9.3.2. A Monte Carlo estimate of the distribution function of $G_Y(\tilde{Y})$ is shown in Figure 9.1(b). Apparently, $G_Y(\tilde{Y})$ is Unif$(0,1)$, so the plausibility function for prediction is valid, by Theorem 9.1.

For further comparison, we performed a simulation study similar to the one presented in [21]. We considered three values for μ (2, 3, 10), six values for σ^2 (0.0625, 0.2, 0.5, 1, 2, 10), five values for n (5, 10, 20, 30, 100), and three values for m (1, 5, 10). For each combination, we evaluated the coverage probability of both the lower and upper 90% prediction limits. In all cases, the coverage probability equals the nominal level, up to Monte Carlo error; these estimates are based on 10,000 Monte Carlo samples. Unlike the Gram–Charlier approximation method in [21], our IM-based interval method does not need technical tools for derivation, and achieves the nominal coverage probability even when $\sigma^2 > 3$. Moreover, the other two methods reported in [21] do not achieve the nominal coverage probability.

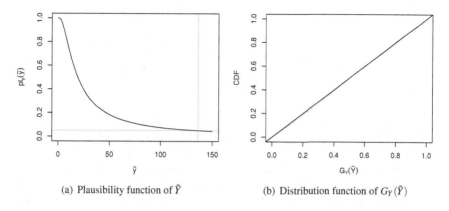

(a) Plausibility function of $\tilde{Y}$ (b) Distribution function of $G_Y(\tilde{Y})$

Figure 9.1 *Taken from [179]. Panel (a): Plausibility function of $\tilde{Y}$ in the log-normal data example. Panel (b): Distribution function of $G_Y(\tilde{Y})$ (gray) compared with that of* $\mathsf{Unif}(0,1)$ *(black) based on Monte Carlo samples from the log-normal distribution with $\mu = 2.173$ and $\sigma^2 = 2.3808$, the maximum likelihood estimates in the example from [21].*

9.3.3 Gamma models

Let $Y = (Y_1, \ldots, Y_n)$ be an iid sample from a gamma distribution with shape parameter $\theta_1 > 0$ and scale parameter $\theta_2 > 0$, both unknown. Gamma models are often used in system reliability applications. Following the calculations in Section 6.4.3, a conditional association for (θ_1, θ_2) based on sufficient statistics is given by

$$T_1 = \theta_2 \Gamma_{n\theta_1}^{-1}(U_1) \quad \text{and} \quad T_2 = F_{\theta_1}^{-1}(U_2),$$

where $U = (U_1, U_2)$ are iid $\mathsf{Unif}(0,1)$, $T_1 = \sum_{i=1}^{n} Y_i$, $T_2 = n^{-1} \sum_{i=1}^{n} \log Y_i - \log(n^{-1}T_1)$, Γ_a is the $\mathsf{Gamma}(a,1)$ distribution function, and F_b is a distribution function without a familiar form. First, suppose the goal is to predict the next (independent) observation $\tilde{Y} = Y_{n+1}$, with the following association:

$$\tilde{Y} = \theta_2 \Gamma_{\theta_1}^{-1}(\tilde{U}), \quad \tilde{U} \sim \mathsf{Unif}(0,1). \tag{9.10}$$

Specifically, we want to give a lower prediction limit for $\tilde{Y}$. The general strategy is to solve for $\theta = (\theta_1, \theta_2)$ in the conditional association, and then plug this solution into the association for the new observation. In particular, for a given $U = (U_1, U_2)$, write

$$\theta(Y, U) = \big(\theta_1(Y, U), \theta_2(Y, U)\big) \tag{9.11}$$

for this solution; it depends on Y only through (T_1, T_2). The solution exists and is unique, though there is no closed-form expression. Justification of this claim and details about computing the solution in (9.11) are given below. Plugging (9.11) into the association for $\tilde{Y}$ gives the marginal association

$$\tilde{Y} = \theta_2(Y, U) \Gamma_{n\theta_1(Y,U)}^{-1}(\tilde{U}). \tag{9.12}$$

Table 9.2 *Taken from [179]. Machine first breakdown times, in hours.*

18	23	29	409	24	74	13	62	46	4
57	19	47	13	19	208	119	209	10	188

This association is not of the separable form in Remark 9.4. In any case, if G_Y denotes the distribution function of the quantity on the right-hand side of (9.12), then we can write $\tilde{Y} = G_Y^{-1}(W)$ for $W \sim \text{Unif}(0,1)$. This completes the A-step. Since we are interested in lower prediction limits, we compute the plausibility function in (9.5).

In some system reliability applications, as in [125, 249], interest may be in the largest among a collection of m future observations. In that case, we have an association that looks exactly like (9.12), except that $\tilde{Y} = \max\{Y_{n+1}, \ldots, Y_{n+m}\}$ is a maximum of m future gamma observations and $\tilde{U}$ is the maximum of m independent uniforms, independent of U. This involves the same solution $\theta(Y, U)$ as before, so nothing changes except the distributions being used in the Monte Carlo simulation of the plausibility function.

For illustration, consider the data $Y = (Y_1, \ldots, Y_n)$ on the first breakdown times of $n = 20$ machines given in [125]. These data are reproduced in Table 9.3.3. At the 5% significance level, the Kolmogorov–Smirnov test cannot reject the null hypothesis that these data are gamma, so the goal is to use the IM machinery described above to produce a lower prediction limit for $\tilde{Y}$, the maximum of $m = 5$ future breakdown times. Proceeding as described above, the plausibility function of $\tilde{Y}$ is given by $\text{pl}_y(\tilde{y}) = G_y(\tilde{y})$, which can be easily evaluated via Monte Carlo. A plot of this plausibility function for $A = \{\tilde{Y} \leq \tilde{y}\}$, as a function of $\tilde{y}$, is given in Figure 9.2(a). A one-sided 90% plausibility interval is the set of all $\tilde{y}$ values such that $\text{pl}_y(\tilde{y}) > 0.10$, and the lower bound in this case is 73.53 hours. For comparison, our lower bound is bigger, i.e., more precise, than the Bayesian lower bound (71.8 hours) in [125] and slightly smaller than the fiducial lower bound (74.36 hours) in [249]. The IM bound, per Remark 9.2, also has a clearer interpretation than the Bayesian and fiducial bounds. To assess the performance of the method in problems similar to this one, we simulate 2000 data sets based on the maximum likelihood estimates based on the failure time data. A Monte Carlo estimate of the distribution function of $G_Y(\tilde{Y})$ is shown in Figure 9.2(b). This distribution function is sufficiently close to that of $\text{Unif}(0,1)$, so we can conclude our one-sided IM-based 90% prediction interval has exact coverage.

For further illustration, we consider a simulation experiment, similar to that in [249], with three values of the sample size n (10, 25, 125), four values of the shape parameter θ_1 (0.5, 1, 5, 10), and two values of m (1, 5); we keep the scale parameter θ_2 fixed at 1. For each combination, we evaluated the coverage probability of both the lower and upper 90% prediction intervals based on 10,000 Monte Carlo samples. In all cases, the coverage probability is within an acceptable range of the target 0.90.

To conclude this section, we investigate the questions of (i) existence and uniqueness of the solution $\theta(T, U) = (\theta_1(T, U), \theta_2(T, U))$ above, and (ii) computation of

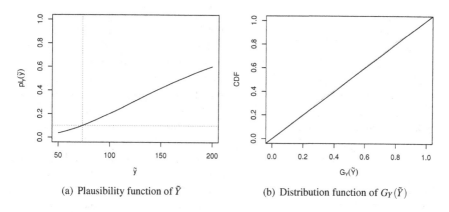

(a) Plausibility function of $\tilde{Y}$

(b) Distribution function of $G_Y(\tilde{Y})$

Figure 9.2 *Taken from [179]. Panel (a): Plausibility function of $\tilde{Y}$ in the gamma data example. Panel (b): Distribution function of $G_Y(\tilde{Y})$ (gray) compared with that of $\mathsf{Unif}(0,1)$ (black) based on Monte Carlo samples from $\mathsf{Gamma}(\hat{\theta}_1, \hat{\theta}_2)$, where $\hat{\theta}_1 = 0.8763$ and $\hat{\theta}_2 = 90.91$ are the maximum likelihood estimates.*

the solution. For uniqueness, the only non-trivial part is the solution θ_1 of equation $F_{\theta_1}(t_2) = u_2$, involving only (t_2, u_2). The challenge is that F_{θ_1} is a non-standard distribution. Glaser [116], in his notation, considers the random variable

$$U^\star = \left\{ \frac{(\prod_{i=1}^n Y_i)^{1/n}}{\frac{1}{n}\sum_{i=1}^n Y_i} \right\}^n,$$

the n-th power of the ratio of geometric and arithmetic means of an iid $\mathsf{Gamma}(\theta_1, 1)$ sample. Then $T_2 = n^{-1}\log(U^\star)$, i.e., our T_2 is a monotone increasing function of Glaser's $U^\star$. Glaser's Corollary 2.2 implies that $U^\star$ is stochastically strictly increasing in θ_1, which implies that $F_{\theta_1}(t_2)$ is a decreasing function of θ_1 for all t_2. Therefore, if a solution exists for θ_1 in (9.11), it must be unique by monotonicity.

Turning to the question of existence, we need to show that, for any t_2, $F_{\theta_1}(t_2)$ spans all of the interval $(0,1)$ for u_2 as θ_1 varies. By monotonicity, it suffices to consider the limits $\theta_1 \to \{0, \infty\}$. Jensen [140] considers the random variable $W = -1/T_2$ and shows, in his Equation (9), that θ_1/W has a limiting distribution as $\theta_1 \to \{0, \infty\}$, which implies the same for $\theta_1 T_2$. It is now clear that $F_{\theta_1}(t_2)$ converges to 1 and 0 as θ_1 converges to 0 and ∞, respectively, for all t_2. Therefore, a solution for θ_1 in (9.11) exists for all (t_2, u_2) pairs, as was to be shown.

Next, we consider computing the solution $\theta(Y,U)$ in (9.11). The only challenging part is solving for θ_1. Suppose t_2 and u_2 are given, and define a function $r(x) = F_x(t_2) - u_2$; the goal is to find the root for r. One can evaluate $r(x)$ by simulating $\mathsf{Gamma}(x, 1)$ variables, giving a Monte Carlo approximation of $F_x(t_2)$, and the root can then be found with any standard method, e.g., bisection. However, this can be fairly expensive computationally. A more efficient alternative approach is available based on large-sample theory. By Theorem 5.2 in [116] and the delta theorem, if n is large, then F_x can be well approximated by a normal distribution function with mean

$\psi(x) - \log(x)$ and variance $n^{-1}\{\psi'(x) - 1/x\}$, where ψ and ψ' are the digamma and trigamma functions, respectively. With this normal approximation, it is easy to evaluate $r(x)$ and find the root numerically. Though this is based on a large-sample approximation, in our experience, there is no significant loss of accuracy.

The normal approximation discussed above is simply a tool to find the solution $\theta(Y, U)$. It also provides some intuition related to the asymptotic argument in Section 9.4.1. When n is large, the variance in the normal approximation is $O(n^{-1})$, so the distribution function F_x will have a steep slope in the neighborhood of the solution to the equation $t_2 = \psi(x) - \log(x)$ and, therefore, the root for $r(x)$ will be in that same neighborhood, no matter the value of u_2. The solution to the equation $t_2 = \psi(x) - \log(x)$ is the maximum likelihood estimator of θ_1 [104], which is consistent. Therefore, when n is large, (9.12) and (9.10) are essentially the same, so the approximate validity of the corresponding prediction plausibility function is clear.

One last modification that we found to be helpful was to modify that normal approximation discussed above by replacing the normal distribution function with a gamma. That is, find solutions for the mean and variance of the normal approximation as before, but then use a gamma distribution function with mean and variance matching those obtained for the normal. See Exercise 9.4.

9.3.4 Binomial models

Let $Y \sim \mathrm{Bin}(n, \theta)$ and $\tilde{Y} \sim \mathrm{Bin}(m, \theta)$ be independent binomial random variables, where n and m are known. The goal is to predict $\tilde{Y}$ based on observing Y. There is considerable literature on this fundamental problem: see, e.g., [250] for frequentist prediction intervals and [239] for Bayesian prediction intervals. The starting point for our IM-based analysis is the following joint association for Y and $\tilde{Y}$,

$$F_{n,\theta}(Y - 1) \le 1 - U < F_{n,\theta}(Y) \quad \text{and} \quad F_{m,\theta}(\tilde{Y} - 1) \le 1 - \tilde{U} < F_{m,\theta}(\tilde{Y}),$$

where U and $\tilde{U}$ are independent uniforms, and $F_{n,\theta}$ is the $\mathrm{Bin}(n, \theta)$ distribution function. To marginalize over θ, we need a known identity linking the binomial and beta distribution functions, i.e., $F_{n,\theta}(y) = 1 - G_{y+1,n-y}(\theta)$, where $G_{a,b}$ is the $\mathrm{Beta}(a, b)$ distribution function; see Exercise 4.6. Now rewrite the first expression in the joint association as a θ-interval:

$$G^{-1}_{Y,n-Y+1}(U) \le \theta < G^{-1}_{Y+1,n-Y}(U). \tag{9.13}$$

Next, rewrite the $\tilde{Y}$ association as

$$F^{-1}_{m,\theta}(1 - \tilde{U}) < \tilde{Y} < F^{-1}_{m,\theta}(1 - \tilde{U}).$$

It is shown in [150] that $F^{-1}_{m,\theta}(v)$ is an increasing function of θ for all v, so we can "plug in" the Y-dependent interval for θ in to this latter inequality, to get

$$F^{-1}_{m,\theta_1(Y,U)}(1 - \tilde{U}) < \tilde{Y} < F^{-1}_{m,\theta_2(Y,U)}(1 - \tilde{U}), \tag{9.14}$$

where $\theta_1(Y, U)$ and $\theta_2(Y, U)$ are, respectively, the left and right endpoints of the interval in (9.13). This completes the A-step. We are interested in two-sided prediction intervals here; see the P- and C-steps for singleton assertions in Section 9.2.3. Note that this association is an interval, compared to the singletons in the previous examples. This is a consequence of the discreteness of the binomial, not a limitation of the IM approach; but see below. Some minor adjustments to the C-step in Section 9.2.3 are needed to handle this discreteness. Since we can easily get a Monte Carlo approximation for the distribution of the two endpoints, constructing a plausibility function for $\tilde{Y}$ is no problem.

In medical applications, it may be desirable to obtain an accurate prediction of the number of future cases of a disease based on the counts in previous years. Section 5 in [250] gives the following example. The total number of newborn babies with permanent hearing loss is $Y = 23$ out of $n = 23061$ normal nursery births over a two-year period. The goal is to predict $\tilde{Y}$, the number of newborns with hearing loss in the following year, based on $m = 12694$ normal births. For a two-sided, IM-based 90% prediction interval for $\tilde{Y}$, we compute the 5th and 95th percentiles of the distribution of the lower and upper endpoints, respectively, in (9.14). The interval obtained is $(6, 21)$, which contains the true $\tilde{Y} = 20$ and is essentially the same as the intervals in [250]; see Remark 9.2.

The plausibility function obtained based on the above construction is a bit conservative. One possible adjustment, based on an idea presented by [249] in the fiducial context, is to eliminate the interval association for $\tilde{Y}$ by first eliminating the interval association (9.13) for θ in terms of the limits $\theta_1(Y, U)$ and $\theta_2(Y, U)$. The idea is to sample a value, $\hat{\theta}(Y, U)$, of θ at random from the interval $(\theta_1(Y, U), \theta_2(Y, U))$. This results in a modified association for $\tilde{Y}$:

$$\tilde{Y} = F^{-1}_{m, \hat{\theta}(Y, U)}(1 - \tilde{U}). \tag{9.15}$$

The intuition is that the uncertainty due to the interval association has been replaced by the uncertainty from sampling. Since the sampled point is "less extreme" than both of the endpoints, this modified association gives a more efficient plausibility function for prediction, which we now demonstrate. Consider binomial samples of size $n = m = 100$ over a range of θ values. Here we compare the coverage probability and average lengths of 95% upper prediction limits based on the modified IM, fiducial, and Jeffreys prior Bayes methods. We simulated 2500 data sets, and each computation of the prediction interval (modified IM, fiducial, and Bayes) used 10,000 Monte Carlo samples. In Figure 9.3, we see that all three methods have coverage slightly above the nominal level over the entire range of θ; this is to be expected, given the discreteness of the binomial model. The modified IM intervals based on (9.15) tend to have slightly higher coverage probability than the others, but with no perceptible difference in length.

9.4   Some further technical details

9.4.1   Asymptotic validity

Outside the separable class in Remark 9.4, or in cases where $\tilde{Y}$ is a non-linear function of several future observables, the theory of prediction validity is more challeng-

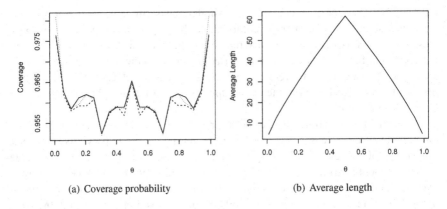

(a) Coverage probability   (b) Average length

Figure 9.3 *Taken from [179]. Coverage probability and average length of the modified IM (solid), fiducial (dashed), and Jeffreys prior Bayes (dotted) upper 95% prediction intervals, as functions of θ; the three curves in panel (b) are indistinguishable. Here n = m = 100 and estimates are based on 2500 simulated data sets.*

ing. However, our examples in Section 9.3 demonstrate that the uniformity assumption of Theorem 9.1 holds at least approximately. Here we give a theoretical argument to explain this phenomenon.

Write $Y^n = (Y_1, \ldots, Y_n)$ for data consisting of n iid components. Suppose that the solution $\theta(Y^n, V)$ converges in probability to θ, as a function of (Y^n, V); this usually is easy to arrange, see the examples in Section 9.3. If $\tilde{Y}$ has a continuous distribution, then, without loss of generality, we can write $G_{Y^n}(\tilde{Y}) = \tilde{F}_{\theta(Y^n,V)}(\tilde{Y})$, where $\tilde{F}_\theta$ is the true distribution of $\tilde{Y}$. Trivially, we have $(\theta(Y^n, V), \tilde{Y}) \to (\theta, \tilde{Y})$ in distribution so, if $(\theta, \tilde{y}) \mapsto \tilde{F}_\theta(\tilde{y})$ is continuous, then the continuous mapping theorem implies that

$$G_{Y^n}(\tilde{Y}) = \tilde{F}_{\theta(Y^n,V)}(\tilde{Y}) \to \tilde{F}_\theta(\tilde{Y}) \sim \mathsf{Unif}(0,1) \quad \text{in distribution.}$$

Therefore, we can generally be sure that the distribution of $G_{Y^n}(\tilde{Y})$ will be approximately $\mathsf{Unif}(0,1)$ when the sample size n is large. This argument holds even if $\tilde{Y}$ is some scalar function of several future observations.

In addition to providing an asymptotic validity result, the argument above is also relevant to prediction accuracy. That is, we have demonstrated that the IM "predictive distribution" for $\tilde{Y}$, which mixes $\tilde{F}_{\theta(Y^n,V)}$ over the distribution of V, converges to $\tilde{F}_\theta$, the true distribution of $\tilde{Y}$. A precise prediction accuracy result requires computing a measure of the distance/divergence of the IM predictive from the truth. We expect that results comparable to those in [155] can be derived, but we leave this as a question to be considered in future work.

9.4.2 Case of dependent τ(U) and η(U)

Recall that, in Section 9.2.2, it was assumed that the original association, $Y = a(\theta, U)$, could be decomposed as $T(Y) = b(\theta, \tau(U))$ and $H(Y) = \eta(U)$, and, furthermore, that $\tau(U)$ and $\eta(U)$ are independent. If $\tau(U)$ and $\eta(U)$ are not independent, which would arise, say, in models that are not regular exponential families, then details of the IM construction are more complicated; see Chapter 6. Here, we describe the three-step construction of an IM for prediction in the case where the model is a location shift of a Student-t distribution.

Let $Y = (Y_1, \ldots, Y_n)^\top$ be iid, with $Y_i = \theta + U_i$ and $U_i \sim t_d$, i.e., θ is a location parameter and the error has a Student-t distribution with known degrees of freedom d. The case with an additional unknown scale parameter can be handled similarly. Let $T(Y)$ be the maximum likelihood estimator for θ, and let $H(Y) = Y - T(Y)1_n$ be the vector of residuals, where 1_n is a (column) n-vector of unity. Since T is equivariant in this example, we have

$$T(Y) = \theta + T(U) \quad \text{and} \quad H(Y) = H(U),$$

so that $\tau = T$ and $\eta = H$. However, $\tau(U)$ and $\eta(U)$ are not independent here. As suggested in Chapter 6 and in Section 9.2.2 above, we want to consider the conditional distribution of $\tau(U)$, given that $\eta(U)$ equals the observed value of $H(Y)$. Let H_0 denote the observed value of $H(Y)$. Write $\mathsf{P}_{V|H_0}$ for the conditional distribution of $V = \tau(U)$, given $\eta(U) = H_0$, so that

$$T(Y) = \theta + V, \quad V \sim \mathsf{P}_{V|H_0}.$$

This can be solved for θ, and a marginal association connecting the observed Y, the next $\tilde{Y}$ to be predicted, and the pair $(V, \tilde{U})$ of auxiliary variables is of the form

$$\tilde{Y} = T(Y) + \tilde{U} - V, \quad \text{where} \quad V \sim \mathsf{P}_{V|H_0}, \; \tilde{U} \sim t_d.$$

Computation with the conditional distributions is more cumbersome, but it is clear that the dependent case is conceptually no different than the independent case. Moreover, using the law of iterated expectation, it can be shown that the conditioning does not affect the validity result in Theorem 9.1.

9.4.3 Multivariate prediction

The focus of this chapter was on the case of predicting a scalar $\tilde{Y}$, which is possibly a scalar-valued function of several future observables. However, in some cases there could be interest in the simultaneous prediction of several future observables. One example is in regression, where the interest may be in predicting the response value corresponding to several values of the predictor variables simultaneously. While the general IM framework is well-equipped to handle the multivariate case, our developments here have employed a few scalar-specific steps. Our goal in this section is simply to highlight those scalar-specific steps, defining a roadmap to extend the present developments to the multivariate case.

First, note that, up to the simplified association formula in (9.3), there is nothing in the developments in Section 9.2.2 specific to the scalar $\tilde{Y}$ case. There, we introduced a distribution G_Y and a probability integral transform, which is only valid in the scalar case. However, (9.2) is well-defined for vector $\tilde{Y}$, and would conclude the

A-step for a multivariate prediction. From here, the P-step proceeds by introducing a predictive random set for the pair $(V, \tilde{U})$. Validity, as usual, would not be a major obstacle, but an efficient choice of predictive random set would be problem specific. For a scalar auxiliary variable, the only reasonable choice of predictive random set is an interval, but in the multivariate case, there are lots of "reasonable" shapes, and the choice among them makes a difference in terms of efficiency.

Second, observe that the discussion of one-sided assertions in Section 9.2.3 is not appropriate in the multivariate setting, where there is no proper ordering. The two most natural assertions would be the singletons and assertions defined via level sets of some scalar-valued function of $\tilde{Y}$, e.g., balls $\{\tilde{y} : \|\tilde{y}\| \leq r\}$ for some fixed radius $r > 0$. The latter case reduces to the scalar prediction problem covered in this paper. For the singleton assertion case, nothing in the present development needs to change, except that the predictive random set, in general, must be specified for the pair $(V, \tilde{U})$ directly. If this predictive random set is valid, then the conclusion of Theorem 9.1 holds. Again, the challenge is that the shape of the predictive random set is directly related to the shape and efficiency of, say, the IM prediction regions for $\tilde{Y}$.

9.5 Concluding remarks

The key to the IM approach in general is the association of data and parameters with unobservable auxiliary variables, and the use of random sets on the auxiliary variable space to construct belief and plausibility functions on the parameter space. In the context of prediction of future observables, all the model parameters are nuisance, and an extreme form of the marginalization technique described in Chapter 7 is required, which allows us to reduce the dimension of the auxiliary variables, increasing efficiency. We give conditions which guarantee that the IM for prediction is valid, and we argue that this notion of IM validity translates to frequentist coverage guarantees for our plausibility intervals for future observables. A sequence of practical examples demonstrates the quality performance of the proposed method, along with its generality and overall simplicity.

The methodology described here covers both discrete and dependent data problems. However, these problems present unique challenges. For example, in the binomial example in Section 9.3.4, our standard IM approach was valid but conservative. A modified and more efficient IM was proposed, and its validity was confirmed numerically, but a theoretical basis for this modification is required. For dependent data problems, marginalization to reduce the dimension of auxiliary variables as described here is possible, but the details would be more challenging. Moreover, as discussed in Section 9.4.3, additional work is needed to properly extend the developments in the present paper to the multivariate prediction problem.

An interesting potential application of the tools for prediction present here is model fitting/checking. That is, the quality of a given model can be assessed by how well it predicts the data Y that is observed. So, our plausibility function for prediction defined herein, evaluated at the observation, can be viewed as a measure of the quality of the model fit. For two competing models, the one which assigns higher plausibility to the value of the observed data can be considered better. At this point, this is only a very rough idea, but does suggest one possibly naive IM approach for the challenging problem of model comparison. More sophisticated techniques, e.g., ones that can also take into consideration the complexity of the model, are currently being explored.

9.6 Exercises

Exercise 9.1. Prove Theorem 9.1.

Exercise 9.2. As mentioned in Remark 9.6, the uniformity condition in Theorem 9.1 can be relaxed, but this slightly weakens the conclusions. For example, suppose $G_Y(\tilde{Y})$ is stochastically no smaller than $\mathsf{Unif}(0,1)$.

   (a) Show that the conclusion of Theorem 9.1 holds for the one-sided predictive random set $\mathcal{S} = [0,W]$, $W \sim \mathsf{Unif}(0,1)$.

   (b) Show that, for assertions $A = \{\tilde{Y} > \tilde{y}\}$, the lower plausibility bounds obtained via (9.8) have the nominal frequentist coverage probability as described in Remark 9.2.

[Similar conclusions hold if $G_Y(\tilde{Y})$ is stochastically no larger than $\mathsf{Unif}(0,1)$, with obvious changes to the predictive random set and assertion.]

Exercise 9.3. Following the calculations in Section 9.3.1, construct a IM for predicting a future response, given a specific value $x_0 \in \mathbb{R}^p$ of the predictor variable, in a normal linear regression model.

Exercise 9.4. For a normal distribution $\mathsf{N}(\mu, \sigma^2)$, with $\mu > 0$, find a gamma distribution approximation by matching the mean and variance.

Exercise 9.5. The multivariate prediction problem was discussed briefly in Section 9.4.3. Consider the special case of predicting the next observation $\tilde{Y} = Y_{n+1}$ after seeing an iid sample $Y = (Y_1, \ldots, Y_n)$ from a multivariate normal distribution $\mathsf{N}_d(\theta, \Sigma)$, where θ is the unknown d-dimensional mean vector and Σ is a known $d \times d$ covariance matrix. Show that, for a ball assertion, $A = \{\tilde{y} : \|\tilde{y}\| \le r\}$, where the radius $r > 0$ is given, the prediction problem can be handled using the machinery discussed here for the scalar prediction problem.

Chapter 10

Simultaneous Inference on Multiple Assertions

10.1 Introduction

Previous chapters have focused on the construction of IMs for prior-free probabilistic inference on a relatively low-dimensional parameter. IMs can handle arbitrary dimensions, so our choice to emphasize the relatively simple problems is just for the sake of understanding. However, one naturally should ask how to apply the IM machinery to challenging high-dimensional problems. We will not address the "high-dimensional" problem here, but we want to present some first ideas in this direction.

In making a transition to more challenging problems, an issue to be considered is *model selection*. Up to now, the particular model—or the set of parameters of interest—has been fixed, and the goal is to produce a meaningful summary of uncertainty about those parameters. More generally, one also has to consider the choice of the model as one with some uncertainty associated with it. The motivating example here is that of variable selection in regression, i.e., to identify which of the coefficients in the linear model are non-zero or, equivalently, to identify the set of "active" predictor variables. In this chapter, we formulate a general problem that involves simultaneous consideration of multiple assertions about the parameter of interest and then specialize this general strategy for variable selection in regression.

Our main technical contribution here is the development of optimal predictive random sets for simultaneous inference on multiple assertions. In particular, we show that the optimal predictive random sets for each individual assertion can be combined to obtain a collection of potentially optimal candidate predictive random sets for simultaneous inference. Under a balance condition, we can identify the unique optimal predictive random set. These optimality considerations, along with the IM's natural insistence on validity, lead to an IM-driven variable procedure for the regression problem which compares favorably with many standard methods, as we demonstrate with simulations in Section 10.5.5.

A problem that has received considerable attention recently is that of model selection bias. When the model is fixed then, as in the regression problem discussed in Chapter 8, there is distribution theory available for constructing hypothesis tests, confidence intervals, etc. However, if the data is used, first, to select a model and, second, for inference about the parameters in the selected model, then the fixed-model inference procedures are not appropriate. In particular, since the selected model likely includes the "most significant" effects, the resulting inference will be overly optimistic, at least in terms of the significance of the selected effects. We believe that this model selection bias likely plays a major role in the recent problems concerning non-reproducibility of significant results in scientific studies. The goal is then to derive procedures for valid "post-selection" inference, i.e., hypothesis tests and confidence intervals that take into account the selection bias. One recent approach is described in [18]. An interesting by-product of our approach in the regression problem here is that the "POSI" method outlined in [18] drops out automatically. This is a consequence of the IM's insistence on validity and the careful construction of the optimal predictive random set for simultaneous assertions.

10.2   Preliminaries

Consider a general setup with association $X = a(\theta, U)$, where $X \in \mathbb{X}$ is the observable data, $\theta \in \Theta$ is the unknown parameter of interest, and U is an unobservable auxiliary variable, with distribution P_U defined on a space $\mathbb{U}$. Then the corresponding sampling model for X, given θ, is denoted by $P_{X|\theta}$. For a given assertion A about θ, an optimal (valid) predictive random set $\mathcal{S}$, if it exists, makes the belief function $\mathrm{bel}_X(A; \mathcal{S})$ stochastically as large as possible (see Exercise 1.18) as a function of $X \sim P_{X|\theta}$, for all $\theta \in A$; motivation for this approach comes from the efficiency principle in Chapter 3. The following definition makes this formal.

Definition 10.1. For a given assertion A, if $\mathcal{S}$ and $\mathcal{S}'$ are two valid predictive random sets, then we say that $\mathcal{S}$ is at least as efficient as $\mathcal{S}'$ (with respect to A) if

$$\mathrm{bel}_X(A; \mathcal{S}) \geq_{\mathrm{st}} \mathrm{bel}_X(A; \mathcal{S}'), \quad \text{as a function of } X \sim P_{X|\theta}, \text{ for } \theta \in A. \qquad (10.1)$$

A predictive random set $\mathcal{S}$ is called optimal (with respect to A) if (10.1) holds for all valid predictive random sets $\mathcal{S}'$.

In this chapter, we will make considerable use of the notion of admissible predictive random sets as discussed in Chapters 4–5. In particular, by Theorem 4.3, we only need to consider predictive random sets which are nested. Moreover, for a given collection $\mathbb{S}$ of (closed) nested subsets of $\mathbb{U}$, there is a "natural" predictive random set $\mathcal{S}$, whose distribution $P_{\mathcal{S}}$ is called the *natural measure*, defined in Definition 5.1. For the readers' convenience, we recall here the definition of the natural measure $P_{\mathcal{S}}$:

$$P_{\mathcal{S}}\{\mathcal{S} \subset K\} = \sup_{S \in \mathbb{S}: S \subset K} P_U(S), \quad K \subseteq \mathbb{U}. \qquad (10.2)$$

That there exists a predictive random set $\mathcal{S}$ with the natural measure $P_{\mathcal{S}}$ satisfying (10.2) was discussed in Chapter 5.

Here we will also introduce the concept of *a-event*, the set of all $u \in \mathbb{U}$ consistent with the association, a particular data point x, and an assertion A. Specifically, the a-event is defined as

$$\mathbb{U}_A(x) = \{u \in \mathbb{U} : \Theta_x(u) \subseteq A\}. \qquad (10.3)$$

We briefly encountered a-events previously, in Section 4.5.2, but these will be essential to the developments in this chapter.

For our last bit of notation and terminology, we should mention some technical (measurability) conditions about the a-events that we will silently assume throughout this chapter. Recall that the auxiliary variable space $\mathbb{U}$ is equipped with a σ-algebra $\mathscr{U}$ of subsets, assumed to contain all the closed sets. Next, equip the sample space $\mathbb{X}$ with a σ-algebra $\mathscr{X}$ and a σ-finite measure μ, where $P_{X|\theta}$ is absolutely continuous with respect to μ (in the measure-theoretic sense) for each θ. Finally, fix a collection $\mathbb{S}$ of (closed) nested subsets in $\mathscr{U}$, and each $S \in \mathbb{S}$ is called a focal element. Then the two key measurability assumptions are:

- $\mathbb{U}_A(x) \in \mathscr{U}$ for all relevant A and for μ-almost all x, and
- $\{x : \mathbb{U}_A(x) \supset S\} \in \mathscr{X}$ for all focal elements $S \in \mathbb{S}$.

General sufficient conditions can be given based on properties of the association mapping $x = a(\theta, u)$ and of the assertion A. It is relatively easy to check the above conditions directly in a particular example and, moreover, these regularity conditions can fail only in non-standard problems.

10.3   Classification of assertions

10.3.1   Simple assertions

When the assertion A is simple, in a sense to be described, it is possible to find an optimal predictive random set as in Definition 10.1. We say that an assertion A is *simple* if the collection of a-events $\mathbb{U}_A \equiv \{\mathbb{U}_A(x) : x \in \mathbb{X}\}$ is nested; otherwise, the assertion is called *complex*; see Section 10.3.2. The following proposition shows that, if A is simple, then the optimal predictive random set (relative to A) obtains by taking the support $\mathbb{S} = \mathbb{U}_A$ and the distribution $P_{\mathcal{S}}$ to satisfy (10.2).

Theorem 10.1. *For a simple assertion A, the optimal predictive random set $\mathcal{S}$ with respect to A is supported on the nested collection $\mathbb{U}_A$ with distribution satisfying (10.2).*

Proof. Proposition 4.1 shows that $\text{bel}_x(A; \mathcal{S}) \leq P_U\{\mathbb{U}_A(x)\}$, for any admissible $\mathcal{S}$ and any x. By assumption, the collection $\{\mathbb{U}_A(x) : x \in \mathbb{X}\}$ is nested, so it can be taken as the support of an admissible predictive random set. In this case, since $P_{\mathcal{S}}$ satisfies (10.2), we have that $\text{bel}_x(A; \mathcal{S}) = P_U\{\mathbb{U}_A(x)\}$ for all x. Therefore, the belief function attains its upper bound for each x, hence, it is optimal. $\qquad\square$

As an example, consider $X \sim N(\theta, 1)$, with association $X = \theta + U$, $U \sim N(0, 1)$. The assertion $A = \{\theta > 0\}$ is simple. To see this, write the a-event:

$$\mathbb{U}_A(x) = \{u : x = \theta + u \text{ for some } \theta > 0\} = (-\infty, x).$$

Then, clearly, if $x < x'$, then $\mathbb{U}_A(x) \subset \mathbb{U}_A(x')$; so, the a-events are nested. The belief function based on the optimal predictive random set $\mathcal{S}$ in Theorem 10.1 is

$$\text{bel}_x(A; \mathcal{S}) = P_{\mathcal{S}}\{\mathcal{S} \subset \mathbb{U}_x(A)\} = P_U\{\mathbb{U}_A(x)\} = \Phi(x),$$

where Φ is the standard normal distribution function. Given the optimality result, it is not a coincidence that $\mathrm{pl}_x(A^c;\mathcal{S}) = 1 - \Phi(x)$ is the p-value for the uniformly most powerful test of $H_0 : \theta \leq 0$ versus $H_1 : \theta > 0$; see [181].

Unfortunately, simple assertions are insufficient for practical applications. For example, in the normal mean example above, we might be interested in the assertion $A = \{\theta \neq 0\}$ or, more generally, we might be interested in several assertions simultaneously, each of which might be simple or not. Practical problems, such as the regression problem discussed in Chapter 8 and below, feature in all of these cases. (More generally, even if there is only one assertion A, IM efficiency considerations require that one think about A and A^c simultaneously, so an understanding of how to handle multiple assertions is fundamental.) The next two sections discuss how to extend the basic optimality results to these important more general cases.

10.3.2   Complex assertions

To motivate our developments, reconsider the normal mean example above. This time, suppose we are interested in the assertion $A = \{\theta \neq 0\}$. This is a union of two disjoint simple assertions, namely, $A_1 = \{\theta < 0\}$ and $A_2 = \{\theta > 0\}$. The optimal predictive random set with respect to A_1 is efficient for A_1 but very inefficient for A_2; likewise, for the optimal predictive random set with respect to A_2. Since we are interested in $A_1 \cup A_2$ we require a predictive random set that is as efficient as possible for both A_1 and A_2. For the general case of a complex assertion written as a union of two simple assertions, an intuitively reasonable strategy is to consider a predictive random set whose focal elements are intersections of the focal elements of the optimal predictive random sets with respect to the two individual simple assertions. The following result says that the optimal predictive random set for the complex assertion must be of this form. (The results below extend to more than two simple component assertions, see Exercise 10.2.)

Theorem 10.2. *Let $A = A_1 \cup A_2$ be a complex assertion, where A_1 and A_2 are simple assertions. Let $\mathcal{S}_1$ and $\mathcal{S}_2$ be the optimal predictive random sets for A_1 and A_2, respectively, as in Theorem 10.1. For any predictive random set $\mathcal{T}$, there exists an $\mathcal{S}$, whose focal elements are intersections of the focal elements of $\mathcal{S}_1$ and $\mathcal{S}_2$, such that $\mathcal{S}$ is at least as efficient as $\mathcal{T}$ with respect to A.*

Proof. Since A_1 and A_2 are simple assertions, the respective optimal predictive random sets $\mathcal{S}_1$ and $\mathcal{S}_2$ have focal elements given by $\mathbb{U}_{A_1}(x)$ and $\mathbb{U}_{A_2}(x)$, as x ranges over $\mathbb{X}$. Without loss of generality, assume that the predictive random set $\mathcal{T}$ for $A = A_1 \cup A_2$ is admissible. That is, assume $\mathcal{T}$ has a nested support $\mathbb{T}$ and that $\mathsf{P}_{\mathcal{T}}$ satisfies (10.2). Define

$$S_j(T) = \mathrm{closure}\left\{ \bigcap_{y:\mathbb{U}_{A_j}(y) \supset T} \mathbb{U}_{A_j}(y) \right\} \quad j = 1, 2, \quad T \in \mathbb{T}.$$

Collect intersections of these sets, $\mathbb{S} = \{S_1(T) \cap S_2(T) : T \in \mathbb{T}\}$. Since $\mathbb{T}$ is nested and the function $T \mapsto S_1(T) \cap S_2(T)$ is monotone, the collection $\mathbb{S}$ is also nested. Define a new predictive random set $\mathcal{S}$, supported on $\mathbb{S}$, with the natural measure $\mathsf{P}_{\mathcal{S}}$

as in (10.2). This predictive random set satisfies the conditions stated in the theorem, i.e., admissible and has focal elements as intersections of the respective optimal predictive random set focal elements. We need to show that $T \subset \mathbb{U}_A(x)$ if and only if $S(T) \subset \mathbb{U}_A(x)$, where $S(T) = S_1(T) \cap S_2(T)$. One direction is easy, since it is clear that $T \subset S(T)$. For the other direction, we want to show that $T \subset \mathbb{U}_A(x)$ implies $S(T) \subset \mathbb{U}_A(x)$. Since $\mathbb{U}_{A_1}(x)$ and $\mathbb{U}_{A_2}(x)$ are disjoint, and the union is $\mathbb{U}_A(x)$, $T \subset \mathbb{U}_A(x)$ implies that either $T \subset \mathbb{U}_{A_1}(x)$ or $T \subset \mathbb{U}_{A_2}(x)$. By definition of $S_1(T)$ and $S_2(T)$, it follows that

$$S_1(T) \subset \mathbb{U}_{A_1}(x) \quad \text{or} \quad S_2(T) \subset \mathbb{U}_{A_2}(x).$$

In either case, $S(T) = S_1(T) \cap S_2(T)$ is contained in $\mathbb{U}_A(x) = \mathbb{U}_{A_1}(x) \cup \mathbb{U}_{A_2}(x)$, which completes the argument that $\{T \subset \mathbb{U}_A(x)\}$ and $\{S(T) \subset \mathbb{U}_A(x)\}$ are equivalent. Finally, since $T \subset S(T)$, we get $\mathsf{P}_U(T) \le \mathsf{P}_U(S(T))$ for all $T \in \mathbb{T}$ and, therefore,

$$
\begin{aligned}
\mathsf{P}_\mathcal{T}\{T \subset \mathbb{U}_A(x)\} &= \sup_{T:T \subset \mathbb{U}_A(x)} \mathsf{P}_U(T) \\
&\le \sup_{T:S(T) \subset \mathbb{U}_A(x)} \mathsf{P}_U(S(T)) = \mathsf{P}_\mathcal{S}\{S \subset \mathbb{U}_A(x)\}.
\end{aligned}
$$

The left-hand side is $\mathrm{bel}_x(A; \mathcal{T})$ and the right-hand side is $\mathrm{bel}_x(A; \mathcal{S})$, and the inequality holds for all x, completing the proof.   □

This result simplifies the search for optimal predictive random sets with respect to a complex assertion. It does not completely resolve the problem, however, since there are many choices of predictive random sets with intersection focal elements. In the normal mean problem, for example, the focal elements for the optimal predictive random sets with respect to A_1 and A_2 are one-sided intervals. Therefore, the focal elements of the optimal predictive random set for $A = A_1 \cup A_2$ must be nested intervals, but it is not clear if the intervals should be symmetric or asymmetric. This ambiguity will be addressed in Section 10.4.2.

10.4   Optimality for a collection of assertions

10.4.1   Multiple complex assertions

Suppose there are multiple assertions, $\{A_j : j \in J\}$, to be considered simultaneously, where $A_j = A_{j1} \cup A_{j2}$, $j \in J$, decomposes as a union of two disjoint simple assertions. Each simple component has an optimal predictive random set according to Theorem 10.1, and the corresponding optimal predictive random set, $\mathcal{S}_j$, for the union A_j has intersection focal elements according to Theorem 10.2. As before, intuition suggests that the optimal predictive random set for $\{A_j : j \in J\}$ would be supported on intersections of the focal elements for the individually optimal $\mathcal{S}_j$, $j \in J$. To justify this intuition, we need a way to measure the efficiency of a predictive random set in this multiple-assertion context.

Definition 10.2. An assertion A is generated by $\{A_j : j \in J\}$ if A can be written as a union of some or all of the A_js. Then a predictive random set $\mathcal{S}$ is at least as efficient as $\mathcal{S}'$ with respect to $\{A_j : j \in J\}$ if (10.1) holds for all A generated by $\{A_j : j \in J\}$.

The following result, a generalization of Theorem 10.2, shows that restricting our consideration to predictive random sets supported on intersection focal elements results in no loss of efficiency.

Theorem 10.3. *Let $\{A_j : j \in J\}$ be a collection of assertions, where $A_j = A_{j1} \cup A_{j2}$ partitions as a union of disjoint simple assertions. Let S_j be the optimal predictive random set for A_j, $j \in J$. Then, for any predictive random set $\mathcal{T}$, there exists an $\mathcal{S}$, whose focal elements are intersections of the focal elements of the S_js, such that $\mathcal{S}$ is at least as efficient as $\mathcal{T}$ with respect to $\{A_j : j \in J\}$.*

Proof. The proof here is similar to that of Theorem 10.2. Consider collection $\{A_j : j \in J\}$ of complex assertions, where each A_j can be written as a union of disjoint simple assertions, i.e., $A_j = A_{j1} \cup A_{j2}$, where $A_{j1} \cap A_{j2} = \varnothing$ and the a-events $\mathbb{U}_{A_{j1}}(\cdot)$ and $\mathbb{U}_{A_{j2}}(\cdot)$ are nested. By Theorem 10.2, we know that the optimal predictive random sets for A_j, $j \in J$, have focal elements $\mathbb{S}_j = \{S_j(v) : v \in V\}$, indexed by a set V, which are intersections of a-events. That is, $S_j(v)$ is (the closure of) $\mathbb{U}_{A_{j1}}(x_{j1}(v)) \cap \mathbb{U}_{A_{j2}}(x_{j2}(v))$ for some $x_{j1}(v)$ and $x_{j2}(v)$.

Without loss of generality, assume that the candidate predictive random set $\mathcal{T}$ for the assertion A generated by $\{A_j : j \in J\}$ is admissible in the sense of Definition 5.1. Given a focal element $T \in \mathbb{T}$ of $\mathcal{T}$, define

$$\tilde{S}_j(T) = \bigcap_{v : S_j(v) \supset T} S_j(v), \quad j \in J.$$

Next, set $\tilde{S}(T) = \bigcap_{j \in J} \tilde{S}_j(T)$ and define $\mathbb{S} = \{\tilde{S}(T) : T \in \mathbb{T}\}$. Now take $\mathcal{S}$ to have support $\mathbb{S}$ and natural measure $\mathsf{P}_{\mathcal{S}}$ as in (10.2); this $\mathcal{S}$ satisfies the conditions of the theorem. It remains to show that $\mathcal{S}$ is more efficient than $\mathcal{T}$.

As in the proof of Theorem 10.2, we need to show that $T \subset \mathbb{U}_A(x)$ if and only if $\tilde{S}(T) \subset \mathbb{U}_A(x)$ for each x and $T \in \mathbb{T}$. By construction, $T \subset \tilde{S}(T)$, so one direction is handled. For the other direction, assume that $T \subset \mathbb{U}_A(x)$. Then $T \subset \mathbb{U}_{A_j}(x)$ for some $j \in J$ and, since A_j splits as a disjoint union of simple assertions, we get that $T \subset \mathbb{U}_{A_{j1}}(x)$ or $T \subset \mathbb{U}_{A_{j2}}(x)$. Then $\tilde{S}_j(T) \subset \mathbb{U}_{A_{j1}} \cap \mathbb{U}_{A_{j2}}(x)$ and, consequently, the no-bigger $\tilde{S}(T)$ must be a subset of $\mathbb{U}_A(x)$. Since $T \subset \mathbb{U}_A(x)$ if and only if $\tilde{S}(T) \subset \mathbb{U}_A(x)$, the claimed superiority of $\mathcal{S}$ to $\mathcal{T}$ follows, just as in the last part of the proof of Theorem 10.2. □

As in the previous section, this result simplifies the search for an optimal predictive random set but does not completely resolve the problem. More on optimality for the multiple-assertion case is given in Section 10.4.2, and some specific details are given for the variable selection application in sections in 10.5.

10.4.2   Balance and optimality

To resolve the ambiguity about the shape of the optimal focal elements, here we will make use of some special structure in the problem. Suppose that there exists transformations of X that do not fundamentally change the inference problem. As a simple example, if $X \sim N(\theta, 1)$ and the association of interest is $A = \{\theta \neq 0\}$, then it is clear that changing the sign of X should not have an impact on how much support there is for A. In other words, this normal mean problem is invariant to sign changes. More generally, let $\mathscr{G}$ be a group of bijections g from $\mathbb{X}$ to itself; as is customary, we shall write gx for the image of x under g, rather than $g(x)$. Practically, the transformations in $\mathscr{G}$ represent symmetries of the inference problem, i.e., nothing fundamental about the problem changes if gX is observed instead of X. The key technical assumption here is that each g commutes with the association mapping a in the following specific way:

$$ga(\theta, u) = a(g\theta, gu), \quad \forall (\theta, u), \quad \forall g \in \mathscr{G}. \tag{10.4}$$

Here we are implicitly assuming that $\mathscr{G}$ also acts upon the parameter space Θ and the auxiliary variable space $\mathbb{U}$. This can be relaxed by introducing groups acting on Θ and $\mathbb{U}$, respectively, different from (but homomorphic to) $\mathscr{G}$ but, for our two applications, the relevant group acts on both Θ and $\mathbb{U}$ directly, so we will not need this extra notational complexity. The above condition has g acting on u, which is different from that which defines the usual group transformation models where g only acts on θ. In fact, our variable selection application below will not fit the usual group transformation structure unless the design is orthogonal.

We shall also require that the assertions respect these symmetries. Let A be an assertion generated the collection $\{A_j : j \in J\}$ of complex assertions. Consider the subgroup of $\mathscr{G}$ to which A is invariant, and write $\mathscr{G}_A = \{g \in \mathscr{G} : gA = A\}$. The intuition is that the inference problem for θ, at least as it concerns the assertion A, is unchanged if the problem is transformed by $g \in \mathscr{G}_A$.

Before proceeding, it may help to see a simple example. Consider, again, the normal mean problem, with $X = \theta + U$, $U \sim \mathrm{N}(0, 1)$, and assertion $A = \{\theta \neq 0\}$. Then changing the sign of X will not affect the problem. So, we can take $\mathscr{G}$ to consist of the identity mapping and $x \mapsto -x$, and clearly (10.4) holds; also, $\mathscr{G}_A = \mathscr{G}$.

Moving on, recall the a-event $\mathbb{U}_A(x) = \{u : x = a(\theta, u) \text{ for some } \theta \in A\}$. In this case, an equivariance property follows (see Exercise 10.3) from (10.4):

$$\mathbb{U}_A(gx) = g\mathbb{U}_A(x), \quad \forall g \in \mathscr{G}_A. \tag{10.5}$$

That is, given A, transforming $x \to gx$, for $g \in \mathscr{G}_A$, and then solving for u is equivalent to solving for u with the given x and then transforming $u \to gu$.

We now have the necessary structure to help specify an optimal predictive random set. It suffices to focus on predictive random sets which are admissible. So, let $\mathcal{S}$ be admissible and, for simplicity, suppose we can write the collection of closed nested focal elements as $\mathbb{S} = \{S_r : r \in [0, 1]\}$, where r corresponds to the set's P_U-probability, i.e., $\mathsf{P}_U(S_r) = 1 - r$. Then we have the following useful representation.

Lemma 10.1. $\mathsf{P}_{X|\theta}\{\mathrm{bel}_X(A; \mathcal{S}) > 1 - r\} = \mathsf{P}_{X|\theta}\{\mathbb{U}_A(X) \supset S_r\}$.

Proof. Recall that $\mathrm{bel}_x(A; \mathcal{S})$ is defined as $\mathsf{P}_{\mathcal{S}}\{\mathcal{S} \subset \mathbb{U}_A(x)\}$. Since $\mathcal{S}$ has the natural measure (10.2), we can write, for any $b \in [0,1]$,

$$
\begin{aligned}
\mathrm{bel}_x(A;\mathcal{S}) > b &\iff P_{\mathcal{S}}\{\mathcal{S} \subset \mathbb{U}_A(x)\} > b \\
&\iff \sup_{r:S_r \subset \mathbb{U}_A(x)} P_U(S_r) > b \\
&\iff \sup_{r:S_r \subset \mathbb{U}_A(x)} (1-r) > b \\
&\iff \mathbb{U}_A(x) \supset S_{1-b}.
\end{aligned}
$$

Therefore, we have that

$$
P_{X|\theta}\{\mathrm{bel}_x(A;\mathcal{S}) > b\} = P_{X|\theta}\{\mathbb{U}_A(X) \supset S_{1-b}\}, \quad \forall b \in [0,1],
$$

which proves the claim, with $r = 1 - b$. □

From Lemma 10.1 and the equivariance property above, if $g \in \mathscr{G}_A$, then

$$
P_{X|\theta}\{\mathrm{bel}_{gX}(A;\mathcal{S}) > 1 - r\} \equiv P_{X|\theta}\{\mathbb{U}_A(gX) \supset S_r\} = P_{X|\theta}\{\mathbb{U}_A(X) \supset g^{-1}S_r\}.
$$

Since the understanding is that the inference problem, at least as it concerns the assertion A, is unchanged by transformations $X \to gX$ for $g \in \mathscr{G}_A$, it is reasonable to require that the distribution of $\mathrm{bel}_X(A;\mathcal{S})$ be invariant to $\mathscr{G}_A$. The previous display reveals that the way to achieve belief function invariance is to require the focal elements of the predictive random set to be invariant to $\mathscr{G}$. This leads to the following notion of *balance*.

Definition 10.3. The predictive random set $\mathcal{S}$ is said to be *balanced with respect to* A if each focal element $S \in \mathbb{S}$ satisfies $gS = S$ for all $g \in \mathscr{G}_A$. Moreover, $\mathcal{S}$ is said to be *balanced* if the aforementioned invariance holds for all $g \in \mathscr{G}$.

Balance itself is a reasonable property, given that the transformations are, by definition, irrelevant to the inference problem. It is also interesting and practically beneficial that balance can be checked without doing any probability calculations; however, the focal elements $\mathbb{S}$ and the transformations $\mathscr{G}$ depend on the model.

Beyond the intuitive appeal of balance, we claim that balance leads to a particular form of optimality. Recall that the goal of IM efficiency is to choose the predictive random set to make the belief function stochastically large when the assertion is true. With this in mind, we propose the following notion of *maximin optimality*.

Definition 10.4. A predictive random set $\mathcal{S}^\star$, with focal elements $\{S_r^\star : r \in [0,1]\}$, is *maximin optimal* if it maximizes

$$
\min_{g \in \mathscr{G}_A} P_{X|\theta}\{\mathrm{bel}_{gX}(A;\mathcal{S}) > 1 - r\} \equiv \min_{g \in \mathscr{G}_A} P_{X|\theta}\{\mathbb{U}_A(X) \supset g^{-1}S_r\}
$$

over all admissible $\mathcal{S}$ with focal elements $\{S_r : r \in [0,1]\}$, uniformly for all A generated by $\{A_j : j \in J\}$, for all $\theta \in A$, and for all $r \in [0,1]$.

Theorem 10.4. *If a predictive random set $\mathcal{S}$ is balanced in the sense of Definition 10.3, then it is maximin optimal in the sense of Definition 10.4.*

The main idea in the proof is the notion of the "core" S° of a given focal element S, defined as $S^\circ = \bigcap_{g \in \mathcal{G}_A} gS$. The proof below relies on the fact that it is both balanced and contained in each gS.

Proof of Theorem 10.4. Take any predictive random set $\mathcal{S}$ as in Theorem 10.3, and let S be a generic focal element. Take any assertion A generated by $\{A_j : j \in J\}$. Define the core of S as $S^\circ = \bigcap_{g \in \mathcal{G}_A} gS$; Exercise 10.4 invites the reader to show that S° is balanced and satisfies $S^\circ \subset gS$ for all $g \in \mathcal{G}_A$. Then,

$$\mathsf{P}_{X|\theta}\{\mathbb{U}_A(X) \supset S^\circ\} \geq \mathsf{P}_{X|\theta}\{\mathbb{U}_A(X) \supset gS\}, \quad \forall g \in \mathcal{G}_A, \quad \forall \theta \in A,$$

with strict inequality in general. Maximin optimality requires that we choose the focal elements to maximize the minimum (over g) of the right-hand side of the above display. However, we can clearly attain the upper bound above by taking the focal element S equal to its core, i.e., balanced. Therefore, balance implies maximin optimality, completing the proof.   □

10.5 Optimal IMs for variable selection

10.5.1 Motivation and problem formulation

Linear regression is arguably one of the most widely used statistical tools in scientific applications. It is standard practice to include as many variables as possible in the planning stages, though likely only a small subset of these variables will be useful in explaining variation in the response variable. Since the inclusion of unimportant explanatory variables in the model generally increases prediction error and makes interpretation of the results more difficult, the variable selection problem—selecting a good subset of explanatory variables—is of fundamental importance. [143] go so far as to say that variable selection is "central to the pursuit of science in general."

Given the importance of the variable selection problem, it is no surprise that there are many proposed methods to solve it. In general, difficulty arises because comparison of models, or subsets of variables, based on likelihood alone will always suggest selection of all the variables. To overcome this limitation, it is now standard to adjust the likelihood by adding some kind of penalty term that depends on the number of variables included in the model. The most well-known of these methods are the Akaike information criterion (AIC) [1] and the Bayesian information criterion (BIC) [216], with the latter imposing a more severe penalty on the number of variables. These methods allow for a ranking of candidate models and, if a single model is required, one can naturally choose the one with highest rank based on AIC, BIC, etc. Despite the certain large-sample variable selection consistency properties of these methods, a shortcoming remains in finite-samples, namely, that the rankings these criteria provide have no inferential meaning. For example, if Model 1 has higher AIC or BIC ranking than Model 2, then we cannot conclude that Model 1 being the correct model is any more plausible than Model 2 being the correct model. In other words, these selection criteria provide no inferentially meaningful measure of the uncertainty that the selected model is correct. Lasso [236, 237] and its variants, including the adaptive lasso [272] and the elastic net [273], all nicely summarized in [133], are useful tools for variable selection, but they too provide no inferentially meaningful measure of uncertainty. And significance tests based on the lasso estimator, e.g., [169], do not resolve the problem.

Bayesian methods, on the other hand, are able to produce measures of uncertainty concerning the candidate models; see, for example, [47]. With the introduction of a prior distribution for the set of possible models and a conditional prior distribution of the model parameters, Markov Chain Monte Carlo methods are available to search the model space for those with high posterior probability. Unfortunately, prior specification and posterior computation remain a challenging problem, especially when the model space is large. Furthermore, the posterior model probability estimates may become less reliable for large p [135], in which case, it is not clear if the most probable model has been identified.

In this section, we present an IM-based alternative. The key is the formulation of the variable selection problem as one involving several simultaneous complex assertions. Then the results on optimal IMs in the previous section can be directly applied. The resulting IM approach for variable selection has some nice features, in particular, that the inferential output corresponding to a particular model, a particular subcollection of the assertions, is valid and, therefore, has a meaningful interpretation. Throughout the book, since the IM output is valid, it is straightforward to derive a variable selection procedure, based on thresholding plausibilities, that has automatic frequentist error rate control. We make connections to a scientifically important issue concerning "post-selection inference," i.e., inference on parameters after data has been used to select a particular model.

Recall the work on the linear regression model in Chapter 8. Starting from the basic (full-rank) regression model $Y = X\beta + \sigma Z$, after carrying out the IM conditioning and marginalization steps, we arrived at the following dimension-reduced association:

$$\hat{\beta} = \beta + \hat{\sigma}W, \quad W \sim \mathsf{P}_W = \mathsf{t}_p(0, M; n - p - 1), \tag{10.6}$$

where $(\hat{\beta}, \hat{\sigma})$ are the usual least-squares estimators of (β, σ), and the auxiliary variable distribution is a p-dimensional Student-t, with $n - p - 1$ degrees of freedom, centered at the origin, with scale matrix $M = (X^\top X)^{-1}$. As discussed in Chapter 8, the key point here is that the original association with a n-dimensional auxiliary variable Z has been re-expressed in terms of a p-dimensional auxiliary variable W. Of interest here is the p-vector β; in particular, we want to identify which of the coordinates of β are non-zero. Without loss of generality, we have assumed that the Y and the columns of X have been centered, so that there is no intercept in the model.

It will be convenient to rewrite the (marginal) association (8.4) once more. Let D be a diagonal $p \times p$ matrix with the same diagonal as M. Then consider the association

$$\hat{\theta} = \theta + \hat{\sigma}U, \tag{10.7}$$

where $\theta = D^{-1/2}\beta$, $\hat{\theta} = D^{-1/2}\hat{\beta}$, and $U = D^{-1/2}W$. After this transformation, the new auxiliary variable U has a $\mathsf{t}_p(0, L, n - p - 1)$ distribution, where $L =$

$D^{-1/2}MD^{-1/2}$. Note, in particular, that $\theta_j = 0$ if and only if $\beta_j = 0$, so the variable selection problem has not been changed; furthermore, the matrix L has all ones on the diagonal.

10.5.2 Variable selection assertions

Recall the dimension-reduced association $\hat{\theta} = \theta + \hat{\sigma}U$ in (10.7), where $\theta = (\theta_1, \ldots, \theta_p)^\top$ and U is a p-vector distributed as $\mathsf{t}_p(0, L, n - p - 1)$, and L is a matrix with ones on the diagonal. The goal is identify which of $\theta_1, \ldots, \theta_p$ are non-zero.

Consider the collection of complex assertions $A_j = \{\theta : \theta_j \neq 0\}$, $j = 1, \ldots, p$. Consider first a particular A_j. This can be written as $A_j = A_{j1} \cup A_{j2}$, where $A_{j1} = \{\theta : \theta_j < 0\}$ and $A_{j2} = \{\theta : \theta_j > 0\}$. We claim that these sub-assertions are both simple in the sense of Section 10.3.1. Take A_{j1}, for example. Then the corresponding a-event is

$$\mathbb{U}_{A_{j1}}(y) = \{u : \hat{\theta} - \hat{\sigma}u \in A_{j1}\} = \{u : u_j > T_j\},$$

where $T_j = \hat{\theta}_j/\hat{\sigma}$. This a-event is nested because it shrinks and expands monotonically as a function of T_j. The same is true for A_{j2} and for all the other $j = 1, \ldots, p$. Therefore, by Theorem 10.1, the optimal predictive random sets for the individual sub-assertions A_{j1} and A_{j2} are each supported on collections of half hyper-planes. Next, it follows from Theorem 10.2 that the optimal predictive random set S_j for the complex assertion A_j is supported on intersections of half hyper-planes, i.e., cylinders $\{u : a_j \leq u_j \leq b_j\}$. If we are considering $\{A_j : j = 1, \ldots, p\}$ simultaneously, then it follows from Theorem 10.3 that the optimal predictive random set is supported on boxes—intersections of p marginal cylinders—in $\mathbb{R}^p$. The remaining question is what shape the boxes should be.

Toward optimality, we need to consider what transformations leave the variable selection problem invariant in the sense of Section 10.4.2. As in the simple normal mean example discussed previously, sign changes to coordinates of $\hat{\theta}$ are irrelevant. In addition, the labeling of the variables $j = 1, \ldots, p$ is arbitrary, so permutations of the variable labels are also irrelevant. This suggests we consider the group $\mathscr{G}$ of *signed permutations*; that is, each $g \in \mathscr{G}$ acts on a p-vector x by matrix multiplication (on the left), where the matrix factors as a product of a diagonal matrix with ± 1 on the diagonal and a permutation matrix. It is clear that the association commutes with $\mathscr{G}$ in the sense of (10.4). [To be precise, one should first take the group as an action on the model for $(\hat{\theta}, \hat{\sigma})$, and then do the marginalization steps discussed in Chapter 8.] With the group $\mathscr{G}$ specified, it is also clear what shape of boxes the optimal predictive random set should be supported on. According to Definition 10.3, a balanced predictive random set should have focal elements—in this case, shaped like boxes in $\mathbb{R}^p$—that are invariant to the transformations in $\mathscr{G}$. The only such boxes are hyper-cubes centered at the origin.

Corollary 10.1. *The admissible hyper-cube predictive random set S, given by*

$$S = \{u \in \mathbb{R}^p : \|u\|_\infty \leq \|U\|_\infty\}, \quad U \sim \mathsf{P}_U = \mathsf{t}_p(0, L; n - p - 1), \qquad (10.8)$$

is balanced in the sense of Definition 10.3 and maximin optimal for $\{A_j : j = 1, \ldots, p\}$ in the sense of Definition 10.4.

Besides the optimality properties, there are some computational advantages to the use of a hyper-cube predictive random set. Specifically, since the sides are parallel to the coordinate axes, it is straightforward to compute the belief function for various assertions related to variable selection. For example,

$$\mathsf{bel}_Y(A_j; \mathcal{S}) = \mathsf{P}_U\{\|U\|_\infty < |T_j|\} = F(|T_j|),$$

where F is the distribution function of $\|U\|_\infty$, with $U \sim \mathsf{t}_p(0, L; n - p - 1)$.

10.5.3 On post-selection inference

Using the balanced hyper-cube predictive random set $\mathcal{S}$ in Corollary 10.1, we can construct a plausibility function for singletons $\{\theta\}$:

$$\mathsf{pl}_Y(\{\theta\}; \mathcal{S}) = 1 - F\left(\|\hat{\sigma}^{-1}(\hat{\theta} - \theta)\|_\infty\right).$$

Henceforth, we will drop the braces and write $\mathsf{pl}_Y(\theta; \mathcal{S})$ for $\mathsf{pl}_Y(\{\theta\}; \mathcal{S})$. This plausibility function satisfies $\mathsf{pl}_Y(\hat{\theta}; \mathcal{S}) = 1$ and is decreasing in θ away from $\hat{\theta}$ with hyper-cube shaped contours. Consequently, by thresholding the plausibility function at level $\alpha \in (0, 1)$, we obtain the plausibility region

$$\{\theta : \mathsf{pl}_Y(\theta; \mathcal{S}) > \alpha\} = \{\theta : F(\|\hat{\sigma}^{-1}(\hat{\theta} - \theta)\|_\infty) < 1 - \alpha\}.$$

Since the predictive random set $\mathcal{S}$ is admissible, it follows that the plausibility region above has nominal frequentist coverage $1 - \alpha$. Moreover, the shape of these plausibility regions is a hyper-cube, with side lengths characterized by quantiles of the ℓ_∞-norm of multivariate Student-t random vectors.

A consequence of our focus on validity simultaneously across a collection of assertions is that a naive projection of these plausibility regions to any sub-model $I \subseteq \{1, 2, \ldots, p\}$ gives a new hyper-cube plausibility region that also has nominal frequentist coverage. Since this conclusion holds for all I, it also holds if I is chosen based on data. Such considerations are relevant in the context of post-selection inference. As [18] argue, when data is used to select a model, then one cannot use the model-specific distribution theory for valid inference. This leads to the fundamental question of how to achieve valid inference when the model is chosen based on data. A "POSI" procedure for valid post-selection inference is proposed in [18], and it turns out that theirs is identical to that obtained by naive projection of the above plausibility region to a sub-model selected by data. The take-away message is that, by insisting on valid probabilistic inference, IMs can automatically handle the challenging post-selection inference problem, further demonstrating their promise.

10.5.4 An IM-driven variable selection procedure

Section 10.5.2 shows how to optimally summarize the uncertainty in data concerning variable selection assertions. From this, it is possible to construct an IM-driven variable selection procedure with good properties.

To start, let $\mathcal{I} \subseteq \{1, 2, \ldots, p\}$ be the collection of indices corresponding to the truly non-zero coefficients, i.e., $\theta_i \neq 0$ for all $i \in \mathcal{I}$. Consider the assertion

$$B_I = \{\mathcal{I} \subseteq I\} = \{\theta : \theta_{I^c} = 0\} = \{\theta : \theta_i = 0, \ \forall \, i \notin I\}.$$

Note that $B_I^c = \bigcup_{i \in I^c} A_i$, so this apparently new kind of assertion is still covered by the variable selection assertions $\{A_j : j = 1, \ldots, p\}$ from before. Using the optimal hyper-cube predictive random set $\mathcal{S}$ in (10.8), we have

$$\mathsf{pl}_T(B_I; \mathcal{S}) = \mathsf{P}_U\{\|U\|_\infty > \|T_{I^c}\|_\infty\} = 1 - F(\|T_{I^c}\|_\infty),$$

where $\|T_{I^c}\|_\infty = \max_{i \in I^c} |T_i|$. It is clear that, if $\theta_{I^c} = 0$, then $\|U\|_\infty$ is stochastically no smaller than $\|T_{I^c}\|_\infty$, which leads to the following basic calibration result.

Lemma 10.2. *If B_I is true, i.e., if $\theta_{I^c} = 0$, then $\mathsf{pl}_T(B_I; \mathcal{S})$, which depends on T only through T_{I^c}, is stochastically no smaller than $\mathsf{Unif}(0, 1)$.*

Intuitively, a model or collection of variables I has some support from data if $\mathsf{pl}_T(B_I; \mathcal{S})$ is not too small. In other words, good models are at least those which are "sufficiently plausible" given data, relative to the optimal IM. Following this idea, an IM-driven approach for variable selection would be to fix $\alpha \in (0, 1)$ and then pick the smallest collection I such that $\mathsf{pl}_T(B_I; \mathcal{S}) > \alpha$. That is, define

$$\hat{I}_\alpha(T) = \text{smallest set } I \text{ such that } \mathsf{pl}_T(B_I; \mathcal{S}) > \alpha. \tag{10.9}$$

We claim that this IM-driven procedure controls the family-wise error rate at level α or, equivalently, it satisfies a certain "selection validity" property:

$$\mathsf{P}_{T|\mathcal{I}}\{\hat{I}_\alpha(T) \subseteq \mathcal{I}\} \geq 1 - \alpha, \quad \forall \, \mathcal{I} \subseteq \{1, 2, \ldots, p\}. \tag{10.10}$$

Theorem 10.5. *The procedure* (10.9) *has the selection validity property* (10.10).

Proof. $\mathsf{pl}_T(B_\mathcal{I}; \mathcal{S}) > \alpha$ implies $\hat{I}_\alpha(T) \subseteq \mathcal{I}$, so apply Lemma 10.2. □

To implement the above procedure, it is not necessary to evaluate the plausibility function at B_I for each $I \subseteq \{1, 2, \ldots, p\}$. In fact, the plausibility depends on the value of $\|T_{I^c}\|_\infty$, so we only need to look at p different models, based on a sorting of the t-statistics by their magnitude. Let π be a permutation that ranks the T values according to their magnitudes, i.e., $|T_{\pi(1)}| < |T_{\pi(2)}| < \cdots < |T_{\pi(p)}|$, and then compute

$$\eta(j) = 1 - F(|T_{\pi(j)}|), \quad j = 1, \ldots, p. \tag{10.11}$$

Each $\eta(j)$ represents a value of $\mathsf{pl}_T(B_I; \mathcal{S})$, but it is important to be clear about which model I it corresponds to. To start, take $\eta(p)$, corresponding to the largest value of $|T|$. If $\eta(p)$ is small, which means $|T_{\pi(p)}|$ is rather large, then there is little evidence to support the null model with no variables. So, $\eta(p)$ is the plausibility $\mathsf{pl}_T(B_I; \mathcal{S})$ for the null model $I = \varnothing$. In general, $\eta(j)$ is the plausibility corresponding to the model that includes only the variables $\pi^{-1}(\{j + 1, \ldots, p\})$, $j = 1, \ldots, p$; it is clear, both intuitively and from the formula, that the the full model, with all variables included, will be assigned plausibility 1. Table 10.1 helps to make this clear. With this

Table 10.1 *IM results for variable selection in the prostate cancer example; here* $T = \hat{\theta}/\hat{\sigma}$ *and "Plausibility" corresponds to the quantity* $\eta(\cdot)$ *in (10.11), ordered by* π.

Order, $\pi(j)$	Variable	$|T|$	Plausibility, $\eta(j)$
1	gleason	0.288	1.0000
2	pgg45	1.029	0.9193
3	lcp	1.165	0.8585
4	age	1.768	0.4502
5	lbph	1.842	0.4000
6	lweight	2.688	0.0646
7	svi	3.154	0.0177
8	lcavol	6.715	0.0000

understanding, we can define the IM-based variable selection procedure as follows: set $j^\star = \max\{j : \eta(j) > \alpha\}$, and then get $\hat{I}_\alpha(T)$ using the rule

$$\hat{I}_\alpha(T) = \pi^{-1}(\{j^\star + 1, \ldots, p\}). \tag{10.12}$$

Theorem 10.5 implies that this procedure will have the selection validity property, i.e., that it will control the family-wise error rate; see the bottom left panels in Figures 10.1–10.4.

10.5.5 Numerical results

10.5.5.1 Real data analysis

Consider the prostate cancer data analyzed previously by [236] and others. This study examined the association between the prostate specific antigen (PSA) level and some clinical measures among men who were about to receive a radical prostatectomy. There were $n = 97$ men with $p = 8$ predictors, four of which, including cancer volume (lcavol), prostate weight (lweight), capsular penetration (lcp), and benign prostatic hyperplasia amount (lbph), were log transformed. The other four predictors were age, seminal vestical invasion (svi), Gleason score (gleason), and percentage Gleason scores 4 or 5 (pgg45). The response variable is the log transformed PSA level (lpsa).

Using the optimal predictive random set S in (10.8), we only need to compute the plausibility (10.11) for $p = 8$ models, which is shown in Table 10.1. Using the method described in Section 10.5.4, at the $\alpha = 0.05$ level, we see that $j^\star = 6$ so, according to (10.12), variables lcavol and svi are selected; using the lasso, [236] also selects the variable lweight. So, the IM selection is more conservative than lasso in this case, but lasso is known to be too aggressive in many cases, such as in Section 10.5.5.2.

10.5.5.2   Simulation studies

For further illustration, here we present the results of several simulation studies. The basic model is $Y \sim N_n(X\beta, \sigma^2 I_n)$, where the rows of the predictor variable matrix are draws from $N_p(0, \Omega)$ with an autoregressive correlation structure, i.e., $\Omega_{jk} = \rho^{|j-k|}$ for all j, k. We consider six different scenarios with varying p, β, and ρ.

Scenario 1. $p = 7$, $\beta = (3, 1.5, 0, 0, 2, 0, 0)^\top$, and $\rho = 0.5$;

Scenario 2. $p = 7$, $\beta = (3, 1.5, 0, 0, 2, 0, 0)^\top$, and $\rho = 0.8$;

Scenario 3. $p = 7$, $\beta = (0.85, 0.85, 0.85, 0, 0, 0, 0)^\top$, and $\rho = 0.5$;

Scenario 4. $p = 7$, $\beta = (0.85, 0.85, 0.85, 0, 0, 0, 0)^\top$, and $\rho = 0.8$;

Scenario 5. $p = 20$, $\beta = (0.85 \, 1_{10}^\top, 0_{10}^\top)^\top$, and $\rho = 0.5$;

Scenario 6. $p = 20$, $\beta = (0.85 \, 1_{10}^\top, 0_{10}^\top)^\top$, and $\rho = 0.8$.

The results of the IM-based variable selection, described in Section 10.5.4, with $\alpha = 0.05$, based on the hypercube predictive random set, for the four scenarios are displayed in Figures 10.1–10.6. Specifically, these figures plot the percentage of true, parsimonious, true or parsimonious, and inclusive models selected, as a function of the sample size n. These are compared with several standard model selection procedures, namely, those based on AIC, BIC, the lasso (with ten-fold cross-validation to select the tuning parameter), and the adaptive lasso; the least angle regression [79] was also used, but those results are not displayed because they closely match the lasso's. The comparisons are based on 1000 simulated data sets in each configuration, with the sample size, n, ranging from 50 to 5000.

In each figure, the quantity being plotted in the "true or parsimonious" panel corresponds to the probability of including in the model no irrelevant variable. Given the theoretical results on IM validity, in particular, Theorem 10.5, we are not surprised to see the IM-based curves sitting at or above the 95% line. The other methods, except for BIC and the adaptive lasso are generally far off the mark in this panel. That BIC and the adaptive lasso can reach above 95% in the "true or parsimonious" panel is due to their asymptotic variable selection consistency property, not because they are properly calibrated at any fixed n. In addition to IM's fixed-n calibration, it performs as well and, in some cases, even better than the classical methods in terms of how frequently it selects the true model. This can be attributed to the optimality considerations in Section 10.5. So, the take-away message is that the IM's fixed-n calibration property does not come with the cost of lower efficiency.

10.6   Concluding remarks

This chapter introduces the concept of multiple simultaneous assertions in the IM framework, and develops a theory of optimal predictive random sets. These general principles are applied, in particular, to the variable selection problem in regression, which leads to simultaneously valid assessments of the uncertainty across all possible models. An important consequence of the simultaneous validity is that it holds even if a model is selected based on a data-driven procedure. That is, the proposed IM leads naturally to valid post-selection inference and, in particular, a naive projection of the

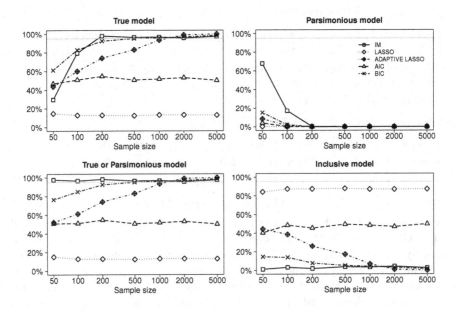

Figure 10.1 *Percentage of true, parsimonious, true or parsimonious, and inclusive models selected, as a function of sample size, for various methods under Scenario 1.*

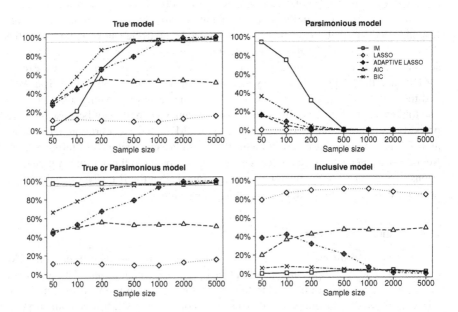

Figure 10.2 *Percentage of true, parsimonious, true or parsimonious, and inclusive models selected, as a function of sample size, for various methods under Scenario 2.*

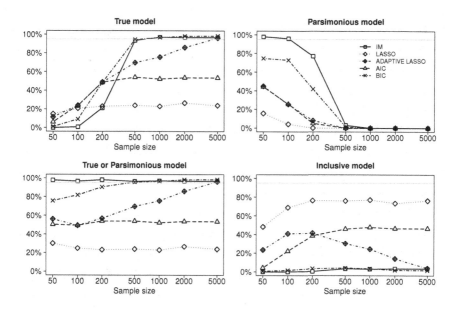

Figure 10.3 *Percentage of true, parsimonious, true or parsimonious, and inclusive models selected, as a function of sample size, for various methods under Scenario 3.*

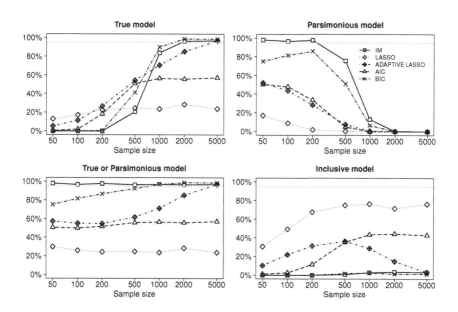

Figure 10.4 *Percentage of true, parsimonious, true or parsimonious, and inclusive models selected, as a function of sample size, for various methods under Scenario 4.*

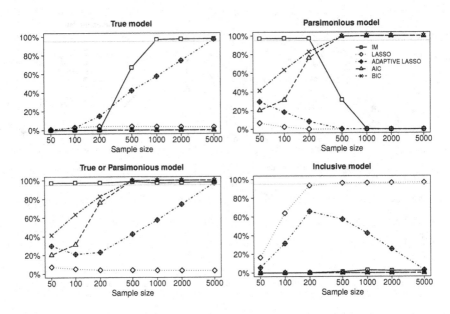

Figure 10.5 *Percentage of true, parsimonious, true or parsimonious, and inclusive models selected, as a function of sample size, for various methods under Scenario 5.*

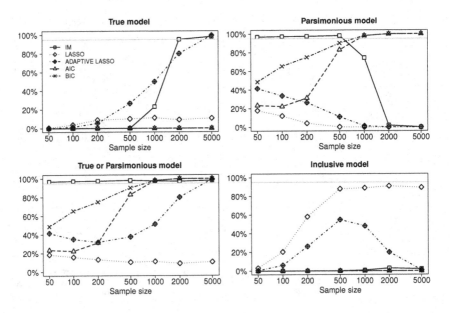

Figure 10.6 *Percentage of true, parsimonious, true or parsimonious, and inclusive models selected, as a function of sample size, for various methods under Scenario 6.*

IM plausibility region based on the optimal hyper-cube predictive random set gives exactly the recently developed "POSI" procedure in [18]. In addition, an IM-driven variable selection procedure is developed which is based on the notion of picking the smallest sub-model which is sufficiently plausible given the data. And, by its connection to the valid IM, this procedure is properly calibrated in the sense that it controls the family-wise error rate. Moreover, our simulation results suggest that the IM's emphasis on inferentially meaningful probabilistic summaries does not come at the cost of decreased efficiency. Extending the developments here for application in other models, such as generalized linear models, is a focus of future research.

The IM approach has already been applied to other problems that involve multiplicity, such as large-scale multinomial inference in genome wide association studies [166] and multiple testing problems [167]. We expect that the optimality considerations here, applied to those problems, would lead to some overall improvements. Perhaps the most important question is how to extend the developments in this paper to the high-dimensional $p \gg n$ context. The $p \gg n$ problem is of great practical interest, and has stimulated considerable research in the last decade; see, for example, [37, 138, 139, 141, 240, 251, 252, 267]. The basic principles of optimal predictive random sets developed here for simultaneous complex assertions are not specific to the $p < n$ case, but the initial dimension reduction steps do not carry over directly to the $p \gg n$ case. We expect that once the A-step can be completed for $p \gg n$ case, ideas similar to those presented here can be successfully applied. This is a topic of ongoing work.

10.7   Exercises

Exercise 10.1. Consider a normal mean model, $X \sim N(\theta, 1)$, with assertion $A = (-\infty, \theta_0)$, for fixed θ_0.

(a) Describe the shape of focal elements for the optimal predictive random set for the collection $\{A, A^c\}$ of simultaneous assertions.

(b) Under a balance condition, find the optimal predictive random set.

(c) Compare the corresponding IM for simultaneous inference on $\{A, A^c\}$ to some classical methods, such as uniformly most powerful unbiased tests, etc.

Exercise 10.2. Extend the result in Theorem 10.2 to the case where the complex assertion A of interest can be decomposed as $A = \bigcap_{j=1}^{J} A_j$, where each of $A_1, \ldots, A_J$ is a simple assertion.

Exercise 10.3. Prove the equivariance property (10.5).

Exercise 10.4. For a set $S \subset \mathbb{U}$ (a focal element of some predictive random set $\mathcal{S}$) and $\mathcal{G}$ a group acting on $\mathbb{U}$, define the core of S as $S^\circ = \bigcup_{g \in \mathcal{G}} gS$. Show that the core is balanced in the sense of Definition 10.3 and satisfies $S^\circ \subset gS$ for all $g \in \mathcal{G}$.

Exercise 10.5. For the variable selection problem discussed in Section 10.5, proposed an alternative predictive random set, e.g., a suitable ellipsoid rather than a hyper-cube. Compare computation of the relevant belief/plausibility functions based on ellipse versus hyper-cube predictive random sets.

Exercise 10.6. Verify the formula for $\hat{I}_\alpha(T)$ in (10.12).

Exercise 10.7. For some of the simulation settings described in Section 10.5.5.2, we find that methods based on BIC and/or the adaptive lasso are able to pick the true model a bit more frequently than the IM-based method when the sample size is large.

(a) Review some of the available literature on these non-IM methods and provide some theoretical justification for this observation.

(b) Explain the benefit of the IM-based method compared to these alternatives despite the conclusion in Part (a).

Chapter 11

Generalized Inferential Models

Portions of the material in this chapter are from R. Martin, "Plausibility functions and exact frequentist inference," *Journal of the American Statistical Association*, to appear, 2015, reprinted with permission by the American Statistical Association, www.amstat.org, and from C. Liu and J. Xie, "Large-scale two-sample multinomial inferences and its applications in genome wide association studies," *International Journal of Approximate Reasoning* **55**, 330–340, 2014, reprinted with permission by Elsevier.

11.1 Introduction

In previous chapters, we have demonstrated that the IM framework provides valid and efficient probabilistic inference without the need for a prior distribution. In this sense, IMs bridge the gap between the classical Bayesian and frequentist schools in a way that no previous attempt at synergy has been able to accomplish. The price that is paid for the IM's desirable properties is that its construction is, at least at first glance, more complicated than the construction of, say, a Bayesian posterior. We believe that the benefits of the IM approach far outweigh the additional costs, and in previous chapters we have demonstrated that the IM construction is quite doable, at least for relatively simple "standard" problems. However, we must also acknowledge that more work is needed to develop efficient IMs for use in "non-standard" problems, such as high-dimensional non-linear models.

The main challenge in the IM construction is the A-step, where an association linking the observable data to the unknown parameter and an unobservable auxiliary variable must be specified. Till now, we have required that this association determine the sampling model of the observable data and, outside a relatively simple class of problems, it may be difficult to identify such an association and/or justify the choice of any particular association. An interesting question then is if we can relax the conditions on the association, allowing for something that stops short of fully characterizing the sampling distribution of the observable data. In this chapter, we show that it is possible to construct a valid IM based on such a relaxed association—we refer to these as *generalized IMs*. The tools to be used in this construction were developed in [176], and these are described in detail in the upcoming sections.

There are a number of interesting by-products of this alternative viewpoint. First, one can do an almost automatic auxiliary variable dimension reduction. In fact, in the approach presented below, one can work with a scalar auxiliary variable, no matter the dimension of the parameter of interest. Second, since there is little-to-no concern about dimension reduction, one can view the proposed generalized IM approach as a sort of default IM analysis, like a IM-based black-box that one can use to crank out valid prior-free probabilistic inference without investing too much effort at the A-step. While a proper A-step that carefully considers the special features of the problem and reduces the auxiliary variable dimension as much as possible, together with the corresponding P-step that invokes an optimal predictive random set, would naturally lead to more efficient inference, the gain in efficiency might not be worth the investment. Therefore, a sort of default IM has some utility. Third, the generalized IM's guaranteed validity property makes it possible to construct exact frequentist procedures whose justification does not require asymptotic theory [176]. Despite the frequentist's desire to control error rates, it seems there is no general strategy besides the one proposed here for constructing exact procedures in finite-samples.

In this chapter we discuss the construction of a generalized IM and its properties. A general strategy is presented but the focus is on a special case that involves the likelihood function. Some examples are presented to compare the results to those of existing frequentist and Bayesian methods. Section 11.4 is devoted to the construction of generalized marginal IMs in the case where nuisance parameters are present. As an application, in Section 11.6, we consider the analysis of a large-scale multinomial model, as in [166], which is a challenging inference problem for all schools of thought. Our generalized IM approach is applied to data coming from a genome-wide association study, with the goal to identify association between genetic variants and rheumatoid arthritis.

11.2   Generalized associations

Despite the simplicity of the three-step IM construction outlined in previous chapters, there is a disadvantage in that, in practice, it requires a special formulation of the sampling model which, for complex models, makes the A- and P-step somewhat challenging. Our goal in this chapter is to relax the requirement that the association fully characterize the sampling model. Below we define exactly what we mean by "relaxed association" and in subsequent sections we show that the corresponding IM approach is easier to apply in complex problems while, at the same time, retaining those desirable properties discussed previously.

The association, as described in previous chapters, plays two main roles. The first is to identify the auxiliary variable U, its space $\mathbb{U}$, and its corresponding distribution P_U. The second is to define the mapping through which the uncertainty about U, for given data Y, is propagated to the parameter θ. So, a generalized association is one that just explicitly describes these two things.

Definition 11.1. Let $Y \in \mathbb{Y}$ be observable data and $\theta \in \Theta$ the parameter of interest. A generalized association specifies an auxiliary variable $U \in \mathbb{U}$, with distribution P_U, and mapping

$$\mathbb{Y} \times \mathbb{U} \ni (y,u) \mapsto \Theta_y(u) \in 2^\Theta. \qquad (11.1)$$

Given a generalized association, the P- and C-steps can be applied just as they are before; see Section 11.3.1. That is, the P-step specified as predictive random set $\mathcal{S}$ for the unobservable U, and then the C-step says to combine $\mathcal{S}$ and Θ_y, for observed $Y = y$, to get the enlarged random set

$$\Theta_y(\mathcal{S}) = \bigcup_{u \in \mathcal{S}} \Theta_y(u).$$

Then the belief and plausibility functions can be defined directly as the distribution of this random set in 2^Θ. The two obvious advantages of this approach are (i) there is no direct concern for writing the sampling model in a suitable form, and (b) since the auxiliary variable U is not tied directly to the observable data Y, potentially we can define U to be of sufficiently low dimension, avoiding the tedious dimension-reduction steps described in previous chapters.

On the other hand, this more general formulation raises some new questions and concerns. First, the generalized association surely cannot be completely arbitrary, i.e., without some connection to the sampling model, the output might not be meaningful. In the next section, we add enough additional structure so that some desirable properties can be demonstrated, in particular, we construct a generalized association using a suitably chosen function of (Y, θ); some further generalizations are discussed in Remark 11.2. Second, since there may be a variety of ways to specify a generalized association, there is a question about efficiency. More work on this is needed, but some supporting results results based on asymptotic theory and numerical example are presented below.

11.3   A generalized IM

11.3.1   Construction

Let Y be a sample from distribution P_θ on $\mathbb{Y}$, where θ is an unknown parameter taking values in Θ, a separable space; here Y could be, say, a sample of size n from a product measure P_θ^n, but we have suppressed the dependence on n in the notation.

To start, let $\ell : \mathbb{Y} \times \Theta \to [0, \infty)$ be a loss function, i.e., a function such that small values of $\ell(y, \theta)$ indicate that the model with parameter θ fits data y reasonably well. This loss function $\ell(y, \theta)$ could be a sort of residual sum-of-squares or the negative log-likelihood. There have been some recent efforts to use loss functions to derive Bayes models in problems without an available likelihood [26, 233], so the appearance of a loss function here may not be a total surprise. Assume that there is a minimizer $\hat{\theta} = \hat{\theta}(y)$ of the loss function $\ell(y, \theta)$ for each y. Next define the function

$$T_{y,\theta} = \exp[-\{\ell(y, \theta) - \ell(y, \hat{\theta})\}], \tag{11.2}$$

We focus here on the case where the loss function is the negative log-likelihood, so the function $T_{y,\theta}$ in (11.2) is the relative likelihood

$$T_{y,\theta} = L_y(\theta) / L_y(\hat{\theta}), \tag{11.3}$$

where $L_y(\theta)$ is the likelihood function, assumed to be bounded, and $\hat{\theta}$ is a maximum likelihood estimator. Other choices of $T_{y,\theta}$ are possible (see Remark 11.1) but the use of likelihood is reasonable, since it conveniently summarizes all information in y concerning θ. Let F_θ be the distribution function of $T_{Y,\theta}$ when $Y \sim P_\theta$, i.e.,

$$F_\theta(t) = P_\theta(T_{Y,\theta} \leq t), \quad t \in \mathbb{R}. \tag{11.4}$$

Often $F_\theta(t)$ will be a smooth function of t for each θ, but the discontinuous case is also possible. To avoid measurability difficulties, we assume throughout that $F_\theta(t)$ is a continuous function in θ for each t. Then the proposed generalized association is

$$F_\theta(T_{Y,\theta}) = U, \quad U \sim P_U = \text{Unif}(0,1), \tag{11.5}$$

which defines the mapping

$$\Theta_y(u) = \{\theta : F_\theta(T_{y,\theta}) = u\}, \quad (y,u) \in \mathbb{Y} \times \mathbb{U}, \quad \mathbb{U} = (0,1).$$

This completes the generalized A-step. For the P-step, we need only specify a (nested) predictive random set for a scalar auxiliary variable, which we are comfortable with by now. However, the particular choice of a predictive random set should depend on the relationship between U and (Y, θ) in the generalized association (11.5). Indeed, since we specified that $T_{y,\theta}$ is large when y and θ agree, it makes sense that the value $u = 1$ should always reside in the predictive random set. That is, the predictive random set should be nested and right-sided, and a natural choice is the random interval $S = [U, 1]$ for $U \sim \text{Unif}(0,1)$. We know from the general discussion in Chapter 4 that this S is valid, so this completes the P-step. Finally, the C-step is exactly as before, yielding the enlarged random set

$$\Theta_y(S) = \bigcup_{u \in S} \Theta_y(u) = \{\theta : F_\theta(T_{y,\theta}) \geq U\}, \quad U \sim \text{Unif}(0,1).$$

An easy calculation (see Exercise 11.1) yields the belief and plausibility functions

$$\text{bel}_y(A) = 1 - \sup_{\theta \notin A} F_\theta(T_{y,\theta}) \quad \text{and} \quad \text{pl}_y(A) = \sup_{\theta \in A} F_\theta(T_{y,\theta}), \tag{11.6}$$

for any $A \in 2^\Theta$, which can be interpreted just as in Chapter 4; and see below. As usual, if we are considering a singleton assertion $A = \{\theta_0\}$, then we write $\text{pl}_y(\theta_0)$ instead of the more tedious $\text{pl}_y(\{\theta_0\})$.

11.3.2 Validity of the generalized IM

Here we demonstrate validity of the generalized IM. The specific goal is to show that $\text{pl}_Y(A)$, as a function of $Y \sim P_\theta$, for a fixed A, has a certain calibration property. Besides providing frequentist error rate control for the IM-based procedures, described below, this calibration guarantees that the belief and plausibility function output has a known objective scale for interpretation. One technical point: continuity of F_θ in θ and separability of Θ ensure that $\text{pl}_y(A)$ is a measurable function in y for each A, so the following probability statements make sense.

Theorem 11.1. *Let A be a subset of Θ. For any $\theta \in A$, if $Y \sim \mathsf{P}_\theta$, then $\mathsf{pl}_Y(A)$ is stochastically larger than uniform. That is, for any $\alpha \in (0,1)$,*

$$\sup_{\theta \in A} \mathsf{P}_\theta \{ \mathsf{pl}_Y(A) \le \alpha \} \le \alpha. \tag{11.7}$$

Proof. Take any $\alpha \in (0,1)$ and any $\theta \in A$. Then, by definition of $\mathsf{pl}_y(A)$ and monotonicity of the probability measure P_θ, we get

$$\mathsf{P}_\theta \{ \mathsf{pl}_Y(A) \le \alpha \} = \mathsf{P}_\theta \{ \sup_{\theta \in A} F_\theta(T_{Y,\theta}) \le \alpha \} \le \mathsf{P}_\theta \{ F_\theta(T_{Y,\theta}) \le \alpha \}. \tag{11.8}$$

The random variable $T_{Y,\theta}$, as a function $Y \sim \mathsf{P}_\theta$, may be continuous or not. In the continuous case, F_θ is a smooth distribution function and $F_\theta(T_{Y,\theta})$ is uniformly distributed. In the discontinuous case, F_θ has jump discontinuities, but it is well-known that $F_\theta(T_{Y,\theta})$ is stochastically larger than uniform. In either case, the latter term in (11.8) can be bounded above by α. Taking supremum over $\theta \in A$ throughout (11.8) gives the result in (11.7). $\qquad\square$

The case where A is a singleton set is an important special case for point estimation and plausibility region construction. The result in Theorem 11.1 specializes nicely to this case.

Theorem 11.2. (i) *If $T_{Y,\theta}$ is a continuous random variable as a function of $Y \sim \mathsf{P}_\theta$, then $\mathsf{pl}_Y(\theta)$ is uniformly distributed.* (ii) *If $T_{Y,\theta}$ is a discrete random variable when $Y \sim \mathsf{P}_\theta$, then $\mathsf{pl}_Y(\theta)$ is stochastically larger than uniform.*

Proof. In this case, $\mathsf{pl}_Y(\theta) = F_\theta(T_{Y,\theta})$ so no optimization is required compared to the general A case. Therefore, the "$\le$" between the second and third terms in (11.8) becomes an "$=$." The rest of the proof goes exactly like that of Theorem 11.1. $\qquad\square$

11.3.3 Derived frequentist procedures and their properties

The IM can, as discussed in previous chapters, be used to construct frequentist decision procedures. What is particularly interesting is that, despite the importance of error rate control in the frequentist program, there seems to be no general strategy for constructing exact procedures. However, the validity theorem for the generalized IM makes it possible to construct frequentist procedures with exact error rate control, independent of the model or sample size. Compare this to those standard methods based on first-order asymptotic theory, those with higher-order asymptotic accuracy [30, 202], and even those based on bootstrap [54, 80], all depending on asymptotic justification. Tastes of what is presented below have appeared previously in the literature but the results are scattered and there seems to be no unified presentation. For example, the use of p-values for hypothesis testing and confidence intervals is certainly not new, nor is the idea of using Monte Carlo methods to approximate critical regions and confidence bounds [20, 27, 107, 130]. Also, the relative likelihood version of the plausibility region below appears in [85, 228, 270]. Each of these papers has a different focus, so the point that there is a simple, useful, and very general method underlying these developments apparently has yet to be made.

First, consider a hypothesis testing problem, $H_0 : \theta \in \Theta_0$ versus $H_1 : \theta \notin \Theta_0$. Define a plausibility function-based test as follows:

$$\text{reject } H_0 \text{ if and only if } \mathsf{pl}_y(\Theta_0) \leq \alpha. \tag{11.9}$$

The intuition is that if Θ_0 is not sufficiently plausible, given $Y = y$, then one should conclude that the true θ is outside Θ_0. An immediate consequence of Theorem 11.1 is that this test controls the probability of a Type I error at level α.

Corollary 11.1. *For any $\alpha \in (0,1)$, the size of the test* (11.9) *is no more than α. That is, $\sup_{\theta \in \Theta_0} \mathsf{P}_\theta\{\mathsf{pl}_Y(\Theta_0) \leq \alpha\} \leq \alpha$. Moreover, if H_0 is a point-null, so that Θ_0 is a singleton, and $T_{Y,\theta}$ is a continuous random variable when $Y \sim \mathsf{P}_\theta$, then the size is exactly α.*

Proof. Exercise 11.2. □

The plausibility function $\mathsf{pl}_y(\theta)$ can also be used to construct confidence regions. Specifically, for any $\alpha \in (0,1)$, define the $100(1-\alpha)\%$ plausibility region

$$\Pi_y(\alpha) = \{\theta : \mathsf{pl}_y(\theta) > \alpha\}. \tag{11.10}$$

The intuition is that θ values which are sufficiently plausible, given $Y = y$, are good guesses for the true parameter value. The size result above for the test (11.9), along with the well-known connection between confidence regions and hypothesis tests, shows that the plausibility regions (11.10) have coverage at the nominal $1 - \alpha$ level.

Corollary 11.2. *For any $\alpha \in (0,1)$, the plausibility region $\Pi_Y(\alpha)$ has the nominal frequentist coverage probability; that is, $\mathsf{P}_\theta\{\Pi_Y(\alpha) \ni \theta\} \geq 1 - \alpha$. Furthermore, the coverage probability is exactly $1 - \alpha$ in case* (i) *of Theorem 11.2.*

Proof. Exercise 11.3. □

Before we discuss sampling distribution properties of the plausibility function in the next section, we consider two important fixed-y properties. These properties motivated the "unified approach" developed by [85] and further studied by [270].

- Since S always contains 1, the minimizer $\hat{\theta}$ of the loss $\ell(y,\theta)$ satisfies $\mathsf{pl}_y(\hat{\theta}) = 1$, i.e., the plausibility region is never empty. In particular, in the case where ℓ is the negative log-likelihood, and $T_{y,\theta}$ is the relative likelihood (11.3), the maximum likelihood estimator is contained in the plausibility region.

- The plausibility function is defined only on Θ; more precisely, $\mathsf{pl}_y(\theta) \equiv 0$ for any θ outside Θ. So, if Θ involves some non-trivial constraints, then only parameter values that satisfy the constraint can be assigned positive plausibility. This implies that the plausibility region cannot extend beyond the effective parameter space. Compare this to the standard "$\hat{\theta} \pm$ something" confidence intervals or those based on asymptotic normality.

One could also ask if the plausibility region is connected or, perhaps, even convex. Unfortunately, like Bayesian highest posterior density regions, the plausibility regions are, in general, neither convex nor connected. An example of non-convexity

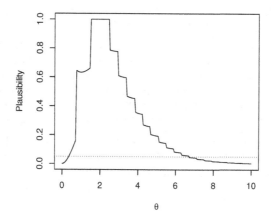

Figure 11.1 *Taken from [176]. Plausibility function for θ based on a single Poisson sample Y = 2.*

can be seen in Figure 11.3. That connectedness might fail is unexpected. Figure 11.1 shows the plausibility function based on a single Poisson sample $Y = 2$ using the relative likelihood (11.3); the small convex portion around $\theta = 1$ shows that disconnected plausibility regions are possible. A better understanding of the complicated dual way that the plausibility function depends on θ—through F_θ and through $T_{y,\theta}$—is needed to properly explain this phenomenon. Discreteness of the distribution F_θ also plays a role, as we have not seen this local convexity in cases where F_θ is continuous. However, suppose that $\theta \mapsto \ell(y, \theta)$ is convex and $F_\theta \equiv F$ does not depend on θ; see Section 11.3.4. In this case, the plausibility region takes the form $\{\theta : T_{y,\theta} > F^{-1}(\alpha)\}$, so convexity and connectedness hold by the quasi-concavity of $\theta \mapsto T_{y,\theta}$. For $T_{Y,\theta}$ the relative likelihood (11.3), under standard conditions, $-2\log T_{Y,\theta}$ is asymptotically chi-square under P_θ for any fixed θ, so the limiting distribution function of $T_{Y,\theta}$ is, indeed, free of θ. Therefore, one could expect convexity of the plausibility region when the sample size is sufficiently large; see Figure 11.4.

The above results demonstrate that the proposed method is valid for any model, any problem, and any sample size. Compare this to the bootstrap or analytical approximations whose theoretical validity holds only asymptotically for suitably regular problems. One could also ask about the asymptotic behavior of the plausibility function. Such a question is relevant because exactness without efficiency may not be particularly useful, i.e., it is desirable that the method can efficiently detect wrong θ values. Take, for example, the case where $T_{y,\theta}$ is the relative likelihood. If $Y = (Y_1, \ldots, Y_n)$ are iid P_θ, then $-2\log T_{Y,\theta}$ is, under mild conditions, approximately chi-square distributed. This means that the dependence of F_θ on θ disappears, asymptotically, so , for large n, the plausibility region (11.10) is similar to

$$\{\theta : -2\log T_{y,\theta} < \chi^2(\alpha)\},$$

where $\chi^2(\alpha)$ is the 100α percentile of the appropriate chi-square. Figure 11.4 displays both of these regions and the similarity is evident. Since the approximate plausibility region in the above display has asymptotic coverage $1 - \alpha$ and is efficient in terms of volume, the efficiency conclusion carries over to the plausibility region.

For a more precise description of the asymptotic behavior of the plausibility function, we present a simple but general and rigorous result. Again, since the plausibility function-based methods are valid for all fixed sample sizes, the motivation for this asymptotic investigation is efficiency. Let $Y = (Y_1, \ldots, Y_n)$ be iid P_{θ^*}. Suppose that the loss function is additive, i.e., $\ell(y, \theta) = \sum_{i=1}^{n} h(y_i, \theta)$, and the function h is such that $H(\theta) = \mathsf{E}_{\theta^*}\{h(Y_1, \theta)\}$ exists, is finite for all θ, and has a unique minimum at $\theta = \theta^*$. Also assume that, for sufficiently large n, the distribution of $T_{Y,\theta}$ under P_θ has no atom at zero. Write pl_n for the plausibility function $\mathsf{pl}_Y = \mathsf{pl}_{(Y_1, \ldots, Y_n)}$.

Theorem 11.3. *Under the conditions in the previous paragraph, $\mathsf{pl}_n(\theta) \to 0$ with P_{θ^*}-probability 1 for any $\theta \neq \theta^*$.*

Proof. Let $\hat{\theta} = \hat{\theta}(Y)$ denote the loss minimizer. Then $\ell(Y, \theta^*) \geq \ell(Y, \hat{\theta})$, so

$$
\begin{aligned}
T_{Y,\theta} &= \exp[-\{\ell(Y, \theta) - \ell(Y, \hat{\theta})\}] \\
&\leq \exp[-\{\ell(Y, \theta) - \ell(Y, \theta^*)\}] \\
&= \exp[-n\{H_n(\theta) - H_n(\theta^*)\}],
\end{aligned}
$$

where $H_n(\theta) = n^{-1}\sum_{i=1}^{n} h(Y_i, \theta)$ is the empirical version of $H(\theta)$. Since $F_\theta(\cdot)$ is non-decreasing,

$$
\mathsf{pl}_n(\theta) = F_\theta(T_{Y,\theta}) \leq F_\theta(e^{-n\{H_n(\theta) - H_n(\theta^*)\}}).
$$

By the assumptions on the loss and the law of large numbers, with P_{θ^*}-probability 1, there exists N such that $H_n(\theta) - H_n(\theta^*) > 0$ for all $n \geq N$. Therefore, the exponential term in the above display vanishes with P_{θ^*}-probability 1. Since $T_{Y,\theta}$ has no atom at zero under P_θ, the distribution function $F_\theta(t)$ is continuous at $t = 0$ and satisfies $F_\theta(0) = 0$. It follows that $\mathsf{pl}_n(\theta) \to 0$ with P_{θ^*}-probability 1. $\square$

The conclusion of Theorem 11.3 is that the plausibility function will correctly distinguish between the true θ^* and any $\theta \neq \theta^*$ with probability 1 for large n. In other words, if n is large, then the plausibility region will not contain points too far from θ^*, hence efficiency. For the case of the relative likelihood, the difference $H_n(\theta) - H_n(\theta^*)$ in the proof converges to the Kullback–Leibler divergence of P_θ from P_{θ^*}, which is strictly positive under the uniqueness condition on θ^*, i.e., identifiability.

It is possible to strengthen the convergence result in Theorem 11.3, at least for the relative likelihood case, with the use of tools from the theory of empirical processes, as in [261]. However, it seems that this approach also requires some uniform control on the small quantiles of the distribution F_θ for θ away from θ^*. These quantiles are difficult to analyze, so more work is needed here.

11.3.4 Computation

Evaluation of $F_\theta(T_{y,\theta})$ is crucial to the proposed methodology. In some problems, it may be possible to derive the distribution F_θ in either closed-form or in terms of some functions that can be readily evaluated, but such problems are rare. So, numerical methods are needed to evaluate the plausibility function and, here, we present a simple Monte Carlo approximation of F_θ. See, also, Remark 11.3 in Section 11.5.

To approximate $F_\theta(T_{y,\theta})$, where $Y = y$ is the observed sample, first choose a large number M; unless otherwise stated, the examples herein use $M = 50,000$, which is conservative. Then construct the following root-M consistent estimate of $F_\theta(T_{y,\theta})$:

$$\widehat{F}_\theta(T_{y,\theta}) = \frac{1}{M} \sum_{m=1}^{M} I\{T_{Y^{(m)},\theta} \le T_{y,\theta}\}, \quad Y^{(1)}, \ldots, Y^{(M)} \overset{\text{iid}}{\sim} \mathsf{P}_\theta. \quad (11.11)$$

This strategy can be performed for any choice of θ, so we may consider $\widehat{F}_\theta(T_{y,\theta})$ as a function of θ. If necessary, the supremum over a set $A \subset \Theta$ can be evaluated using a standard optimization package; in our experience, the optim function in R works well. To compute a plausibility interval, solutions to the equation $\widehat{F}_\theta(T_{y,\theta}) = \alpha$ are required. These can be obtained using, for example, standard bisection or stochastic approximation [107].

An interesting question is if, and under what conditions, the distribution function F_θ does not depend on θ. Indeed, if F_θ is free of θ, then there is no need to simulate new $Y^{(m)}$'s for different θ's—the same Monte Carlo sample can be used for all θ—which amounts to substantial computational savings. Next we describe a general context where this θ-independence can be discussed.

Let $\mathscr{G}$ be a group of transformations $g : \mathbb{Y} \to \mathbb{Y}$, and let $\bar{\mathscr{G}}$ be a corresponding group of transformations $\bar{g} : \Theta \to \Theta$ defined by the invariance condition:

$$\text{if } Y \sim \mathsf{P}_\theta, \text{ then } gY \sim \mathsf{P}_{\bar{g}\theta}, \quad (11.12)$$

where, e.g., gy denotes the image of y under transformation g. Note that g and $\bar{g}$ are tied together by the relation (11.12). Models that satisfy (11.12) are called group transformation models. The next result, similar to Corollary 1 in [228], shows that F_θ is free of θ in group models when $T_{y,\theta}$ has a certain invariance property.

Theorem 11.4. *Suppose* (11.12) *holds for groups $\mathscr{G}$ and $\bar{\mathscr{G}}$ as described above. If $\bar{\mathscr{G}}$ is transitive on Θ and $T_{y,\theta}$ satisfies*

$$T_{gy,\bar{g}\theta} = T_{y,\theta} \quad \text{for all } y \in \mathbb{Y} \text{ and } g \in \mathscr{G}, \quad (11.13)$$

then the distribution function F_θ in (11.4) *does not depend on θ.*

Proof. For $Y \sim \mathsf{P}_\theta$, pick any fixed θ_0 and choose corresponding $g, \bar{g}$ such that $\theta = \bar{g}\theta_0$; such a choice is possible by transitivity. Let $Y_0 \sim \mathsf{P}_{\theta_0}$, so that gY_0 also has distribution P_θ. Since $T_{gY_0,\bar{g}\theta_0} = T_{Y_0,\theta_0}$, by (11.13), it follows that T_{Y_0,θ_0} has distribution free of θ. □

For a given function $T_{y,\theta}$, condition (11.13) needs to be checked. For the loss function-based description of $T_{y,\theta}$, if $\ell(y,\theta)$ is invariant with respect to $\mathscr{G}$, i.e.,

$$\ell(gy, \bar{g}\theta) = \ell(y,\theta), \quad \forall (g,\bar{g}), \quad \forall (y,\theta),$$

and if the loss minimizer $\hat{\theta} = \hat{\theta}(y)$ is equivariant, i.e.,

$$\hat{\theta}(gy) = \bar{g}\hat{\theta}(y), \quad \forall\,(g,\bar{g}), \quad \forall\,y,$$

then (11.13) holds. For the special case where $\ell(y,\theta)$ is the negative log-likelihood, so that $T_{y,\theta}$ is the relative likelihood (11.3), we have the following result.

Corollary 11.3. *Suppose the model has a dominating measure relatively invariant with respect to $\mathscr{G}$. Then (11.12) holds, and the relative likelihood $T_{Y,\theta}$, in (11.3), satisfies (11.13). Therefore, if $\bar{\mathscr{G}}$ is transitive, the distribution function F_θ in (11.4) does not depend on θ.*

Proof. See Theorem 3.1 and the discussion on pages 46–47 in [72]. $\qquad\qquad\square$

11.3.5 Examples

Example 11.1. Inference on the success probability θ based on a sample Y from a binomial distribution is a fundamental problem in statistics. For this problem, [32, 33] showed that the widely-used Wald confidence interval often suffers from strikingly poor frequentist coverage properties, and that other intervals can be substantially better in terms of coverage. In the present context, the relative likelihood is given by

$$T_{y,\theta} = \left(\frac{n\theta}{y}\right)^y \left(\frac{n - n\theta}{n - y}\right)^{n-y}.$$

For given (n,y), one can exactly evaluate $\mathrm{pl}_y(\theta) = P_\theta(T_{Y,\theta} \le T_{y,\theta})$, where $Y \sim \mathrm{Bin}(n,\theta)$, numerically, using the binomial mass function. Given $\mathrm{pl}_y(\theta)$, the $100(1 - \alpha)\%$ plausibility interval for θ can be found by solving the equation $\mathrm{pl}_y(\theta) = \alpha$ numerically. Figure 11.2(a) shows a plot of the plausibility function for data $(n,y) = (25,15)$. As expected, at $\hat{\theta} = 15/25 = 0.6$, the plausibility function is unity. The steps in the plausibility function are caused by the discreteness of the underlying binomial distribution. The figure also shows (in gray) an approximation of the plausibility function obtained by Monte Carlo sampling ($M = 1000$) from the binomial distribution, as in (11.11), and the exact and the approximation plausibility functions are almost indistinguishable. By the general theory above, this $100(1 - \alpha)\%$ plausibility interval has guaranteed coverage probability $1 - \alpha$. However, the discreteness of the problem implies that the coverage is conservative. A plot of the coverage probability, as a function of θ, for $n = 50$, is shown in Figure 11.2(b), confirming the claimed conservativeness (up to simulation error). Of the intervals considered in [32], only the Clopper–Pearson interval has guaranteed 0.95 coverage probability. The plausibility interval here is clearly more efficient, particularly for θ near 0.5.

Example 11.2. Let $Y_1,\dots,Y_n$ be independent samples from a distribution with density $p_\theta(y) = \theta^2(\theta + 1)^{-1}(y + 1)e^{-\theta y}$, for $y,\theta > 0$. This non-standard distribution, a mixture of a gamma and an exponential density, appears in [162]. In this case, $\hat{\theta} = \{1 - \bar{y} + (\bar{y}^2 + 6\bar{y} + 1)^{1/2}\}/2\bar{y}$, and the relative likelihood is

$$T_{y,\theta} = (\theta/\hat{\theta})^{2n}\{(\hat{\theta} + 1)/(\theta + 1)\}^n e^{n\bar{y}(\hat{\theta} - \theta)}.$$

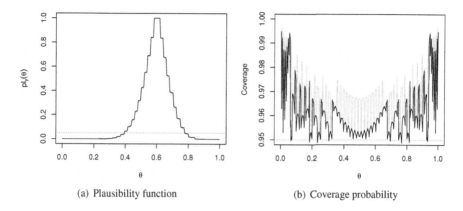

Figure 11.2 *Taken from [176]. Panel (a): Exact (black) and approximate (gray) plausibility functions for a data set with $(n,y) = (25,15)$. Panel (b): Coverage probability of the 95% plausibility (black) and Clopper–Pearson (gray) intervals, as a function of θ, for $n = 50$.*

For illustration, I compare the coverage probability of the 95% plausibility interval versus those based on standard asymptotic normality of $\hat\theta$ and a corresponding parametric bootstrap. With 1000 random samples of size $n = 50$ from the distribution above, with $\theta = 1$, the estimated coverage probabilities are 0.949, 0.911, and 0.942 for plausibility, asymptotic normality, and bootstrap, respectively. The plausibility interval hits the desired coverage probability on the nose, while the other two, especially the asymptotic normality interval, fall a bit short.

Example 11.3. Consider an iid sample $Y_1,\ldots,Y_n$ from a gamma distribution with unknown shape θ_1 and scale θ_2. Maximum likelihood estimation of (θ_1,θ_2) in the gamma problem has an extensive body of literature, e.g., [28, 121, 131]. In this case, the maximum likelihood estimate has no closed-form expression, but the relative likelihood can be readily evaluated numerically and the plausibility function can be found via (11.11). For illustration, consider the data presented in [104] on the survival times of $n = 20$ rats exposed to a certain amount of radiation. A plot of the 90% plausibility region for $\theta = (\theta_1,\theta_2)$ is shown in Figure 11.3. A Bayesian posterior sample is also shown, based on the Jeffreys prior, along with a plot of the 90% confidence ellipse based on asymptotic normality of the maximum likelihood estimate. Since n is relatively small, the shape the Bayes posterior is non-elliptical. The plausibility region captures the non-elliptical shape, and has roughly the same center and size as the maximum likelihood region. Moreover, the plausibility region has exact coverage.

Example 11.4. Consider a binary response variable Y that depends on a set of covariates $x = (1,x_1,\ldots,x_p)^\top \in \mathbb{R}^{p+1}$. An important special case is the probit regression model, with likelihood $L_y(\theta) = \prod_{i=1}^n \Phi(x_i^\top \theta)^{y_i}\{1 - \Phi(x_i^\top \theta)\}^{1-y_i}$, where $y_1,\ldots,y_n$ are the observed binary response variables, $x_i = (1,x_{i1},\ldots,x_{ip})^\top$ is a vector of covariates associated with y_i, Φ is the standard Gaussian distribution function, and

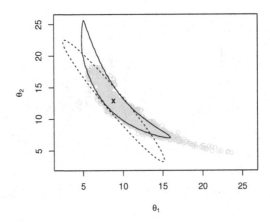

Figure 11.3 *Taken from [176]. Solid line gives the 90% plausibility region for $\theta = (\theta_1, \theta_2)$ in the gamma example; X marks the maximum likelihood estimate; the dashed line gives the 90% confidence region for θ based on asymptotic normality of the maximum likelihood estimator; the gray points are samples from the (Jeffreys prior) Bayesian posterior distribution.*

$\theta = (\theta_0, \theta_1, \ldots, \theta_p)^\top \in \Theta$ is an unknown coefficient vector. This likelihood function can be maximized to obtain the maximum likelihood estimate $\hat{\theta}$ and, hence, the relative likelihood $T_{y,\theta}$ in (11.3). Then $\mathrm{pl}_y(\theta)$ can be evaluated as in (11.11).

For illustration, consider a real data set with a single covariate ($p = 1$). The data, taken from Table 8.4 in [111], concerns the relationship between exposure to choleric acid and the death of mice. In particular, the covariate x is the acid dosage and $y = 1$ if the exposed mice dies and $y = 0$ otherwise. Here a total of $n = 120$ mice are exposed, ten at each of the twelve dosage levels. Figure 11.4 shows the 90% plausibility region for $\theta = (\theta_0, \theta_1)^\top$. For comparison, the 90% confidence region based on the asymptotic normality of $\hat{\theta}$ is also given. In this case, the plausibility and confidence regions are almost indistinguishable, likely because n is relatively large. The 0.9 coverage probability of the plausibility region is, however, guaranteed and its similarity to the classical region suggests that it is also efficient.

11.4   A generalized marginal IM

11.4.1   Construction

In many cases, θ can be partitioned as $\theta = (\psi, \lambda) \in \Psi \times \Lambda$ where ψ is the interest parameter and λ is a nuisance parameter. For example, ψ could be just a component of the parameter vector θ or, more generally, ψ is some function of θ. In such a case, the approach described above can be applied with special kinds of sets, e.g., $A = \{\theta = (\psi, \lambda) : \lambda \in \Lambda\}$, to obtain marginal inference for ψ. However, it may be easier to interpret a redefined *marginal* plausibility function. The natural extension to the methodology presented above is to consider some loss function $\ell(y, \psi)$

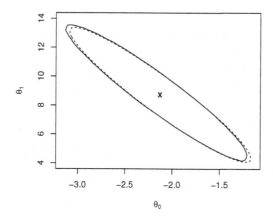

Figure 11.4 *Taken from [176]. Solid line gives the 90% plausibility region for $\theta = (\theta_0, \theta_1)$ in the binary regression example; dashed line gives the 90% confidence region for θ based on asymptotic normality of the maximum likelihood estimator.*

that does not directly involve the nuisance parameter λ, and construct the function $T_{y,\psi}$, depending only on the interest parameter ψ, like before. For example, taking $\ell(y, \psi) = -\sup_\lambda \log L(\psi, \lambda)$ to be the negative profile likelihood corresponds to replacing the relative likelihood (11.3) with the relative profile likelihood

$$T_{y,\psi} = L_y(\psi, \hat{\lambda}_\psi) / L_y(\hat{\psi}, \hat{\lambda}), \qquad (11.14)$$

where $\hat{\lambda}_\psi$ is the conditional maximum likelihood estimate of λ when ψ is fixed, and $(\hat{\psi}, \hat{\lambda})$ is a global maximizer of the likelihood. As before, other choices of $T_{y,\psi}$ are possible, but (11.14) is an obvious choice and shall be my focus in what follows.

   If the distribution of $T_{Y,\psi}$, as a function of $Y \sim P_{\psi,\lambda}$, does not depend on λ, then the development in the previous section carries over without a hitch. That is, one can define the distribution function F_ψ of $T_{Y,\psi}$ and construct a marginal plausibility function just as before:

$$\mathsf{mpl}_y(A) = \sup_{\psi \in A} F_\psi(T_{y,\psi}), \quad A \subseteq \Psi, \qquad (11.15)$$

This function can, in turn, be used exactly as in Section 11.3.3 for inference on the parameter ψ of interest, e.g., a $100(1-\alpha)\%$ marginal plausibility region for ψ is

$$\{\psi : \mathsf{mpl}_y(\psi) > \alpha\}. \qquad (11.16)$$

The distribution function F_ψ can be approximated via Monte Carlo just as in Section 11.3.4; see (11.17) below. Unfortunately, checking that F_ψ does not depend on the nuisance parameter λ seems to be a difficult task; see below.

11.4.2 Theoretical considerations

It is straightforward to verify that the validity properties (Theorems 11.1–11.2) carry over exactly in this more general case, provided that $T_{Y,\psi}$ in (11.14) has distribution free of λ, as a function of $Y \sim P_{\psi,\lambda}$. Consequently, the basic properties of the plausibility regions and tests (Corollaries 11.1–11.2) also hold in this case. It is rare, however, that $T_{Y,\psi}$ can be written in closed-form, so checking if its distribution depends on λ can be challenging.

Following the ideas in Corollary 11.3, it is natural to consider models having a special structure. The particular structure of interest here is that where, for each fixed ψ, $P_{\psi,\lambda}$ is a transformation model with respect to λ. That is, there exist associated groups of transformations, namely, $\mathcal{G}$ and $\bar{\mathcal{G}}$, such that

$$\text{if } Y \sim P_{\psi,\lambda}, \text{ then } gY \sim P_{\psi,\bar{g}\lambda}, \quad \text{for all } \psi.$$

This is called a composite transformation model; see [11] and Example 11.7 below. For such models, it follows from the argument in the proof of Theorem 11.4 that, if the loss $\ell(y, \psi)$ is invariant to the group action, i.e., if $\ell(gy, \psi) = \ell(y, \psi)$ for all y and g, then the corresponding $T_{Y,\psi}$ has distribution that does not depend on λ. Therefore, inference based on the marginal plausibility function mpl_y is exact in these composite transformation models. See Examples 11.5 and 11.7.

What if the problem is not a composite transformation model? In some cases, it is possible to show directly that the distribution of $T_{Y,\psi}$ does not depend on λ (see Examples 11.5–11.7 and 11.9) but, in general, this seems difficult. Large-sample theory can, however, provide some guidance. For example, if $Y = (Y_1, \ldots, Y_n)$ is a vector of iid samples from $P_{\psi,\lambda}$, then it can be shown, under certain standard regularity conditions, that $-2 \log T_{Y,\psi}$ is asymptotically chi-square, *for all values of* λ [23, 191]. Similar conclusions can be reached for the case where $T_{Y,\psi}$ is a conditional likelihood [3]. This suggests that, at least for large n, λ has a relatively weak effect on the sampling distribution of $T_{Y,\psi}$. This, in turn, suggests the following intuition: since λ has only a minimal effect, construct a marginal plausibility function for ψ, by fixing λ to be at some convenient value λ_0. In particular, a Monte Carlo approximation of F_ψ is as follows:

$$\widehat{F}_\psi(T_{y,\psi}) = \frac{1}{M} \sum_{m=1}^{M} I\{T_{Y^{(m)},\psi} \leq T_{y,\psi}\}, \quad Y^{(1)}, \ldots, Y^{(M)} \stackrel{\text{iid}}{\sim} P_{\psi,\lambda_0}. \tag{11.17}$$

Numerical justification for this approximation is provided in Example 11.8.

One could also consider different choices of $T_{Y,\psi}$ that might be less sensitive to the choice of λ. For example, the Bartlett correction to the likelihood ratio or the signed likelihood root often have faster convergence to a limiting distribution, suggesting less dependence on λ [9, 10, 227]. Such quantities have also been used in conjunction with bootstrap/Monte Carlo schemes that avoid use of the approximate limiting distribution [70, 156]. These adjustments did not appear to be necessary in the examples considered below. However, further work is needed along these lines, particularly in the case of high-dimensional λ.

11.4.3   Examples

Example 11.5. For a simple illustrative example, let $Y_1,\ldots,Y_n$ independent with distribution $N(\psi,\lambda)$, where $\theta = (\psi,\lambda)$ is completely unknown, but only the mean ψ is of interest. In this case, the relative profile likelihood is

$$T_{Y,\psi} = \left\{1 + n(\bar{Y} - \psi)^2/S^2\right\}^{-n/2},$$

where $S^2 = \sum_{i=1}^{n}(Y_i - \bar{Y})^2$ is the usual residual sum-of-squares. Since $T_{Y,\psi}$ is a monotone decreasing function of the squared t-statistic, it is easy to see that the marginal plausibility interval (11.16) for ψ is exactly the textbook t-interval; see Exercise 11.4. Exactness and efficiency of the marginal plausibility interval follow from the well-known results for the t-interval.

Example 11.6. Suppose that $Y_1,\ldots,Y_n$ are independent real-valued observations from an unknown distribution P, a nonparametric problem. Consider the so-called empirical likelihood ratio, given by $n^n \prod_{i=1}^{n} P(\{Y_i\})$, where P ranges over all probability measures on $\mathbb{R}$ [197]. Here interest is in a functional $\psi = \psi(P)$, namely the $100p$th quantile of P, where $p \in (0,1)$ is fixed. Theorem 5 in [253] shows that

$$T_{Y,\psi} = \left(\frac{p}{r}\right)^r \left(\frac{1-p}{n-r}\right)^{n-r}, \quad \text{where} \quad r = \begin{cases} \#\{i : Y_i \leq \psi\} & \text{if } \psi < \hat{\psi} \\ np & \text{if } \psi = \hat{\psi} \\ \#\{i : Y_i < \psi\} & \text{if } \psi > \hat{\psi}, \end{cases}$$

and $\hat{\psi}$ is the $100p$th sample quantile. The distribution of $T_{Y,\psi}$ depends on P only through ψ, so the marginal plausibility function is readily obtained via basic Monte Carlo. In fact, it is now essentially a binomial problem, like that in Example 11.1. Regarding efficiency, we report a small simulation study. For standard normal, exponential, and Student-t (df $= 3$) distributions, 1000 samples of size $n = 100$ are taken and, for each, a 95% plausibility interval for the 75th percentile is calculated. The estimated coverage probabilities are 0.957, 0.962, and 0.959, respectively. This suggests, as expected, that the plausibility intervals are conservative. Similar results are seen for different percentiles and for smaller/larger samples sizes.

Example 11.7. Consider a bivariate Gaussian distribution with all five parameters unknown. That is, the unknown parameter is $\theta = (\psi,\lambda)$, where the correlation coefficient ψ is the parameter of interest, and $\lambda = (\mu_1,\mu_2,\sigma_1,\sigma_2)$ is the nuisance parameter. From the calculations in [232], the relative profile likelihood is

$$T_{Y,\psi} = \left\{(1-\psi^2)^{1/2}(1-\hat{\psi}^2)^{1/2}/(1-\psi\hat{\psi})\right\}^n,$$

where $\hat{\psi}$ is the sample correlation coefficient. It is clear from the previous display and basic properties of $\hat{\psi}$ that the distribution of $T_{Y,\psi}$ is free of λ; this fact could also have been deduced directly from the problem's composite transformation structure. Therefore, for the Monte Carlo approximation in (11.11), data can be simulated from the bivariate Gaussian distribution with any convenient choice of λ.

For illustration, we replicate a simulation study in [232]. Here 10,000 samples of size $n = 10$ from a bivariate Gaussian distribution with $\lambda = (1,2,1,3)$ and various ψ

Table 11.1 *Taken from [176]. Estimated coverage probabilities of 95% intervals for ψ in the Example 11.7 simulation. First three rows are taken from Table 1 in [232]. Last two rows correspond to the parametric bootstrap and marginal plausibility intervals, respectively.*

	Correlation, ψ				
Method	-0.9	-0.5	0.0	0.5	0.9
z	0.9527	0.9525	0.9500	0.9517	0.9542
z_4	0.9499	0.9509	0.9494	0.9502	0.9516
R^*	0.9488	0.9500	0.9517	0.9508	0.9492
PB	0.9385	0.9425	0.9453	0.9438	0.9411
MPL	0.9492	0.9496	0.9502	0.9505	0.9509

values. Coverage probabilities of the 95% plausibility intervals (11.16) are displayed in Table 11.4.3. For comparison, several other methods are considered:

- Fisher's interval, based on approximate normality of $z = \frac{1}{2}\log\{(1 + \hat{\psi})/(1 - \hat{\psi})\}$;

- A modification of Fisher's z, due to [136], based on approximate normality of

$$z_4 = z - \frac{3z + \hat{\psi}}{4(n-1)} - \frac{23z + 33\hat{\psi} - 5\hat{\psi}^2}{96(n-1)^2};$$

- Third-order approximate normality of $R^* = R - R^{-1}\log(RQ^{-1})$, where R is a signed log-likelihood root and Q is a measure of maximum likelihood departure, with expressions for R and Q worked out in [232];

- Standard parametric bootstrap percentile confidence intervals based on the sample correlation coefficient, with 5000 bootstrap samples.

In this case, based on the first three digits, z, z_4, and R^* perform reasonably well, but the parametric bootstrap intervals suffer from under-coverage near ± 1. The plausibility intervals are quite accurate across the range of ψ values.

Example 11.8. Consider a gamma distribution with mean ψ and shape λ; that is, the density is $p_\theta(y) = \Gamma(\lambda)^{-1}(\lambda/\psi)^\lambda y^{\lambda-1}e^{-\lambda y/\psi}$, where $\theta = (\psi, \lambda)$. The goal is to make inference on the mean ψ. Likelihood-based solutions to this problem are presented in [102, 104, 122]. In the present context, it is straightforward to evaluate the relative profile likelihood $T_{Y,\psi}$ in (11.14). However, it is difficult to check if the distribution function F_ψ of $T_{Y,\psi}$ depends on nuisance shape parameter λ. So, following the general intuition above, we assume that it has a negligible effect and fix $\lambda \equiv 1$ in the Monte Carlo step. That is, the Monte Carlo samples, $Y^{(1)}, \ldots, Y^{(M)}$, in (11.17), are each iid samples of size n taken from a gamma distribution with mean ψ and shape $\lambda_0 = 1$. That the results are robust to fixing $\lambda_0 = 1$ in large samples is quite reasonable, but what about in small samples? It would be comforting if the distribution of the relative profile likelihood $T_{Y,\psi}$ were not sensitive to the underlying value of the shape parameter λ. Monte Carlo estimates of the distribution function of $T_{Y,\psi}$ are shown in Figure 11.5 for $n = 10$, $\psi = 1$, and a range of λ values. It is clear

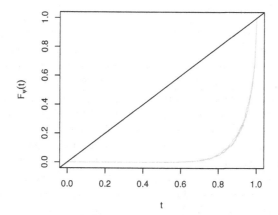

Figure 11.5 *Taken from [176]. Distribution functions (gray) of the relative profile likelihood $T_{Y,\psi}$ for the gamma mean problem in Example 11.8, for mean $\psi = 1$ and a range of shapes λ from 0.1 to 10.*

Table 11.2 *Taken from [176]. Interval estimates for the gamma mean ψ in Example 11.8. First three rows are taken from [104]; last row gives the marginal plausibility interval.*

	95% intervals for ψ		
Method	Lower	Upper	Length
classical	96.7	130.9	34.2
[260]	97.0	134.7	37.7
[104]	97.2	134.2	37.0
MPL	97.1	133.6	36.5

that the distribution is not particularly sensitive to the value of λ, which provides comfort in fixing $\lambda_0 = 1$. Similar comparisons hold for ψ different from unity. For further justification for fixing $\lambda_0 = 1$, we computed the coverage probability for the 95% marginal plausibility interval for ψ, based on fixed $\lambda_0 = 1$ in (11.17), over a range of true (ψ, λ) values; in all cases, the coverage is within an acceptable range of 0.95.

Reconsider the data on survival times in Example 11.3 above. Table 11.4.3 shows 95% intervals for ψ based on four different methods: the classical first-order accurate approximation; the second-order accurate parameter-averaging approximation of [260]; the best of the two third-order accurate approximations in [104]; and the marginal plausibility interval. In this case, the marginal plausibility interval is shorter than both the second- and third-order accurate confidence intervals.

11.4.4   Comparison with parametric bootstrap

The plausibility function-based method described above allows for the construction of exact frequentist methods in many cases, which is particularly useful in problems where an exact sampling distribution is not available. An alternative method for such problems is the parametric bootstrap, where the unknown P_θ is replaced by the estimate $P_{\hat\theta}$, and the sampling distribution is approximated by simulating from $P_{\hat\theta}$. This approach and variations thereof have been carefully studied [70, 156] and have many desirable properties. The proposed plausibility function method is, at least superficially, quite similar to the parametric bootstrap, so it is interesting to see how the two methods compare. In many cases, plausibility functions and the parametric bootstrap give similar answers, such as the bivariate normal correlation example above. Here I show one simple example where the former clearly outperforms the latter.

Example 11.9. Consider a simple Gaussian random effects model, i.e., $Y_1,\ldots,Y_n$ are independently distributed, with $Y_i \sim N(\mu_i,\sigma_i^2)$, $i = 1,\ldots,n$, where the means $\mu_1,\ldots,\mu_n$ are unknown, but the variances $\sigma_1^2,\ldots,\sigma_n^2$ are known. The Gaussian random effects portion comes from the assumption that the individual means are an independent $N(\lambda,\psi^2)$ sample, where $\theta = (\psi,\lambda)$ is unknown. Here $\psi \geq 0$ is the parameter of interest, and the overall mean λ is a nuisance parameter.

Using well known properties of Gaussian convolutions, it is possible to recast this hierarchical model in a non-hierarchical form. That is, $Y_1,\ldots,Y_n$ are independent, with $Y_i \sim N(\lambda,\sigma_i^2 + \psi^2)$, $i = 1,\ldots,n$. Here the conditional maximum likelihood estimate of λ, given ψ, is the weighted average $\hat\lambda_\psi = \sum_{i=1}^n w_i(\psi)Y_i / \sum_{i=1}^n w_i(\psi)$, where $w_i(\psi) = 1/(\sigma_i^2 + \psi^2)$, $i = 1,\ldots,n$. From here it is straightforward to write down the relative profile likelihood $T_{Y,\psi}$ in (11.14); see Exercise 11.5. Moreover, since the model is of the composite transformation form, the distribution of $T_{Y,\psi}$ is free of λ, so any choice of λ (e.g., $\lambda = 0$) will suffice in the Monte Carlo step (11.17). One can then readily compute plausibility intervals for ψ.

For interval estimation of ψ, a parametric bootstrap is a natural choice. But it is known that bootstrap methods tends to have difficulties when the true parameter is at or near the boundary. The following simulation study will show that the marginal plausibility interval outperforms the parametric bootstrap in the important case of ψ near the boundary, i.e., $\psi \approx 0$. Since the two-sided bootstrap interval cannot catch a parameter exactly on the boundary, we consider true values of ψ getting closer to the boundary as the sample size increases. In particular, for various n, we take the true $\psi = n^{-1/2}$ and compare interval estimates based on coverage probability and mean length. Here data are simulated independently, according to the model $Y_i \sim N(0,\sigma_i^2 + \psi^2)$, $i = 1,\ldots,n$, where ψ is as above and the $\sigma_1,\ldots,\sigma_n$ are iid samples from an exponential distribution with mean 2. Table 11.4.4 shows the results of 1000 replications of this process for four sample sizes. Observe that the plausibility intervals hit the target coverage probability on the nose for each n, while the bootstrap suffers from drastic under-coverage for all n.

Table 11.3 *Taken from [176]. Estimated coverage probabilities and expected lengths of the 95% interval estimates for ψ in Example 11.9*

	MPL		PB	
n	Coverage	Length	Coverage	Length
50	0.952	0.267	0.758	0.183
100	0.946	0.162	0.767	0.138
250	0.948	0.079	0.795	0.079
500	0.950	0.041	0.874	0.039

11.4.5   Variable selection in Gaussian linear regression

This example demonstrates the proposed method's ability to handle non-standard inference problems, in this case, variable selection in regression. Here we give a relatively simple generalized IM-based approach.

Consider the full-rank Gaussian linear model as discussed in Chapters 8 and 10. In particular, we have a set of p predictor variables, stacked in a full-rank $n \times p$ matrix X, with unknown regression coefficients $\beta = (\beta_1, \ldots, \beta_p)^\top$ and unknown error variance σ^2. The goal is to select a subset of variables (columns of X) that suitably describe the variation in the response Y. Toward this, let $\gamma \in \{0,1\}^p$ index the collection of sub-models, with β_γ the corresponding sub-vector of β. For any given γ, evaluation of the marginal plausibility function for β_γ is immediate, i.e., no Monte Carlo methods are required; see Exercise 11.7. Now consider the assertion $A_\gamma = \{\beta_{1-\gamma} = 0\}$, i.e., that model γ is not wrong. Evidence in the observed y for model γ can, therefore, be measured by

$$\mathsf{mpl}_y(A_\gamma) = F_{[\beta_{1-\gamma}=0]}\left(T_{y,[\beta_{1-\gamma}=0]}\right),$$

with the subscript "$[\beta_{1-\gamma} = 0]$" emphasizing the fact that the calculation is based on the marginal plausibility function for $\beta_{1-\gamma}$, evaluated at zero. As A_γ corresponds to a singleton assertion with respect to $\beta_{1-\gamma}$, the frequentist calibration in Theorem 11.2 holds. For inference on the number of non-zero coefficients in β, one can consider

$$\mathsf{mpl}_y(k) = \max_{\gamma:|\gamma|\leq k} \mathsf{mpl}_y(A_\gamma), \tag{11.18}$$

the plausibility that the model has at most k non-zero coefficients. It follows from the calibration property quoted above that a rule that rejects the claim of at most k non-zero coefficients when $\mathsf{mpl}_y(k) \leq \alpha$ with control the frequentist Type I error rate at α. Therefore, a reasonable model selection strategy is to choose $k^\star$ variables, where $k^\star$ is the smallest k such that $\mathsf{mpl}_y(k) \leq \alpha$.

For a numerical illustration, consider the diabetes data set analyzed in [79]. These data consist of observations for $n = 442$ diabetes patients with $p = 10$ covariates: age, sex, body mass index (bmi), average blood pressure (map), and six blood serum measurements (tc, ldl, hdl, tch, ltg, and glu). In this case, there are $2^{10} = 1024$ possible models; that $\mathsf{mpl}_y(A_\gamma)$ can be evaluated in closed-form, without the Monte Carlo,

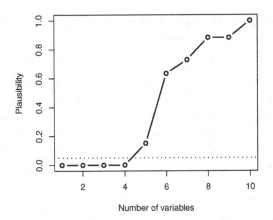

Figure 11.6 *Plot of the marginal plausibility function* $\mathsf{mpl}_y(k)$ *in* (11.18) *versus* k *for the diabetes data regression problem.*

makes the computations very fast, even for much larger p. A plot of $\mathsf{mpl}_y(k)$ versus k for these data is shown in Figure 11.6. Here, with $\alpha = 0.05$, the method selects $k^\star = 5$ variables. Of all models γ with five variables, the one with largest $\mathsf{mpl}_y(A_\gamma)$ consists of sex, bmi, map, hdl, and ltg. For comparison, these five variables are a subset of the seven selected by lasso [79] and a superset of the four selected by the Bayesian lasso [198].

11.5   Remarks on generalized IMs

Remark 11.1. We mentioned previously that the relative likelihood (11.3) is not the only possible choice for $T_{y,\theta}$. For example, if the likelihood is unbounded, then one might consider $T_{y,\theta} = L_y(\theta)$. A penalized version of the likelihood might also be appropriate in some cases, i.e., $T_{y,\theta} = L_y(\theta)\pi(\theta,y)$, where π is something like a Bayesian prior (although it could depend on y too). This could be potentially useful in high-dimensional problems. Another interesting class of $T_{y,\theta}$ quantities is motivated by higher-order asymptotics, as in [202] and the references therein. Choosing $T_{y,\theta}$ to be Barndorff-Nielsen's $r^\star = r^\star(\theta,y)$ quantity, or some variation thereof, could potentially give better results, particularly in the marginal inference problem involving $\theta = (\psi,\lambda)$. However, the possible gain in efficiency comes at the cost of additional analytical computations, and, based on my empirical results, it is unclear if these refinements would lead to any noticeable improvements. Also, recently, composite likelihoods [241] have been considered in problems where a genuine likelihood is either not available or is too complicated to compute. There are also possibilities besides those directly connected to likelihood, see, Section S:gims:lsnultinomial. In general, the method proposed here seems like a promising alternative to the methods more commonly used, but further investigation is needed.

Remark 11.2. More general versions of the generalized associations are possible.

Indeed, if $(y, u) \mapsto \Theta_y(u)$ is suitably compatible with the sampling model $Y \sim \mathsf{P}_{Y|\theta}$, then the generalized IM can be shown to be valid, provided that the predictive random set for U is valid. The challenge is in defining a meaningful notion of "compatibility" in this context. More work is needed along these lines, but we expect that the developments will shed light on various model-free inference, such as M-estimation, but from an IM perspective.

Remark 11.3. Using Monte Carlo approximations (11.11) and (11.17) to construct exact frequentist inferential procedures is, to our knowledge, new. Despite its novelty, the method is surprisingly simple and general. On the other hand, there is a computational price to pay for this simplicity and generality. Specifically, determination of plausibility intervals requires evaluation of the Monte Carlo estimate of $F_\theta(T_{y,\theta})$ for several θ values. This can be potentially time-consuming, but running the Monte Carlo simulations for different θ in parallel can help reduce this cost. The proposed method works—in theory and in principle—in high-dimensional problems, but there the computational cost is further exaggerated. An important question is if some special techniques can be developed for problems where only the nuisance parameter is high- or infinite-dimensional. Clever marginalization can reduce the dimension to something manageable within the proposed framework, making exact inference in semiparametric problems possible.

11.6 Application: Large-scale multinomial inference

11.6.1 Problem background

Consider the problem of comparing the frequencies of single nucleotide polymorphisms (SNPs) across cases and controls. We scan the whole genome sequence using blocks of SNPs, where the block size is ten SNPs. For a given block, there are two independent multinomial distributions corresponding to the SNP distributions in cases and controls. The two multinomial distributions can be derived by a $2 \times K$ table of independent Poisson counts, where K is the total number of SNP genotypes in the block. More specifically, let N_{tk} denote the number of occurrences of genotype k in group t, where $N_{tk} \sim \mathsf{Pois}(\lambda_{tk})$, for t taking values 0 (control) and 1 (case), and $k = 1, \ldots, K$.

By conditioning on $N_t = \sum_{k=1}^{K} N_{tk}$, for $t = 0, 1$, the observed $\{N_{tk}\}$ follow two independent multinomial models:

$$(N_{t1}, \ldots, N_{tK}) \sim \mathsf{Mult}(N_t; \eta_{t1}, \ldots, \eta_{tK}), \quad t = 0, 1,$$

where $\eta_{tk} = \lambda_{tk} / \sum_{k=1}^{K} \lambda_{tk}$, $k = 1, \ldots, K$, are the SNP frequencies. The problem of interest is to determine of the η_{tk}s across $t = 0, 1$ are the same. In terms of the original λ_{tk}s, the problem is that $\lambda_{0k} \propto \lambda_{1k}$ for all $k = 1, \ldots, K$. An alternative characterization, the one we will take here, is as follows. By conditioning on each column of the $2 \times K$ table of Poisson counts, we obtain K binomial distributions. Set $\theta_k = \lambda_{1k}/(\lambda_{0k} + \lambda_{1k})$, $n_k = N_{0k} + N_{1k}$, and $Y_k = N_{1k}$, $k = 1, \ldots, K$. Then

$$(Y_k \mid n_k) \sim \mathsf{Bin}(n_k, \theta_k), \quad k = 1, \ldots, K, \quad \text{independent.} \tag{11.19}$$

Then the assertion A of interest, concerning the presence of a case effect, in terms of the binomial parameters, is

$$A = \{\theta = (\theta_1, \ldots, \theta_K) : \theta_1 = \cdots = \theta_K\}. \tag{11.20}$$

We discuss here a generalized IM approach based on ideas presented in [166].

11.6.2   A generalized IM

To construct a valid generalized IM for inference on θ, we need to identify a function of the data $Y = (Y_1, \ldots, Y_K)$, possibly depending on θ, similar to the relative likelihood in the previous sections. If we let $n = \sum_{k=1}^K n_k$ be the total sample size, then we consider the statistic

$$T_Y = \sum_{k=1}^K \frac{w_k}{n_k(n - n_k)} \left(Y_k - \frac{n_k}{n} \sum_{j=1}^K Y_j \right)^2,$$

where $w_k = (n_k - 1)/(n_k + 1)$ down-weights observations with a small column total. In this case, note that the function $y \mapsto T_y$ does not depend on θ itself; however, its distribution does depend on θ. Write F_θ for the conditional distribution function of T_Y, given $\sum_{k=1}^K Y_k$; the formula for F_θ can be derived from the fact that, given $\sum_{k=1}^K Y_k$, the vector $(Y_1, \ldots, Y_K)$ has a multivariate hypergeometric distribution. In addition, F_θ depends on θ only up to the relative magnitudes of $\theta_1, \ldots, \theta_K$, i.e., it only depends on $\theta_k / \sum_{j=1}^K \theta_j$. For a generalized association, we can then take

$$F_\theta(T_Y) = U, \quad U \sim \mathsf{Unif}(0,1).$$

This completes the A-step. For the P-step, we want a nested predictive random set $\mathcal{S}$ for U. In this case, since small values of $F_\theta(T_Y)$ indicate that θ matches the observed data well, we take a nested left-sided predictive random set: $\mathcal{S} = [0, U]$, with $U \sim \mathsf{Unif}(0,1)$. For the C-step, we construct the enlarged random set

$$\Theta_y(\mathcal{S}) = \{\theta : F_\theta(T_Y) \leq U\}, \quad U \sim \mathsf{Unif}(0,1).$$

Since the assertion A of interest is a lower-dimensional subset of the full K-dimensional parameter space Θ, the belief in A is necessarily 0. The plausibility function, however, is generally non-zero and given by

$$\mathsf{pl}_y(A) = 1 - \inf_{\theta \in A} F_\theta(T_y);$$

the appearance of "inf" instead of "sup" as in previous sections is due to the use of a left- instead of right-sided predictive random set. However, as we discuss next, no minimization is needed.

11.6.3   Methodology and computation

For $\theta \in A$, all the components $\theta_1, \ldots, \theta_K$ are equal; therefore, there is only one unspecified value. Moreover, F_θ only depends on the relative magnitudes of the θ_ks so, there is only one value of $F_\theta(T_Y)$ for θ restricted to A. This distribution function is computed by using a scaled chi-square distribution, with scale and degrees of freedom estimated from a Monte Carlo sample via the method-of-moments. In particular, we carry out the following four steps:

1. Simulate M independent Monte Carlo samples $Y^{(1)}, \ldots, Y^{(M)}$ of K-vectors Y from their conditional distribution, given $\sum_{k=1}^K Y_k$.

2. Compute $T^m = T_{Y^{(m)}}, m = 1, \ldots, M$.

3. Calculate sample mean and variance of the sampled Y:

$$\bar{Y} = \frac{1}{M} \sum_{m=1}^M Y^{(m)} \quad \text{and} \quad S_Y^2 = \frac{1}{M-1} \sum_{m=1}^M (Y^{(m)} - \bar{Y})^2.$$

4. Find the degrees of freedom v and scale parameter γ of the chi-square approximation by matching its first two theoretical moments to the two sample moments from Step 3, i.e.,

$$\gamma = S_Y^2/(2\bar{Y}) \quad \text{and} \quad v = \bar{Y}/\gamma.$$

Steps 2–4 above are straightforward, so we provide details only for Step 1. Note that, given $\tau = \sum_{k=1}^K Y_k$, with $\theta_k \equiv 1$, the sampling distribution of $(Y_1, \ldots, Y_K)$ is the well-known multivariate hypergeometric distribution, with parameters K, $(n_1, \ldots, n_K)$, and τ. This distribution has the attractive feature that both the marginal distributions and conditional distributions are also hypergeometric. This allows for a simple way of generating multivariate hypergeometric random variables by simulating univariate hypergeometric deviates. For example, the marginal distribution of X_1 is the univariate hypergeometric distribution with parameters τ, $\sum_{k=1}^K n_k - \tau$, and $\sum_{k=1}^K n_k$, which, in the familiar context of sampling from an urn without replacement, represent the number of "white balls" in the urn, the number of "black balls" in the urn, and the number of balls selected from the urn, respectively. We then use the R function rhyper to simulate univariate hypergeometric random variables, which are then stacked together in the appropriate way to produce the required multivariate hypergeometric samples in Step 1.

Finally, for the observed $Y = y$, the distribution function $F_\theta(T_y)$ which, for $\theta \in A$, is a value $F(T_y)$ not depending on θ, is approximated by the distribution function of the scaled chi-square distribution identified from the Monte Carlo simulation above. In particular, the plausibility of the assertion A based on data $Y = y$ is simply $F(T_y)$.

11.6.4 Analysis in genome-wide association studies

We apply the generalized IM methodology on the GAW16 (Genetic Analysis Workshop 16) data from the North American Rheumatoid Arthritis Consortium. This genome-wide association study aims at identifying genetic variants, or SNPs, which are associated with Rheumatoid Arthritis. The data consists of 2062 samples—868 are cases and 1194 are controls. For each sample, whole genome SNPs are observed with a total coverage of 545,080 SNPs.

We partition the entire SNP sequence on each chromosome into a sequence of B blocks of consecutive SNPs, each block consisting of, for example, 10 SNPs. For each block, indexed by $b = 1, ..., B$, our proposed analysis of the two-sample multinomial counts produces a plausibility p_b for the assertion A in (11.20), i.e., that "the two samples, cases versus controls, are from the same population." The quantity $q_b = 1 - p_b$, $b = 1, ..., B$, provides evidence against the truthfulness of the assertion.

To assess the performance of our IM method for this "large K, small n" problem, we first consider a simulated study, using the real SNP genotype data to randomly simulate disease phenotypes. A phenotype variable w is generated from a simple additive model $w_i = \sum_j z_{ij} \beta_{ij} + e_i$, where z_{ij} denotes the SNP genotype of subject i at SNP j, $z_{ij} = 0$, 1, or 2 for wild type homozygous, heterozygous, and mutation homozygous, respectively, $i = 1, ..., 2062$, and j from 1 to the number of SNPs for a chromosome. We consider chromosome 14, which contains 17,947 SNPs in the Rheumatoid Arthritis genotype data. The coefficient β_{ij} is the effect of the jth SNP for the ith subject and is set equal to zero except for five SNPs at positions $j = 5000, 5001, ..., 5004$, where β_{ij} for these five SNPs are simulated from independent normal distributions with means of 5 and standard deviation of 1. In addition e_i is the residual effect generated from a normal distribution with mean of 0 and standard deviation of 1. Finally, disease subjects are sampled from the individuals with phenotypes w exceeding a threshold, which is the normal quantile corresponding to the proportion of $868/2062 = 0.42$, and controls are sampled from the remaining individuals. This simulation creates a new case-control data set with disease causal SNPs at positions $j = 5000, 5001, ..., 5004$ in chromosome 14.

Figure 11.7 displays a sequence of the q-value for chromosomes 14 in terms of Z-score, $Z_b = \Phi^{-1}(q_b)$. Positions around 5000 have very large q_b values hence show strong evidences against the assertion that "the two samples, cases versus controls, are from the same population." In other words, the generalized IM provides evidence to support the claim that SNPs around 5000 have different frequencies between disease and control subjects and hence are associated with the disease.

We now apply the IM method to the real data of 868 cases and 1194 controls and compare SNP genotype frequencies between the two groups. One block of 10 SNPs is studied at a time, with the scale of multinomial up to $3^{10} = 59,049$ categories. Figure 11.8 displays sequences of the q-value for chromosomes 6 and 14 in terms of Z-score, $Z_b = \Phi^{-1}(q_b)$. When larger than 8, the values of the Z-scores are replaced with 8 in the plots. Figure 11.9 displays the histograms of the q-values for chromosomes 6 and 14. Large values in Figure 11.8(a) correspond to those on the right tail in Figure 11.9 (a). They indicate that there are some blocks on chromosome 6 potentially associated with Rheumatoid Arthritis. This result is consistent with the known fact that the HLA (human leukocyte antigen) region on chromosome 6 contributes to disease risk. On the other hand, Figures 11.8(b) and 11.9(b) show that there are very few blocks on chromosome 14 that have Z-scores larger than 6 and are considered to associate with Rheumatoid Arthritis. Except for large values, the q-value in Figures 11.9(a,b) have very smooth distributions. This implies that we can specify a null distribution so that SNPs or blocks potentially associated with Rheumatoid Arthritis can be identified.

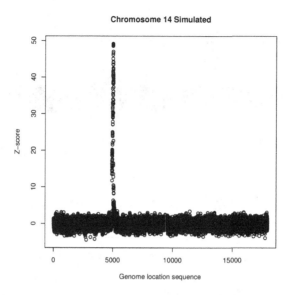

Chromosome 14 Simulated

Figure 11.7 *Taken from [166]. The time-series plots of the Z-scores of the probabilities for the assertion that control and case populations are different, computed based on the inferential model for the simulated data from chromosome 14.*

For this large K and small n problem, with K up to $3^{10} = 59,049$, it is difficult to apply standard frequentist or Bayesian approaches and no such analysis has been done for a block of SNPs. Instead, we conduct a standard approach of chi-square tests for 2×3 contingency tables for a single SNP at a time and compare the results with our previous analysis. The simple chi-square tests of one SNP at a time identify the same HLA region on chromosome 6 with significant association to the disease. However, the chi-square tests also produce many extremely significant SNPs, corresponding to Z-scores larger than 10, on all other chromosomes. This result indicates that the standard method tends to make falsely significant associations whereas our generalized IM method is both simple and more accurate in assessing uncertainty for this challenging problem.

11.7 Exercises

Exercise 11.1. Verify the formulas in (11.6).

Exercise 11.2. Prove Corollary 11.1.

Exercise 11.3. Prove Corollary 11.2.

Exercise 11.4. Show that the marginal plausibility interval for the normal mean in Example 11.5 is identical to the familiar t-interval.

Exercise 11.5. Derive the formula for the conditional maximum likelihood estima-

(a) Chromosome 6

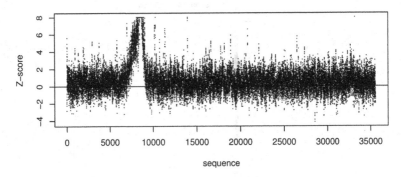

(b) Chromosome 14

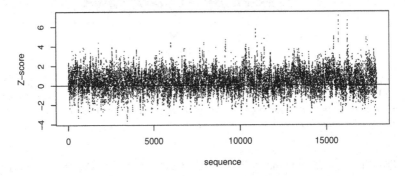

Figure 11.8 *Taken from [166]. The time-series plots of the Z-scores of the probabilities for the assertion that control and case populations are different, computed based on the two-multinomial model for SNPs in blocks of 10 in (a) chromosome 6 and (b) chromosome 14.*

tor of λ, given ψ, as stated in Section 11.4.4. Also write down the relative profile likelihood.

Exercise 11.6. The comparison between the generalized IM and the bootstrap method made in Section 11.4.4 was based on a very simple version of the bootstrap. Apply a more sophisticated version of the bootstrap discussed in [69]. Do the relative comparisons remain the same?

Exercise 11.7. In the regression problem in Section 11.4.5, show that the relative profile likelihood for β_γ has a distribution free of all other parameters. In particular, its distribution is just that of a simple transformation of a F-distribution, .

Exercise 11.8. The method discussed in Section 11.4.5 for the regression problem boils down to one of making inference on the "order" of the model. Another application where inference on the model order is important is in mixture models. Consider a K-component normal mixture model, where K has some fixed upper bound. Can

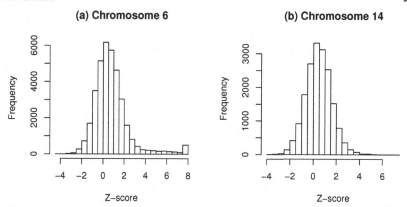

Figure 11.9 *Taken from [166]. Histograms of the Z-scores of the probabilities for the assertion that control and case populations are different, computed based on the two-multinomial model for SNPs in blocks of 10 in (a) chromosome 6 and (b) chromosome 14.*

you work out a generalized marginal IM for inference on K? (*Note:* This is a research problem, not a simple exercise!)

Chapter 12

Future Research Topics

12.1 Introduction

Reid and Cox [203] discuss the importance of statistical theory, in particular, "the foundations of statistical analysis, rather than the theoretical analysis of specific statistical methods." Their essay makes a number of important points; we especially support their claim that calibration, or *validity* (in our IM language), is essential. However, it seems they dismiss the possibility that the inferential output can carry any meaningful probabilistic interpretation and, since a general strategy for constructing default priors remains elusive, they conclude that confidence-based methods are the most appropriate. This book has demonstrated that there is another alternative—the *IM approach*—and that it provides valid, prior-free probabilistic inference. As Brad Efron writes in his discussion of [262]: "Perhaps the most important unresolved problem in statistical inference is the use of Bayes theorem in the absence of prior information." We interpret Efron's comment as a challenge to develop a framework that provides meaningful probabilistic inference in the absence of prior information, thereby bridging the gap between frequentist and Bayesian thinking and solidifying the foundations of statistical inference. We believe that the IM framework is poised to meet this challenge, and we hope that the reader agrees.

Statistics is a broad and profound subject and, as Reid and Cox [203] argue, it would be a drastic oversimplification to claim that the foundations could be fixed once and for all. Our "foundations" must be re-evaluated regularly, or whenever new scientific applications arise. The classical frequentist and Bayesian ideas by now have had many years of fine-tuning and the new IM approach will similarly need such tuning. In fact, we believe there is much to be done, and there are exciting opportunities in efforts to overcome the various challenges. In this last chapter, we want to outline what, in our minds are some of these key challenges, and to suggest some important open problems. Our "top ten" list in Section 12.3 can be viewed as a sort of checklist for moving the IM approach to the mainstream of statistics. Among other things, we include some discussion here about high-dimensional problems, including nonparametric problems.

12.2   New directions to explore

12.2.1   IM-based interpretation of frequentist output

The disagreement between the classical and Bayesian schools of thought has been unhealthy for statistics in general. As an answer to this, many authors have put forth proposals to bridge the gap between the two perspectives; all attempt to provide a sort of synergy. Unfortunately, apparently none of these proposals have been successful because, while the philosophical arguments between the two sides have died down, there certainly is no synergy. In our opinion, the bottom line is that there rarely is any meaningful prior information available, so practitioners are more likely to adopt a procedure that does not require the use of a meaningless non-informative prior in such cases. The IM framework discussed here seems to provide something like the "best of both worlds," but we expect it will not be easy for practitioners to switch over to this new framework, despite its appeal.

The bottleneck is that people are familiar with the existing tools and how to interpret them. To break through this obstacle, it will be important to make connections between the familiar output and the IM output. Progress along these lines was made in [181] where it was shown that, roughly, any standard p-value is a plausibility relative to some IM. In other words, the familiar p-value is just a feature of an IM plausibility function, thus making our plausibility functions more accessible. Toward making the IM output more accessible, it will be important to make further connections with the familiar classical summaries, such as confidence intervals. The ideas presented in Chapter 11, especially its version in [176], are expected to be helpful along these lines, as are the connections to post-selection inference in Chapter 10.

As a last point here, it is worth mentioning that the developments in Chapter 10 already show that family-wise error rates in the multiple testing-type context are natural, that is, IM validity considerations, not a special choice of loss function, identifies the family-wise error rate as the appropriate summary of a test's performance. In that sense, the recent emphasis on false discovery rates may not be well-founded, as those methods that control false discovery rate are motivated only by goal to reject more hypotheses.

12.2.2   More conditioning

Suppose that information about a common parameter θ is coming from several sources. An example is where $X_1, \ldots, X_n$ are iid from a distribution depending on θ. In such cases, to improve efficiency in the P-step, [182] propose a dimension reduction strategy based on conditioning. As discussed in Chapter 6, a fairly general recipe to identify the quantity to condition on is based on solving a suitable partial differential equation.

Though this new approach is interesting, there are two obvious questions that come to mind. The first is a theoretical question: in what class of statistical problems would it be possible to properly formulate this partial differential equation such that a solution is guaranteed to exist? The second question is more of a practical type: how to actually solve the equation when a solution exists? To answer these questions, some clear understanding of both the theory of partial differential equations and the tools used to solve them.

One last point concerning the PDE approach for conditioning and dimension reduction. There are cases where some non-differentiability is present. For example, consider the triangular distribution with density function f_θ, on $[0,1]$, given by

$$f_\theta(x) = \begin{cases} 2x/\theta & \text{if } 0 \le x \le \theta, \\ 2(1-x)/(1-\theta) & \text{if } \theta < x \le 1. \end{cases}$$

The triangular distribution is a sort of benchmark example where inference on θ based on an independent sample $Y_1,\dots,Y_n$ from f_θ is quite challenging. Indeed, the standard likelihood-based theory fails, since the Fisher information is not defined; this also makes defining a suitable non-informative prior difficult. A challenge to the IM approach in Chapter 6 in this case is the fact that $F_\theta(x)$ is not everywhere differentiable in θ. This suggests that the definition of a "solution" to the relevant partial differential equation might need to be relaxed. For this, the notion of "weak derivatives" or "distributions" as solutions should be explored.

12.2.3 More marginalization

For concreteness, consider a (possibly unbalanced) normal linear mixed effect model with two variance components, a model widely used in applications, particularly in genetics studies; see Chapter 8. The standard model is written as $Y = X\beta + Z\alpha + \varepsilon$, where β is a vector of fixed parameters, α is a vector of iid $N(0,\sigma_\alpha^2)$ random variables, and ε is a vector of iid $N(0,\sigma_\varepsilon^2)$ random variables; both X and Z are fixed matrices of predictor variables. The parameter $\theta = (\beta,\sigma_\alpha^2,\sigma_\varepsilon^2)$ is unknown, but suppose that only σ_ε^2 is of interest. It can be shown [195] that the minimal sufficient statistic $(Y_1,\dots,Y_L)$ for $(\sigma_\alpha^2,\sigma_\varepsilon^2)$ has distribution characterized by the equations

$$Y_\ell = (\sigma_\alpha^2 \lambda_\ell + \sigma_\varepsilon^2)U_\ell, \quad \ell = 1,\dots,L,$$

where $U_1,\dots,U_L$ are independent with $U_\ell \sim \text{ChiSq}(r_\ell)$. The constants λ_ℓ, r_ℓ, and L are known and depend on X and Z; the λ_ℓ's are distinct and $\lambda_1 > \cdots > \lambda_L \ge 0$.

Let $\psi = \sigma_\varepsilon^2$ be the parameter of interest, and $\eta = \sigma_\alpha^2/\sigma_\varepsilon^2$ the nuisance parameter. Then the above equation can be rewritten as $Y_\ell = \psi(\eta\lambda_\ell + 1)U_\ell, \ell = 1,\dots,L$. To our knowledge, there is no method available for exact marginal inference on ψ in this case—see page 855 in [71]—so a suitable marginal IM for this problem is expected to be an important contribution. Unfortunately, this model is not regular in the sense of Chapter 7, so marginalization is not straightforward. We expect that a marginalization strategy that works for this case would also work for marginal inference on $\psi = \sigma_\alpha^2$, but further work is needed.

12.2.4 Various kinds of discrete problems

Discrete data problems, such as logistic or Poisson regression, can be challenging for methods which are based on asymptotic justification. But IMs are valid for all models and all sample sizes, so they can be particularly advantageous in discrete problems. One interesting setup is a multinomial problem, where it can be shown that, as in a Bayesian analysis, the Dirichlet distribution plays a key role. For contingency table data, there are some interesting questions to consider. First, marginal inference on the odds ratio is an important problem, especially when sample size/cell counts are small. Another interesting application of the multinomial model is the assessment of agreement [31, 161]. One approach would be to get valid marginal inference on the kappa coefficient [6, 49]. For these and other discrete-data applications, we expect that new marginalization techniques will be needed.

12.2.5   Model comparison

A challenge to all frameworks of statistics is that of model selection or, more simply, model comparison. That is, given two or more candidate models for the observed data, how to compare them, how to select a good one, and, more importantly, how to summarize the uncertainty about the model choice. We discussed the variable selection problem in Chapter 10 and we have a bit more to say about this below, so here let us focus on just the simple case of two models for the observable data X. Interestingly, the question boils down to one of prediction, i.e., where we want to see which model can predict the observed data best. So, the developments in Chapter 9 on predicting a future observation are expected to be useful in this model comparison problem too. The reader will recall that the prediction problem is one of "extreme marginalization," where the full parameter is a nuisance to be marginalized over. In that case, questions about efficient marginalization come to the fore once again. A shortcoming of this over-simplified strategy is that it does not directly account for the complexity of the model; see the next two subsections. The ideal formulation is one where a marginal IM for the model index is derived, so that uncertainty about the model choice can be considered. Something along these lines was presented in Section 11.4.5, but further work is needed.

12.2.6   High-dimensional inference

These days, all the hot topics in statistics research focus around high-dimensional problems. The starting point for our IM developments as described in this book were, naturally, low-dimensional problems. Therefore, extending the IM machinery to handle the popular high-dimensional problems is of utmost importance. In this section we describe a few perspectives on how IMs might be applied for inference on a high-dimensional parameter. For simplicity, we focus on the canonical normal means model,

$$X_i \sim N(\theta_i, 1), \quad i = 1, \ldots, n, \quad \text{independent,}$$

but the ideas would surely extend to other more complicated models.

An important observation is that, because the individual problems are being treated simultaneously, there must be some connection between them, despite the independence assumption. For example, the component θ_i's are related in some way. This connection can be made formal by specifying an aggregate loss function, e.g., sum of squared error loss. In this light, Stein's "paradoxical" result [229] on the inadmissibility of the maximum likelihood estimator is not so surprising, given that it ignores any connection between the problems. Corrections to the maximum likelihood estimator, such as the James–Stein estimator [230], look ad hoc until one formulates the connection between the problems through the model. Since this "connection" can be viewed as some prior information, it makes sense to introduce a Bayes model. To accommodate this connection, it is standard to introduce an exchangeable—or conditionally iid—prior for $(\theta_1, \ldots, \theta_n)$, i.e.,

$$[(\theta_1, \ldots, \theta_n) \mid \alpha] \overset{\text{iid}}{\sim} \pi_\alpha, \quad \text{and} \quad \alpha \sim \nu;$$

here, α is treated as a hyper-parameter. To avoid specifying a prior for α, one can take an empirical Bayes approach, e.g., [38, 205, 206], whereby α is estimated from data by the maximum marginal likelihood estimator, $\hat{\alpha}_n$; as reviewed in [77], the James–Stein estimator and its variants can be given an empirical Bayes interpretation. However, fixing α at an estimator ignores the underlying uncertainty, which can lead to some misleading results [218], unless n is very large. Despite this concern, empirical Bayes methods have been applied successfully across a wide range of high-dimensional problems, and we expect that an IM version of empirical Bayes, one that accounts for the uncertainty in α without requiring a prior distribution, would be a significant contribution.

An approach not unlike the Bayes/empirical Bayes perspective presented above is one that introduces, in addition to a loss function (likelihood), a penalty term on the "structure/complexity" of θ. For example, the connection among the θ_i's is often represented by a length condition on θ, i.e., that the distance $d(\theta, \theta_0)$ of θ from some specified θ_0 is not too large. In that case, it makes sense to consider an estimator obtained by minimizing

$$-\ell_X(\theta) + \alpha d(\theta, \theta_0),$$

where α is a tuning parameter that controls the trade-off between model fit and model complexity. Here, as in the empirical Bayes context, despite the importance of α, it is standard to plug in an estimate of α and ignore the underlying uncertainty. An IM approach that handles, rather than ignores, the uncertainty in α would be an important and novel contribution to the literature on penalized estimation.

In a slightly different direction, in some applications, it is known that many of the θ_i's are exactly zero, but the indices of the zeros are unknown. This is one special case of "sparsity." Bayes and empirical Bayes analysis of this problem proceeds by taking the prior for θ_i to be of a "two-groups" form, e.g., [76, 175, 217]. Perhaps a more direct approach is one that treats the model $M \subseteq \{1, 2, \ldots, n\}$, which highlights the location of, say, the non-zero entries in the mean vector; that is, write $\theta = (M, \theta_M)$. In recent theoretical investigations, in the Bayes and empirical Bayes literature, it has been shown that the prior on M drives the performance, while the prior on θ_M, given M, can only interfere; see, in particular, [185]. Good priors for M are available, but the choice of conditional prior for θ_M is not clear. Unfortunately, the Bayesian approach does not allow the analyst to leave the prior for θ_M unspecified. However, the IM approach has sufficient flexibility to incorporate priors where they are available and work without priors where they are not available. We believe that the IM framework's ability to handle partial prior information will help to push it into the statistical mainstream. We propose to call this general idea *partial-* or *semi-Bayes*.

12.2.7 Nonparametric problems

By nonparametric, here we mean those problems where the parameter of interest is infinite-dimensional, i.e., θ is a function rather than a number or vector. For example, θ could be the density function itself or perhaps a regression function. Research in this direction has focused primarily on producing estimators of the infinite-dimensional parameter and studying the estimator's convergence properties. While there has been some interest in recent literature on the construction of confidence bands for θ, work on valid probabilistic inference for θ is at least scarce, maybe even non-existent. Even the Bayesian work on this area is primarily concerned with asymptotic frequentist properties of their nonparametric posterior distribution.

The task of building up solid foundations for statistical inference for an infinite-dimensional problem is a daunting one. However, we believe that the goal is within reach. At some level, all existing approaches to the nonparametric problem introduce some finite-dimensional representation. For example, finite mixture models often drive the density estimation problem, and finite basis expansions are frequently employed in nonparametric regression. So, since IMs can produce valid prior-free probabilistic inference under very mild conditions, it is reasonable to expect that IMs will yield quality inference in the finite-dimensional representations of the problems, which is a step in the right direction. At this point, we have preliminary results to suggest that the fundamental nonparametric problem—inference on the underlying distribution P—can be handled using a generalization of the Dirichlet process model of Ferguson [88] commonly used in Bayesian nonparametrics. While this result is promising, further work is needed. In particular, there will be challenges in taking essential advantage the underlying smoothness of the target function θ, but we expect that the developments in the previous subsection will provide some insights on this. There will also be computational challenges to overcome.

12.3 Our "top ten list" of open problems

To summarize the above discussion, we collect here our "top ten list" of open problems. Solutions to these problems, in our opinion, will be major breakthroughs in the development of the IM framework, in particular, and statistical inference, in general. We have made some progress on some of these problems, but we hope that the reader will take a shot at solving one or more of them.

- IMs for $p \gg n$ linear regression models, as well as other similar "very-high-dimensional" problems;
- IMs for generalized linear regression models, especially discrete logistic and Poisson regression;
- An IM framework for model comparison;
- A general computational framework for implementing IM-based methods;

- Necessary and sufficient conditions for the existence of regular conditional IMs, and the computational tools needed to solve the corresponding PDEs;

- Optimal non-regular conditional IMs;

- Necessary and sufficient conditions for the existence of regular marginal IMs;

- Optimal non-regular marginal IMs;

- A general framework for partial- or semi-Bayes;

- Nonparametric IMs and necessary extensions of the existing Markov chain Monte Carlo methods.

12.4   Final remarks

For the very last section of the book, we have two very simple goals. First, we want to thank the persistent reader who has made it all this way. As we admitted in the Preface, the difficulty in this book is not the mathematical technicalities but, rather, in the way it challenges the reader to consider a perspective very different from those in essentially every other book. In this way, we think our contribution here is a unique one and we hope the reader agrees. More than anything, we hope the reader has been inspired by our efforts to think carefully about the foundations of statistical inference, even if he/she does not completely agree with our perspective.

Second, we want to take a very brief look back to see if we have accomplished the main goals set in Chapters 1 and 2. Of specific interest is the question about the role of probability in the statistical inference. One clear role that probability plays is in the statement of the sampling models, and we believe that this is a fundamental ingredient of the problem. In particular, the sampling model is what drives key validity/calibration property that we have argued is necessary for scientific inference. Beyond the sampling model, the role of probability is less clear. The Bayesian framework relies on probability throughout the analysis but, as we have argued, this is questionable unless there is a meaningful probability (prior) to start with. Fiducial inference seems to avoid the specification of a prior distribution but sacrifices its meaningfulness during the "continue to regard" operation. The main proposal in this book is to work on a fixed auxiliary variable probability space and to carry out valid prediction of these predictable quantities. In this way, no probabilities are being invented or reinvented and, therefore, we can guarantee validity, not just in an asymptotic limit. The IM framework is based on the use of a predictive random set to carry out this prediction, but there may be other ways to implement the general proposal. In our opinion, this book has clearly demonstrated that the IM framework, if not the only implementation of the general proposal, is certainly a promising one and deserves further investigation in years to come.

To conclude, we recall a well-known Chinese saying (pāo zhuān yǐn yù) which translates as

discard a brick in hopes of acquiring a gem.

We can interpret our efforts here in relation to this saying. That is, our book can be viewed as a modest attempt to encourage others to take up the challenge of developing a solid foundation of statistics and to make more substantial contributions.

Bibliography

[1] H. Akaike. Information theory and an extension of the maximum likelihood principle. In *Second International Symposium on Information Theory (Tsahkadsor, 1971)*, pages 267–281. Akadémiai Kiadó, Budapest, 1973.

[2] D. E. Amos and W. G. Bulgren. Computation of a multivariate *F* distribution. *Math. Comp.*, 26:255–264, 1972.

[3] Erling B. Andersen. The asymptotic distribution of conditional likelihood ratio tests. *J. Amer. Statist. Assoc.*, 66:630–633, 1971.

[4] Astride Aregui and Thierry Denœux. Constructing consonant belief functions from sample data using confidence sets of pignistic probabilities. *Internat. J. Approx. Reason.*, 49(3):575–594, 2008.

[5] Michael S. Balch. Mathematical foundations for a theory of confidence structures. *Internat. J. Approx. Reason.*, 53:1003–1019, 2012.

[6] Mousumi Banerjee, Michelle Capozzoli, Laura McSweeney, and Debajyoti Sinha. Beyond kappa: a review of interrater agreement measures. *Canad. J. Statist.*, 27(1):3–23, 1999.

[7] O. Barndorff-Nielsen. Exponentially decreasing distributions for the logarithm of particle size. *Proc. R. Soc. Lond. A.*, 353(1674):401–419, 1977.

[8] O. Barndorff-Nielsen. On a formula for the distribution of the maximum likelihood estimator. *Biometrika*, 70(2):343–365, 1983.

[9] O. E. Barndorff-Nielsen. Inference on full or partial parameters based on the standardized signed log likelihood ratio. *Biometrika*, 73(2):307–322, 1986.

[10] O. E. Barndorff-Nielsen and Peter Hall. On the level-error after Bartlett adjustment of the likelihood ratio statistic. *Biometrika*, 75(2):374–378, 1988.

[11] Ole E. Barndorff-Nielsen. *Parametric Statistical Models and Likelihood*, volume 50 of *Lecture Notes in Statistics*. Springer-Verlag, New York, 1988.

[12] Rudolf Beran. Calibrating prediction regions. *J. Amer. Statist. Assoc.*, 85(411):715–723, 1990.

[13] James Berger. The case for objective Bayesian analysis. *Bayesian Anal.*, 1(3):385–402, 2006.

[14] James O. Berger and José M. Bernardo. On the development of reference priors. In *Bayesian Statistics, 4 (Peñíscola, 1991)*, pages 35–60. Oxford Univ. Press, New York, 1992.

[15] James O. Berger, José M. Bernardo, and Dongchu Sun. The formal definition

of reference priors. *Ann. Statist.*, 37(2):905–938, 2009.

[16] James O. Berger, Brunero Liseo, and Robert L. Wolpert. Integrated likelihood methods for eliminating nuisance parameters. *Statist. Sci.*, 14(1):1–28, 1999.

[17] James O. Berger and Robert L. Wolpert. *The Likelihood Principle*. Institute of Mathematical Statistics Lecture Notes—Monograph Series, 6. Institute of Mathematical Statistics, Hayward, CA, 1984.

[18] Richard Berk, Lawrence Brown, Andreas Buja, Kai Zhang, and Linda Zhao. Valid post-selection inference. *Ann. Statist.*, 41(2):802–837, 2013.

[19] Jose-M. Bernardo. Reference posterior distributions for Bayesian inference. *J. Roy. Statist. Soc. Ser. B*, 41:113–147, 1979.

[20] Julian Besag and Peter Clifford. Generalized Monte Carlo significance tests. *Biometrika*, 76(4):633–642, 1989.

[21] Dulal Kumar Bhaumik and Robert David Gibbons. An upper prediction limit for the arithmetic mean of a lognormal random variable. *Technometrics*, 46(2):239–248, 2004.

[22] Peter J. Bickel and Kjell A. Doksum. *Mathematical Statistics*. Holden-Day, Inc., San Francisco, Calif.-Düsseldorf-Johannesburg, 1976.

[23] Peter J. Bickel, Chris A. J. Klaassen, Ya'acov Ritov, and John A. Wellner. *Efficient and Adaptive Estimation for Semiparametric Models*. Springer-Verlag, New York, 1998.

[24] Patrick Billingsley. *Probability and Measure*. John Wiley & Sons Inc., New York, third edition, 1995.

[25] Allan Birnbaum. On the foundations of statistical inference. *J. Amer. Statist. Assoc.*, 57:269–326, 1962.

[26] Pier Bissiri, Chris Holmes, and Stephen G. Walker. A general framework for updating belief distributions. Unpublished manuscript, `arxiv:1306.6430`, 2013.

[27] Erik Bølviken and Eva Skovlund. Confidence intervals from Monte Carlo tests. *J. Amer. Statist. Assoc.*, 91(435):1071–1078, 1996.

[28] K. O. Bowman and L. R. Shenton. *Properties of Estimators for the Gamma Distribution*, volume 89 of *Statistics: Textbooks and Monographs*. Marcel Dekker Inc., New York, 1988. With a contribution by Y. C. Patel.

[29] George E. P. Box and George C. Tiao. *Bayesian Inference in Statistical Analysis*. Addison-Wesley Publishing Co., Reading, Mass.-London-Don Mills, Ont., 1973.

[30] A. R. Brazzale, A. C. Davison, and N. Reid. *Applied Asymptotics: Case Studies in Small-Sample Statistics*. Cambridge University Press, Cambridge, 2007.

[31] Lyle D. Broemeling. *Bayesian Methods for Measures of Agreement*. Chapman & Hall/CRC Press, Boca Raton, FL, 2009.

[32] Lawrence D. Brown, T. Tony Cai, and Anirban DasGupta. Interval estimation for a binomial proportion (with discussion). *Statist. Sci.*, 16:101–133, 2001.

[33] Lawrence D. Brown, T. Tony Cai, and Anirban DasGupta. Confidence intervals for a binomial proportion and asymptotic expansions. *Ann. Statist.*, 30:160–201, 2002.

[34] Lawrence D. Brown, T. Tony Cai, and Anirban DasGupta. Interval estimation in exponential families. *Statist. Sinica*, 13(1):19–49, 2003.

[35] Brent D. Burch and Hari K. Iyer. Exact confidence intervals for a variance ratio (or heritability) in a mixed linear model. *Biometrics*, 53(4):1318–1333, 1997.

[36] T. Tony Cai. One-sided confidence intervals in discrete distributions. *J. Statist. Plann. Inference*, 131(1):63–88, 2005.

[37] Emmanuel Candes and Terence Tao. The Dantzig selector: Statistical estimation when *p* is much larger than *n*. *Ann. Statist.*, 35(6):2313–2351, 2007.

[38] Bradley P. Carlin and Thomas A. Louis. *Bayes and Empirical Bayes Methods for Data Analysis*, volume 69 of *Monographs on Statistics and Applied Probability*. Chapman & Hall, London, 1996.

[39] George Casella. Comment on "setting confidence intervals for bounded parameters." *Statist. Sci.*, 17(2):159–160, 2002.

[40] George Casella and Roger L. Berger. *Statistical Inference*. The Wadsworth & Brooks/Cole Statistics/Probability Series. Wadsworth & Brooks/Cole Advanced Books & Software, Pacific Grove, CA, 1990.

[41] Qianshun Cheng, Xu Gao, and Ryan Martin. Exact prior-free probabilistic inference on the heritability coefficient in a linear mixed model. *Electron. J. Stat.*, 8(2):3062–3076, 2014.

[42] Andy K. L. Chiang. A simple general method for constructing confidence intervals for functions of variance components. *Technometrics*, 43(3):356–367, 2001.

[43] Jessi Cisewski and Jan Hannig. Generalized fiducial inference for normal linear mixed models. *Ann. Statist.*, 40(4):2102–2127, 2012.

[44] Bertrand S. Clarke and Andrew R. Barron. Information-theoretic asymptotics of Bayes methods. *IEEE Trans. Inform. Theory*, 36(3):453–471, 1990.

[45] Bertrand S. Clarke and Andrew R. Barron. Jeffreys' prior is asymptotically least favorable under entropy risk. *J. Statist. Plann. Inference*, 41(1):37–60, 1994.

[46] William Cleveland. *Visualizing Data*. Hobart Press, 1993.

[47] Merlise Clyde and Edward I. George. Model uncertainty. *Statist. Sci.*, 19(1):81–94, 2004.

[48] Carlos A. Coelho and João T. Mexia. On the distribution of the product and ratio of independent generalized gamma-ratio random variables. *Sankhyā*, 69(2):221–255, 2007.

[49] J. Cohen. A coefficient of agreement for nominal scales. *Edu. and Psych. Meas.*, 20:37–46, 1960.

[50] D. R. Cox. *Principles of Statistical Inference*. Cambridge University Press, Cambridge, 2006.

[51] Monica A. Creasy. Symposium on interval estimation: Limits for the ratio of means. *J. Roy. Statist. Soc. Ser. B.*, 16:186–194, 1954.

[52] M. N. Das and G. A. Kulkarni. Incomplete block designs for bio-assays. *Biometrics*, 22:706–729, 1966.

[53] Gauri Sankar Datta and Jayanta Kumar Ghosh. On priors providing frequentist validity for Bayesian inference. *Biometrika*, 82(1):37–45, 1995.

[54] A. C. Davison and D. V. Hinkley. *Bootstrap Methods and Their Application*, volume 1. Cambridge University Press, Cambridge, 1997.

[55] A. P. Dawid. Calibration-based empirical probability. *Ann. Statist.*, 13(4):1251–1285, 1985. With discussion.

[56] A. P. Dawid and M. Stone. The functional-model basis of fiducial inference. *Ann. Statist.*, 10(4):1054–1074, 1982. With discussion.

[57] A. P. Dawid, M. Stone, and J. V. Zidek. Marginalization paradoxes in Bayesian and structural inference. *J. Roy. Statist. Soc. Ser. B*, 35:189–233, 1973. With discussion and reply by the authors.

[58] Bruno de Finetti. *Theory of Probability. Vol. 1*. Wiley Classics Library. John Wiley & Sons, Ltd., Chichester, 1990. A critical introductory treatment, Translated from the Italian and with a preface by Antonio Machì and Adrian Smith, With a foreword by D. V. Lindley, Reprint of the 1974 translation.

[59] Bruno de Finetti. *Theory of Probability. Vol. 2*. Wiley Classics Library. John Wiley & Sons, Ltd., Chichester, 1990. A critical introductory treatment, Translated from the Italian and with a preface by Antonio Machì and Adrian Smith, With a foreword by D. V. Lindley, Reprint of the 1975 translation.

[60] A. P. Dempster. Further examples of inconsistencies in the fiducial argument. *Ann. Math. Statist.*, 34:884–891, 1963.

[61] A. P. Dempster. On the difficulities inherent in Fisher's fiducial argument. *J. Amer. Statist. Assoc.*, 59:56–66, 1964.

[62] A. P. Dempster. New methods for reasoning towards posterior distributions based on sample data. *Ann. Math. Statist.*, 37:355–374, 1966.

[63] A. P. Dempster. Upper and lower probabilities induced by a multivalued mapping. *Ann. Math. Statist.*, 38:325–339, 1967.

[64] A. P. Dempster. A generalization of Bayesian inference. (With discussion). *J. Roy. Statist. Soc. Ser. B*, 30:205–247, 1968.

[65] A. P. Dempster. Upper and lower probabilities generated by a random closed interval. *Ann. Math. Statist.*, 39:957–966, 1968.

[66] A. P. Dempster. Upper and lower probability inferences for families of hypotheses with monotone density ratios. *Ann. Math. Statist.*, 40:953–969, 1969.

[67] A. P. Dempster. The Dempster–Shafer calculus for statisticians. *Internat. J. Approx. Reason.*, 48(2):365–377, 2008.

[68] T. Denoeux. Constructing belief functions from sample data using multinomial confidence regions. *Internat. J. of Approx. Reason.*, 42(3):228–252, 2006.

[69] Thomas J. DiCiccio and Bradley Efron. Bootstrap confidence intervals. *Statist. Sci.*, 11(3):189–228, 1996. With comments and a rejoinder by the authors.

[70] Thomas J. DiCiccio, Michael A. Martin, and Steven E. Stern. Simple and accurate one-sided inference from signed roots of likelihood ratios. *Canad. J. Statist.*, 29(1):67–76, 2001.

[71] Lidong E, Jan Hannig, and Hari Iyer. Fiducial intervals for variance components in an unbalanced two-component normal mixed linear model. *J. Amer. Statist. Assoc.*, 103(482):854–865, 2008.

[72] Morris L. Eaton. *Group Invariance Applications in Statistics*. Institute of Mathematical Statistics, Hayward, CA, 1989.

[73] Paul T. Edlefsen, Chuanhai Liu, and Arthur P. Dempster. Estimating limits from poisson counting data using dempster-shafer analysis. *Ann. Appl. Stat.*, 3(2):764–790, 2009.

[74] B. Efron. Bootstrap methods: Another look at the jackknife. *Ann. Statist.*, 7(1):1–26, 1979.

[75] Bradley Efron. R. A. Fisher in the 21st century. *Statist. Sci.*, 13(2):95–122, 1998.

[76] Bradley Efron. Microarrays, empirical Bayes and the two-groups model. *Statist. Sci.*, 23(1):1–22, 2008.

[77] Bradley Efron. *Large-Scale Inference*, volume 1 of *Institute of Mathematical Statistics Monographs*. Cambridge University Press, Cambridge, 2010.

[78] Bradley Efron. Bayes' theorem in the 21st century. *Science*, 340(6137):1177–1178, 2013.

[79] Bradley Efron, Trevor Hastie, Iain Johnstone, and Robert Tibshirani. Least angle regression. *Ann. Statist.*, 32(2):407–499, 2004. With discussion, and a rejoinder by the authors.

[80] Bradley Efron and Robert J. Tibshirani. *An Introduction to the Bootstrap*. Chapman & Hall, New York, 1993.

[81] Duncan Ermini Leaf, Jun Hui, and Chuanhai Liu. Statistical inference with a single observation of $n(\theta, 1)$. *Pak. J. Statist.*, 25:571–586, 2009.

[82] Duncan Ermini Leaf and Chuanhai Liu. Inference about constrained parameters using the elastic belief method. *Internat. J. Approx. Reason.*, 53(5):709–727, 2012.

[83] Michael D. Escobar and Mike West. Bayesian density estimation and inference using mixtures. *J. Amer. Statist. Assoc.*, 90(430):577–588, 1995.

[84] Michael Evans. What does the proof of Birnbaum's theorem prove? *Electron. J. Stat.*, 7:2645–2655, 2013.

[85] G. J. Feldman and R. D. Cousins. Unified approach to the classical statistical analysis of small signals. *Phys. Rev. D*, 57(7):3873–3889, 1998.

[86] William Feller. *An Introduction to Probability Theory and Its Applications. Vol. I.* Third edition. John Wiley & Sons, Inc., New York-London-Sydney, 1968.

[87] Alan P. Fenech and David A. Harville. Exact confidence sets for variance components in unbalanced mixed linear models. *Ann. Statist.*, 19(4):1771–1785, 1991.

[88] Thomas S. Ferguson. A Bayesian analysis of some nonparametric problems. *Ann. Statist.*, 1:209–230, 1973.

[89] K. W. Fertig and N. R. Mann. One-sided prediction intervals for at least p out of m future observations from a normal population. *Technometrics*, 19:167–167, 1977.

[90] E. C. Fieller. Symposium on interval estimation: Some problems in interval estimation. *J. Roy. Statist. Soc. Ser. B.*, 16:175–185, 1954.

[91] Stephen E. Fienberg. When did Bayesian inference become "Bayesian"? *Bayesian Anal.*, 1(1):1–40 (electronic), 2006.

[92] R. A. Fisher. Inverse probability. *Proceedings of the Cambridge Philosophical Society*, 26:528–535, 1930.

[93] R. A. Fisher. The fiducial argument in statistical inference. *Ann. Eugenics*, 6:391–398, 1935.

[94] Ronald A. Fisher. *Statistical Methods and Scientific Inference.* 2nd ed., revised. Hafner Publishing Company, New York, 1959.

[95] Ronald A. Fisher. *Statistical Methods and Scientific Inference.* Hafner Press, New York, 3rd edition, 1973.

[96] Ronald A. Fisher. *Statistical Methods for Research Workers.* Hafner Publishing Co., New York, 1973. Fourteenth edition—revised and enlarged.

[97] Ailana M. Fraser, D. A. S. Fraser, and Ana-Maria Staicu. Second order ancillary: a differential view from continuity. *Bernoulli*, 16(4):1208–1223, 2010.

[98] D. A. S. Fraser. Structural probability and a generalization. *Biometrika*, 53:1–9, 1966.

[99] D. A. S. Fraser. *The Structure of Inference.* John Wiley & Sons Inc., New York, 1968.

[100] D. A. S. Fraser. Tail probabilities from observed likelihoods. *Biometrika*, 77(1):65–76, 1990.

[101] D. A. S. Fraser. Is Bayes posterior just quick and dirty confidence? *Statist. Sci.*, 26(3):299–316, 2011.

[102] D. A. S. Fraser and N. Reid. Adjustments to profile likelihood. *Biometrika*, 76(3):477–488, 1989.

[103] D. A. S. Fraser, N. Reid, E. Marras, and G. Y. Yi. Default priors for Bayesian

and frequentist inference. *J. R. Stat. Soc. Ser. B Stat. Methodol.*, 72(5):631–654, 2010.

[104] D. A. S. Fraser, N. Reid, and A. Wong. Simple and accurate inference for the mean of a gamma model. *Canad. J. Statist.*, 25(1):91–99, 1997.

[105] D. A. S. Fraser, N. Reid, and A. Wong. Inference for bounded parameters. *Phys. Rev. D*, 69(3):033002, 2004.

[106] D. A. S. Fraser, A. Wong, and Y. Sun. Three enigmatic examples and inference from likelihood. *Canad. J. Statist.*, 37(2):161–181, 2009.

[107] Paul H. Garthwaite and Stephen T. Buckland. Generating Monte Carlo confidence intervals by the Robbins-Monro process. *J. Roy. Statist. Soc. Ser. C*, 41(1):159–171, 1992.

[108] Seymour Geisser. *Predictive Inference*, volume 55 of *Monographs on Statistics and Applied Probability*. Chapman & Hall, New York, 1993.

[109] Andrew Gelman. Prior distributions for variance parameters in hierarchical models (comment on article by Browne and Draper). *Bayesian Anal.*, 1(3):515–533, 2006.

[110] Andrew Gelman, John B. Carlin, Hal S. Stern, and Donald B. Rubin. *Bayesian Data Analysis*. Chapman & Hall/CRC, Boca Raton, FL, second edition, 2004.

[111] Jayanta K. Ghosh, Mohan Delampady, and Tapas Samanta. *An Introduction to Bayesian Analysis*. Springer, New York, 2006.

[112] M. Ghosh, N. Reid, and D. A. S. Fraser. Ancillary statistics: A review. *Statist. Sinica*, 20:1309–1332, 2010.

[113] Malay Ghosh. Objective priors: An introduction for frequentists. *Statist. Sci.*, 26(2):187–202, 2011.

[114] Malay Ghosh and Yeong-Hwa Kim. The Behrens-Fisher problem revisited: a Bayes-frequentist synthesis. *Canad. J. Statist.*, 29(1):5–17, 2001.

[115] C. Giunti. New ordering principle for the classical statistical analysis of Poisson processes with background. *Phys. Rev. D*, 59(5):053001, 1999.

[116] Ronald E. Glaser. The ratio of the geometric mean to the arithmetic mean for a random sample from a gamma distribution. *J. Amer. Statist. Assoc.*, 71(354):480–487, 1976.

[117] L. J. Gleser. Comment on "setting confidence intervals for bounded parameters." *Statist. Sci.*, 17(2):161–163, 2002.

[118] Leon Jay Gleser and Jiunn T. Hwang. The nonexistence of $100(1 - \alpha)\%$ confidence sets of finite expected diameter in errors-in-variables and related models. *Ann. Statist.*, 15(4):1351–1362, 1987.

[119] David Golan and Saharon Rosset. Accurate estimation of heritability in genome wide studies using random effects models. *Bioinformatics*, 27:317–323, 2011.

[120] Franklin A. Graybill. *Theory and Application of the Linear Model*. Duxbury Press, North Scituate, Mass., 1976.

[121] J. Arthur Greenwood and David Durand. Aids for fitting the gamma distribution by maximum likelihood. *Technometrics*, 2:55–65, 1960.

[122] John V. Grice and Lee J. Bain. Inferences concerning the mean of the gamma distribution. *J. Amer. Statist. Assoc.*, 75(372):929–933, 1980.

[123] G. J. Hahn and W. Q. Meeker. *Statistical Intervals:A Guide for Practitioners*. Wiley, New York, 1991.

[124] Jaroslav Hájek. *Sampling from a Finite Population*, volume 37. Marcel Dekker, Inc., New York, 1981. Edited by Václav Dupač.

[125] Michael Hamada, Valen Johnson, Leslie M. Moore, and Joanne Wendelberger. Bayesian prediction intervals and their relationship to tolerance intervals. *Technometrics*, 46(4):452–459, 2004.

[126] Jan Hannig. On generalized fiducial inference. *Statist. Sinica*, 19(2):491–544, 2009.

[127] Jan Hannig. Generalized fiducial inference via discretization. *Statist. Sinica*, 23(2):489–514, 2013.

[128] Jan Hannig, Hari Iyer, and Paul Patterson. Fiducial generalized confidence intervals. *J. Amer. Statist. Assoc.*, 101(473):254–269, 2006.

[129] Jan Hannig and Thomas C. M. Lee. Generalized fiducial inference for wavelet regression. *Biometrika*, 96(4):847–860, 2009.

[130] Matthew T. Harrison. Conservative hypothesis tests and confidence intervals using importance sampling. *Biometrika*, 99(1):57–69, 2012.

[131] H. Leon Harter and Albert H. Moore. Maximum-likelihood estimation of the parameters of gamma and Weibull populations from complete and from censored samples. *Technometrics*, 7:639–643, 1965.

[132] David A. Harville and Alan P. Fenech. Confidence intervals for variance ratio, or for heritability, in an unbalanced mixed linear model. *Biometrics*, 41(1):137–152, 1985.

[133] Trevor Hastie, Robert Tibshirani, and Jerome Friedman. *The Elements of Statistical Learning*. Springer-Verlag, New York, 2nd edition, 2009.

[134] David Heath and William Sudderth. On finitely additive priors, coherence, and extended admissibility. *Ann. Statist.*, 6(2):333–345, 1978.

[135] Matthew J. Heaton and James G. Scott. Bayesian computation and the linear model. In Ming-Hui Cheh, Dipak Dey, Peter Müller, Dongchu Sun, and Keying Ye, editors, *Frontiers of Statistical Decision Making and Bayesian Analysis*, pages 527–545. Springer, 2010.

[136] Harold Hotelling. New light on the correlation coefficient and its transforms. *J. Roy. Statist. Soc. Ser. B.*, 15:193–225; discussion, 225–232, 1953.

[137] P. L. Hsu. Contributions to the theory of "student's" *t*-test as applied to the problem of two samples. In *Statistical Research Memoirs*, pages 1–24. University College, London, 1938.

[138] Jian Huang, Joel L. Horowitz, and Fengrong Wei. Variable selection in non-

parametric additive models. *Ann. Statist.*, 38(4):2282–2313, 2010.

[139] Jian Huang, Shuangge Ma, Hongzhe Li, and Cun-Hui Zhang. The sparse Laplacian shrinkage estimator for high-dimensional regression. *Ann. Statist.*, 39(4):2021–2046, 2011.

[140] J. L. Jensen. Inference for the mean of a gamma distribution with unknown shape parameter. *Scand. J. Statist.*, 13(2):135–151, 1986.

[141] Valen E. Johnson and David Rossell. Bayesian model selection in high-dimensional settings. *J. Amer. Statist. Assoc.*, 107(498):649–660, 2012.

[142] Joseph B. Kadane. *Principles of Uncertainty.* Texts in Statistical Science Series. CRC Press, Boca Raton, FL, 2011. http://uncertainty.stat.cmu.edu.

[143] Joseph B. Kadane and Nicole A. Lazar. Methods and criteria for model selection. *J. Amer. Statist. Assoc.*, 99(465):279–290, 2004.

[144] Hyun Min Kang, Jae Hoon Sul, Susan K. Service, Noah A. Zaitlen, Sit-yee Kong, Nelson B. Freimer, Chiara Sabatti, and Eleazar Eskin. Variance component model to account for sample structure in genome-wide association studies. *Nat. Genet.*, 42(4):348–354, 2010.

[145] R. E. Kass and L. Wasserman. The selection of prior distributions by formal rules. *J. Amer. Statist. Assoc.*, 91(435):1343–1370, 1996.

[146] D. G. Kendall. Foundations of a theory of random sets. In *Stochastic Geometry (a tribute to the memory of Rollo Davidson)*, pages 322–376. Wiley, London, 1974.

[147] A. I. Khuri and Hardeo Sahai. Variance components analysis: a selective literature survey. *Internat. Statist. Rev.*, 53(3):279–300, 1985.

[148] Hea-Jung Kim. A Monte Carlo method for estimating prediction limit for the arithmetic mean of lognormal sample. *Comm. Statist. Theory Methods*, 36(9-12):2159–2167, 2007.

[149] Seock-Ho Kim and Allan S. Cohen. On the Behrens-Fisher problem: A review. *Journal of Educational and Behavioral Statistics*, 23(4):356–377, 1998.

[150] A. Klenke and L. Mattner. Stochastic ordering of classical discrete distributions. *Adv. Appl. Probab.*, 42(2):392–410, 2010.

[151] J. Kohlas and P.-A. Monney. An algebraic theory for statistical information based on the theory of hints. *Internat. J. of Approx. Reason.*, 48(2):378–398, 2008.

[152] K. Krishnamoorthy, Thomas Mathew, and Shubhabrata Mukherjee. Normal-based methods for a gamma distribution: prediction and tolerance intervals and stress-strength reliability. *Technometrics*, 50(1):69–78, 2008.

[153] H. V. Kulkarni and S. K. Powar. A new method for interval estimation of the mean of the Gamma distribution. *Lifetime Data Anal.*, 16(3):431–447, 2010.

[154] R. C. S. Lai, J. Hannig, and T. C. M. Lee. Generalized fiducial inference for ultrahigh dimensional regression. *J. Amer. Statist. Assoc.*, to appear,

`arXiv:1304.7847`, 2013.

[155] J. F. Lawless and Marc Fredette. Frequentist prediction intervals and predictive distributions. *Biometrika*, 92(3):529–542, 2005.

[156] Stephen M. S. Lee and G. Alastair Young. Parametric bootstrapping with nuisance parameters. *Statist. Probab. Lett.*, 71(2):143–153, 2005.

[157] Youngjo Lee and Justus Seely. Computing the Wald interval for a variance ratio. *Biometrics*, 52:1486–1491, 1996.

[158] E. L. Lehmann and George Casella. *Theory of Point Estimation*. Springer Texts in Statistics. Springer-Verlag, New York, second edition, 1998.

[159] E. L. Lehmann and Joseph P. Romano. *Testing Statistical Hypotheses*. Springer Texts in Statistics. Springer, New York, third edition, 2005.

[160] Faming Liang, Chuanhai Liu, and Raymond J. Carroll. *Advanced Markov Chain Monte Carlo Methods*. Wiley Series in Computational Statistics. John Wiley & Sons, Ltd., Chichester, 2010. Learning from past samples.

[161] Lawrence Lin, A. S. Hedayat, and Wenting Wu. *Statistical Tools for Measuring Agreement*. Springer, New York, 2012.

[162] D. V. Lindley. Fiducial distributions and Bayes' theorem. *J. Roy. Statist. Soc. Ser. B*, 20:102–107, 1958.

[163] Dennis. V Lindley. A statistical paradox. *Biometrika*, 44:187–192, 1957.

[164] Roderick Little. Calibrated Bayes, for statistics in general, and missing data in particular. *Statist. Sci.*, 26(2):162–174, 2011.

[165] Chuanhai Liu and Ryan Martin. Frameworks for prior-free posterior probabilistic inference. *WIREs Comput. Stat.*, 7(1):77–85, 2015.

[166] Chuanhai Liu and Jun Xie. Large scale two sample multinomial inferences and its applications in genome-wide association studies. *Internat. J. Approx. Reason.*, 55(1, part 3):330–340, 2014.

[167] Chuanhai Liu and Jun Xie. Probabilistic inference for multiple testing. *Internat. J. Approx. Reason.*, 55(2):654–665, 2014.

[168] Ruitao Liu, Arijit Chakrabarti, Tapas Samanta, Jayanta K. Ghosh, and Malay Ghosh. On divergence measures leading to Jeffreys and other reference priors. *Bayesian Anal.*, 9(2):331–369, 2014.

[169] Richard Lockhart, Jonathan Taylor, Ryan J. Tibshirani, and Robert Tibshirani. A significance test for the lasso. *Ann. Statist.*, 42(2):413–468, 2014.

[170] S. MacEachern and P. Müller. Estimating mixture of Dirichlet process models. *J. Comput. Graph. Statist.*, 7:223–238, 1998.

[171] Steven N. MacEachern. Computational methods for mixture of Dirichlet process models. In D. Dey, P. Müller, and D. Sinha, editors, *Practical Nonparametric and Semiparametric Bayesian Statistics*, volume 133 of *Lecture Notes in Statist.*, pages 23–43. Springer, New York, 1998.

[172] M. Mandelkern and J. Schultz. Coverage of confidence intervals based on

conditional probability. *J. High Energy Phys.*, 11:036, 2000.

[173] M. Mandelkern and J. Schultz. The statistical analysis of Gaussian and Poisson signals near physical boundaries. *J. Math. Phys.*, 41(8):5701–5709, 2000.

[174] Mark Mandelkern. Setting confidence intervals for bounded parameters. *Statist. Sci.*, 17(2):149–172, 2002. With comments.

[175] R. Martin and Surya T. Tokdar. A nonparametric empirical Bayes framework for large-scale multiple testing. *Biostatistics*, 13(3):427–439, 2012.

[176] Ryan Martin. Plausibility functions and exact frequentist inference. *J. Amer. Statist. Assoc.*, to appear, arXiv:1203.6665, 2014.

[177] Ryan Martin. Random sets and exact confidence regions. *Sankhyā A*, 76:288–304, 2014.

[178] Ryan Martin, Duncan Ermini Leaf, and Chuanhai Liu. Optimal inferential models for a Poisson mean. Unpublished manuscript, arXiv:1207.0105, 2012.

[179] Ryan Martin and Rama Lingham. Prior-free probabilistic prediction of future observations. *Technometrics*, to appear, arXiv:1403.7589, 2015.

[180] Ryan Martin and Chuanhai Liu. Inferential models: A framework for prior-free posterior probabilistic inference. *J. Amer. Statist. Assoc.*, 108(501):301–313, 2013.

[181] Ryan Martin and Chuanhai Liu. A note on *p*-values interpreted as plausibilities. *Statist. Sinica*, 24:1703–1716, 2014.

[182] Ryan Martin and Chuanhai Liu. Conditional inferential models: Combining information for prior-free probabilistic inference. *J. R. Stat. Soc. Ser. B*, 77(1):195–217, 2015.

[183] Ryan Martin and Chuanhai Liu. Marginal inferential models: Prior-free probabilistic inference on interest parameters. *J. Amer. Statist. Assoc.*, to appear; arXiv:1306.3092, 2015.

[184] Ryan Martin, Raymond Mess, and Stephen G. Walker. Empirical Bayes posterior concentration in sparse high-dimensional linear models. Unpublished manuscript, arXiv:1406.7718, 2014.

[185] Ryan Martin and Stephen G. Walker. Asymptotically minimax empirical Bayes estimation of a sparse normal mean vector. *Electron. J. Stat.*, 8(2):2188–2206, 2014.

[186] Ryan Martin, Jianchun Zhang, and Chuanhai Liu. Dempster–Shafer theory and statistical inference with weak beliefs. *Statist. Sci.*, 25(1):72–87, 2010.

[187] G. Matheron. *Random Sets and Integral Geometry*. John Wiley & Sons, New York-London-Sydney, 1975.

[188] Deborah Mayo. On the Birnbaum argument for the strong likelihood principle. *Statist. Sci.*, 29(2):227–239, 2014.

[189] Ilya Molchanov. *Theory of Random Sets*. Probability and Its Applications (New York). Springer-Verlag London Ltd., London, 2005.

[190] Peter Müller and Fernando A. Quintana. Nonparametric Bayesian data analysis. *Statist. Sci.*, 19(1):95–110, 2004.

[191] S. A. Murphy and A. W. van der Vaart. On profile likelihood. *J. Amer. Statist. Assoc.*, 95(450):449–485, 2000. With discussion.

[192] J. Neyman. Fiducial argument and the theory of confidence intervals. *Biometrika*, 32:128–150, 1941.

[193] Hung T. Nguyen. On random sets and belief functions. *J. Math. Anal. Appl.*, 65(3):531–542, 1978.

[194] Robert E. Odeh. Two-sided prediction intervals to contain at least k out of m future observations from a normal distribution. *Technometrics*, 32(2):203–216, 1990.

[195] Anthony Olsen, Justus Seely, and David Birkes. Invariant quadratic unbiased estimation for two variance components. *Ann. Statist.*, 4(5):878–890, 1976.

[196] W. R. Ott. *Environmental Statistics and Data Analysis*. CRC Press, Boca Raton, FL, 1995.

[197] Art B. Owen. Empirical likelihood ratio confidence intervals for a single functional. *Biometrika*, 75(2):237–249, 1988.

[198] Trevor Park and George Casella. The Bayesian lasso. *J. Amer. Statist. Assoc.*, 103(482):681–686, 2008.

[199] E. J. G. Pitman. Statistics and science. *J. Amer. Statist. Assoc.*, 52:322–330, 1957.

[200] A. D. Polyanin, V. F. Zaitsev, and A. Moussiaux. *Handbook of First Order Partial Differential Equations*, volume 1 of *Differential and Integral Equations and Their Applications*. Taylor & Francis Ltd., London, 2002.

[201] N. Reid. The roles of conditioning in inference. *Statist. Sci.*, 10(2):138–157, 1995.

[202] N. Reid. Asymptotics and the theory of inference. *Ann. Statist.*, 31(6):1695–1731, 2003.

[203] Nancy Reid and David R. Cox. On some principles of statistical inference. *Int. Statist. Rev.*, 83(2):293–308, 2015.

[204] H. E. Robbins. On the measure of a random set. *Ann. Math. Statistics*, 15:70–74, 1944.

[205] Herbert Robbins. An empirical Bayes approach to statistics. In *Proceedings of the Third Berkeley Symposium on Mathematical Statistics and Probability, 1954–1955, vol. I*, pages 157–163, Berkeley and Los Angeles, 1956. University of California Press.

[206] Herbert Robbins. The empirical Bayes approach to statistical decision problems. *Ann. Math. Statist.*, 35:1–20, 1964.

[207] Christian P. Robert. On the Jeffreys–Lindley paradox. *Philos. Sci.*, 81(2):216–232, 2014.

[208] B. P. Roe and M. Woodroofe. Improved probability method for estimating signal in the presence of background. *Phys. Rev. D*, 60(5):053009, 1999.

[209] B. P. Roe and M. B. Woodroofe. Setting confidence belts. *Phys. Rev. D*, 63(1):013009, 2000.

[210] Sheldon Ross. *A First Course in Probability*. Macmillan Co., New York; Collier Macmillan Ltd., London, second edition, 1984.

[211] Donald B. Rubin. Bayesianly justifiable and relevant frequency calculations for the applied statistician. *Ann. Statist.*, 12(4):1151–1172, 1984.

[212] Leonard J. Savage. *The Foundations of Statistics*. Dover Publications, Inc., New York, revised edition, 1972.

[213] Leonard J. Savage. On rereading R. A. Fisher. *Ann. Statist.*, 4(3):441–500, 1976. Edited posthumously by John W. Pratt, With a discussion by B. Efron, Churchill Eisenhart, Bruno de Finetti, D. A. S. Fraser, V. P. Godambe, I. J. Good, O. Kempthorne and Stephen M. Stigler.

[214] Henry Scheffé. Practical solutions of the Behrens–Fisher problem. *J. Amer. Statist. Assoc.*, 65:1501–1508, 1970.

[215] Mark J. Schervish. *Theory of Statistics*. Springer-Verlag, New York, 1995.

[216] Gideon Schwarz. Estimating the dimension of a model. *Ann. Statist.*, 6(2):461–464, 1978.

[217] James G. Scott and James O. Berger. An exploration of aspects of Bayesian multiple testing. *J. Statist. Plann. Inference*, 136(7):2144–2162, 2006.

[218] James G. Scott and James O. Berger. Bayes and empirical-Bayes multiplicity adjustment in the variable-selection problem. *Ann. Statist.*, 38:2587–2619, 2010.

[219] Shayle R. Searle, George Casella, and Charles E. McCulloch. *Variance Components*. Wiley Series in Probability and Mathematical Statistics: Applied Probability and Statistics. John Wiley & Sons, Inc., New York, 1992. A Wiley-Interscience Publication.

[220] Thomas A. Severini. On the relationship between Bayesian and non-Bayesian elimination of nuisance parameters. *Statist. Sinica*, 9(3):713–724, 1999.

[221] Glenn Shafer. *A Mathematical Theory of Evidence*. Princeton University Press, Princeton, N.J., 1976.

[222] Glenn Shafer. Nonadditive probabilities in the work of Bernoulli and Lambert. *Arch. Hist. Exact Sci.*, 19(4):309–370, 1978/79.

[223] Glenn Shafer. Allocations of probability. *Ann. Probab.*, 7(5):827–839, 1979.

[224] Glenn Shafer. Constructive probability. *Synthese*, 48(1):1–60, 1981.

[225] Glenn Shafer. Belief functions and parametric models. *J. Roy. Statist. Soc. Ser. B*, 44(3):322–352, 1982. With discussion.

[226] Glenn Shafer. Belief functions and possibility measures. In James C. Bezdek, editor, *The Analysis of Fuzzy Information, Vol. 1: Mathematics and Logic*,

pages 51–84. CRC, 1987.

[227] Ib M. Skovgaard. Likelihood asymptotics. *Scand. J. Statist.*, 28(1):3–32, 2001.

[228] Emil Spjøtvoll. Unbiasedness of likelihood ratio confidence sets in cases without nuisance parameters. *J. Roy. Statist. Soc. Ser. B*, 34:268–273, 1972.

[229] Charles Stein. Inadmissibility of the usual estimator for the mean of a multivariate normal distribution. In *Proceedings of the Third Berkeley Symposium on Mathematical Statistics and Probability, 1954–1955, vol. I*, pages 197–206, Berkeley and Los Angeles, 1956. University of California Press.

[230] Charles M. Stein. Estimation of the mean of a multivariate normal distribution. *Ann. Statist.*, 9(6):1135–1151, 1981.

[231] Stephen M. Stigler. *The History of Statistics*. The Belknap Press of Harvard University Press, Cambridge, MA, 1990. The measurement of uncertainty before 1900, Reprint of the 1986 original.

[232] Y. Sun and A. C. M. Wong. Interval estimation for the normal correlation coefficient. *Statist. Probab. Lett.*, 77(17):1652–1661, 2007.

[233] Nick Syring and Ryan Martin. Likelihood-free Bayesian inference on the minimum clinically important difference. unpublished manuscript, arXiv:1501.01840, 2015.

[234] Gunnar Taraldsen and Bo Henry Lindqvist. Fiducial theory and optimal inference. *Ann. Statist.*, 41(1):323–341, 2013.

[235] Robert Tibshirani. Noninformative priors for one parameter of many. *Biometrika*, 76(3):604–608, 1989.

[236] Robert Tibshirani. Regression shrinkage and selection via the lasso. *J. Roy. Statist. Soc. Ser. B*, 58(1):267–288, 1996.

[237] Robert Tibshirani. Regression shrinkage and selection via the lasso: a retrospective. *J. R. Stat. Soc. Ser. B Stat. Methodol.*, 73(3):273–282, 2011.

[238] John W. Tukey. *Exploratory Data Analysis*. Pearson, 1977.

[239] Frank Tuyl, Richard Gerlach, and Kerrie Mengersen. Posterior predictive arguments in favor of the Bayes-Laplace prior as the consensus prior for binomial and multinomial parameters. *Bayesian Anal.*, 4(1):151–158, 2009.

[240] Sara van de Geer, Peter Bühlmann, Ya'acov Ritov, and Ruben Dezeure. On asymptotically optimal confidence regions and tests for high-dimensional models. *Ann. Statist.*, 42(3):1166–1202, 2014.

[241] Cristiano Varin, Nancy Reid, and David Firth. An overview of composite likelihood methods. *Statist. Sinica*, 21(1):5–42, 2011.

[242] Peter M. Visscher, William G. Hill, and Naomi R. Wray. Heritability in the genomics era—concepts and misconceptions. *Nat. Rev. Genet.*, 9:255–266, 2008.

[243] Abraham Wald. A note on the analysis of variance with unequal class frequencies. *Ann. Math. Statistics*, 11:96–100, 1940.

[244] Peter Walley. *Statistical Reasoning with Imprecise Probabilities*, volume 42 of *Monographs on Statistics and Applied Probability*. Chapman & Hall Ltd., London, 1991.

[245] Peter Walley. Inferences from multinomial data: Learning about a bag of marbles. *J. Roy. Statist. Soc. Ser. B*, 58(1):3–57, 1996. With discussion and a reply by the author.

[246] Damian V. Wandler and Jan Hannig. Fiducial inference on the largest mean of a multivariate normal distribution. *J. Multivariate Anal.*, 102(1):87–104, 2011.

[247] Damian V. Wandler and Jan Hannig. A fiducial approach to multiple comparisons. *J. Statist. Plann. Inference*, 142(4):878–895, 2012.

[248] Damian V. Wandler and Jan Hannig. Generalized fiducial confidence intervals for extremes. *Extremes*, 15(1):67–87, 2012.

[249] C. M. Wang, Jan Hannig, and Hari K. Iyer. Fiducial prediction intervals. *J. Statist. Plann. Inference*, 142(7):1980–1990, 2012.

[250] Hsiuying Wang. Closed form prediction intervals applied for disease counts. *Amer. Statist.*, 64(3):250–256, 2010.

[251] Lan Wang, Yongdai Kim, and Runze Li. Calibrating nonconvex penalized regression in ultra-high dimension. *Ann. Statist.*, 41(5):2505–2536, 2013.

[252] Larry Wasserman and Kathryn Roeder. High-dimensional variable selection. *Ann. Statist.*, 37(5A):2178–2201, 2009.

[253] Larry A. Wasserman. Belief functions and statistical inference. *Canad. J. Statist.*, 18(3):183–196, 1990.

[254] Larry Alan Wasserman. Prior envelopes based on belief functions. *Ann. Statist.*, 18(1):454–464, 1990.

[255] Samaradasa Weerahandi. Generalized confidence intervals. *J. Amer. Statist. Assoc.*, 88(423):899–905, 1993.

[256] B. L. Welch. The significance of the difference between two means when the population variances are unequal. *Biometrika*, 29:350–362, 1938.

[257] B. L. Welch. The generalization of 'Student's' problem when several different population variances are involved. *Biometrika*, 34:28–35, 1947.

[258] B. L. Welch and H. W. Peers. On formulae for confidence points based on integrals of weighted likelihoods. *J. Roy. Statist. Soc. Ser. B*, 25:318–329, 1963.

[259] R. D. Wolfinger and Robert E. Kass. Nonconjugate Bayesian analysis of variance component models. *Biometrics*, 56(3):768–774, 2000.

[260] Augustine C. M. Wong. A note on inference for the mean parameter of the gamma distribution. *Statist. Probab. Lett.*, 17(1):61–66, 1993.

[261] Wing Hung Wong and Xiaotong Shen. Probability inequalities for likelihood ratios and convergence rates of sieve MLEs. *Ann. Statist.*, 23(2):339–362, 1995.

[262] M. Xie and K. Singh. Confidence distribution, the frequentist distribution of a parameter – a review. *Int. Statist. Rev.*, 81(1):3–39, 2013.

[263] Minge Xie, Kesar Singh, and William E. Strawderman. Confidence distributions and a unifying framework for meta-analysis. *J. Amer. Statist. Assoc.*, 106(493):320–333, 2011.

[264] Ronald Yager and Liping Liu, editors. *Classic Works of the Dempster–Shafer Theory of Belief Functions*, volume 219. Springer, Berlin, 2008.

[265] J. Yang, B. Benyamin, B. P. McEvoy, S. Gordon, A. K. Henders, D. R. Nyholt, P. A. Madden, A. C. Heath, N. G. Martin, G. W. Montgomery, M. E. Goddard, and P. M. Visscher. Common SNPs explain a large proportion of the heritability for human height. *Nat. Genet.*, 42(7):565–569, 2010.

[266] S. L. Zabell. R. A. Fisher and the fiducial argument. *Statist. Sci.*, 7(3):369–387, 1992.

[267] Cun-Hui Zhang and Jian Huang. The sparsity and bias of the LASSO selection in high-dimensional linear regression. *Ann. Statist.*, 36(4):1567–1594, 2008.

[268] Jianchun Zhang. *Statistical inference with weak beliefs*. PhD thesis, Purdue University, West Lafayette, IN, 2010.

[269] Jianchun Zhang and Chuanhai Liu. Dempster–Shafer inference with weak beliefs. *Statist. Sinica*, 21(2):475–494, 2011.

[270] Tonglin Zhang and Michael Woodroofe. Credible and confidence sets for the ratio of variance components in the balanced one-way model. *Sankhyā Ser. A*, 64(3, part 1):545–560, 2002.

[271] Xiang Zhou and Matthew Stephens. Efficient multivariate linear mixed model algorithms for genome-wide association studies. *Nat. Methods*, 11:407–409, 2014.

[272] Hui Zou. The adaptive lasso and its oracle properties. *J. Amer. Statist. Assoc.*, 101(476):1418–1429, 2006.

[273] Hui Zou and Trevor Hastie. Regularization and variable selection via the elastic net. *J. R. Stat. Soc. Ser. B Stat. Methodol.*, 67(2):301–320, 2005.

Index

Printed in the United States
by Baker & Taylor Publisher Services